*Martin Maldovan and
Edwin L. Thomas*

**Periodic Materials and
Interference Lithography**

Related Titles

Sailor, M. J.
Porous Silicon in Practice
Preparation, Characterization and Applications

2009
Softcover
ISBN: 978-3-527-31378-5

Kaskel, S.
Porous Materials
Introduction to Materials Chemistry, Properties, and Applications

2009
Softcover
ISBN: 978-3-527-32035-6

Misawa, H., Juodkazis, S. (eds.)
3D Laser Microfabrication
Principles and Applications

2006
Hardcover
ISBN: 978-3-527-31055-5

*Martin Maldovan and
Edwin L. Thomas*

Periodic Materials and Interference Lithography

for Photonics, Phononics and Mechanics

WILEY-VCH Verlag GmbH & Co. KGaA

The Authors

Dr. Martin Maldovan
MIT
Materials Science & Engineering
77 Massachusetts Avenue
Cambridge, MA 02139
USA

Prof. Edwin L. Thomas
MIT
Materials Science & Engineering
77 Massachusetts Avenue
Cambridge, MA 02139
USA

■ All books published by Wiley-VCH are carefully produced. Nevertheless, authors, editors, and publisher do not warrant the information contained in these books, including this book, to be free of errors. Readers are advised to keep in mind that statements, data, illustrations, procedural details or other items may inadvertently be inaccurate.

Library of Congress Card No.:
applied for

British Library Cataloguing-in-Publication Data
A catalogue record for this book is available from the British Library.

Bibliographic information published by the Deutsche Nationalbibliothek
Die Deutsche Nationalbibliothek lists this publication in the Deutsche Nationalbibliografie; detailed bibliographic data are available on the Internet at <http://dnb.d-nb.de>.

© 2009 WILEY-VCH Verlag GmbH & Co. KGaA, Weinheim

All rights reserved (including those of translation into other languages). No part of this book may be reproduced in any form – by photoprinting, microfilm, or any other means – nor transmitted or translated into a machine language without written permission from the publishers. Registered names, trademarks, etc. used in this book, even when not specifically marked as such, are not to be considered unprotected by law.

Typesetting Laserwords Private Limited, Chennai, India
Printing Strauss GmbH, Mörlenbach
Binding Litges & Dopf GmbH, Heppenheim

Printed in the Federal Republic of Germany
Printed on acid-free paper

ISBN: 978-3-527-31999-2

Contents

Preface *XI*
Introduction *XIII*

Theory *1*

1	**Structural Periodicity** *3*	
1.1	Nonperiodic versus Periodic Structures *4*	
1.2	Two-dimensional Point Lattices *6*	
1.3	Three-dimensional Point Lattices *10*	
1.3.1	Primitive and Nonprimitive Unit Cells *14*	
1.4	Mathematical Description of Periodic Structures *16*	
1.5	Fourier Series *20*	
1.5.1	Fourier Series for Two-dimensional Periodic Functions *20*	
1.5.2	Fourier Series for Three-dimensional Periodic Functions *23*	
1.5.3	Arbitrary Unit Cells *25*	
	Further Reading *26*	
	Problems *26*	
2	**Periodic Functions and Structures** *29*	
2.1	Introduction *30*	
2.2	Creating Simple Periodic Functions in Two Dimensions *31*	
2.2.1	The Square Lattice *31*	
2.2.2	The Triangular Lattice *38*	
2.3	Creating Simple Periodic Functions in Three Dimensions *41*	
2.3.1	The Simple Cubic Lattice *44*	
2.3.2	The Face-centered-cubic Lattice *47*	
2.3.3	The Body-centered-cubic Lattice *51*	
2.4	Combination of Simple Periodic Functions *59*	
	Problems *61*	
3	**Interference of Waves and Interference Lithography** *63*	
3.1	Electromagnetic Waves *64*	

Periodic Materials and Interference Lithography. M. Maldovan and E. Thomas
Copyright © 2009 WILEY-VCH Verlag GmbH & Co. KGaA, Weinheim
ISBN: 978-3-527-31999-2

3.2	The Wave Equation	65
3.3	Electromagnetic Plane Waves	68
3.4	The Transverse Character of Electromagnetic Plane Waves	69
3.5	Polarization	72
3.5.1	Linearly Polarized Electromagnetic Plane Waves	73
3.5.2	Circularly Polarized Electromagnetic Plane Waves	74
3.5.3	Elliptically Polarized Electromagnetic Plane Waves	75
3.6	Electromagnetic Energy	75
3.6.1	Energy Density and Energy Flux for Electromagnetic Plane Waves	77
3.6.2	Time-averaged Values	77
3.6.3	Intensity	80
3.7	Interference of Electromagnetic Plane Waves	81
3.7.1	Three-dimensional Interference Patterns	86
3.8	Interference Lithography	89
3.8.1	Photoresist Materials	89
3.8.2	The Interference Lithography Technique	92
3.8.3	Designing Periodic Structures	93
	Further Reading	94
	Problems	94
4	**Periodic Structures and Interference Lithography**	**97**
4.1	The Connection between the Interference of Plane Waves and Fourier Series	98
4.2	Simple Periodic Structures in Two Dimensions Via Interference Lithography	100
4.3	Simple Periodic Structures in Three Dimensions Via Interference Lithography	104
	Further Reading	110
	Problems	111

Experimental 113

5	**Fabrication of Periodic Structures**	**115**
5.1	Introduction	116
5.2	Light Beams	116
5.3	Multiple Gratings and the Registration Challenge	118
5.4	Beam Configuration	119
5.4.1	Using Four Beams	119
5.4.2	Using a Single Beam (Phase Mask Lithography)	120
5.5	Pattern Transfer: Material Platforms and Photoresists	122
5.5.1	Negative Photoresists	124
5.5.2	Positive Photoresists	126
5.5.3	Organic–Inorganic Hybrids Resists	128
5.6	Practical Considerations for Interference Lithography	128

5.6.1	Preserving Polarizations and Directions	*128*
5.6.2	Contrast	*131*
5.6.3	Drying	*132*
5.6.4	Shrinkage	*133*
5.6.5	Backfilling – Creating Inverse Periodic Structures	*133*
5.6.6	Volume Fraction Control	*134*
5.7	Closing Remarks	*135*
	Further Reading	*136*

Applications *139*

6	**Photonic Crystals**	*141*
6.1	Introduction	*142*
6.2	One-dimensional Photonic Crystals	*143*
6.2.1	Finite Periodic Structures	*143*
6.2.2	Infinite Periodic Structures	*147*
6.2.3	Finite versus Infinite Periodic Structures	*150*
6.3	Two-dimensional Photonic Crystals	*151*
6.3.1	Reciprocal Lattices and Brillouin Zones in Two Dimensions	*152*
6.3.2	Band Diagrams and Photonic Band Gaps in Two Dimensions	*157*
6.3.3	Photonic Band Gaps in Two-dimensional Simple Periodic Structures	*160*
6.4	Three-dimensional Photonic Crystals	*162*
6.4.1	Reciprocal Lattices and Brillouin Zones in Three Dimensions	*164*
6.4.2	Band Diagrams and Photonic Band Gaps in Three Dimensions	*168*
6.4.3	Photonic Band Gaps in Three-dimensional Simple Periodic Structures	*170*
	Further Reading	*176*
	Problems	*179*

7	**Phononic Crystals**	*183*
7.1	Introduction	*184*
7.1.1	Elastic Waves in Homogeneous Solid Materials	*184*
7.1.2	Acoustic Waves in Homogeneous Fluid Materials	*187*
7.2	Phononic Crystals	*188*
7.3	One-dimensional Phononic Crystals	*190*
7.3.1	Finite Periodic Structures	*190*
7.3.2	Infinite Periodic Structures	*194*
7.4	Two-dimensional Phononic Crystals	*198*
7.4.1	Vacuum Cylinders in a Solid Background	*198*
7.4.2	Solid Cylinders in Air	*202*
7.4.3	Phononic Band Gaps in Two-dimensional Simple Periodic Structures	*205*
7.5	Three-dimensional Phononic Crystals	*207*

7.5.1	Solid Spheres in a Solid Background Material	208
	Further Reading	210
	Problems	213

8 Periodic Cellular Solids 215
8.1 Introduction 216
8.2 One-dimensional Hooke's Law 218
8.3 The Stress Tensor 219
8.4 The Strain Tensor 221
8.4.1 Expansion 225
8.4.2 General Deformation 226
8.4.3 Resolving a General Deformation as Strain Plus Rotation 227
8.5 Stress–Strain Relationship: The Generalized Hooke's Law 229
8.6 The Generalized Hooke's Law in Matrix Notation 230
8.7 The Elastic Constants of Cubic Crystals 232
8.7.1 Young's Modulus and Poisson's Ratio 233
8.7.2 The Shear Modulus 235
8.7.3 The Bulk Modulus 237
8.8 Topological Design of Periodic Cellular Solids 238
8.9 Finite Element Program to Calculate Linear Elastic Mechanical Properties 243
8.10 Linear Elastic Mechanical Properties of Periodic Cellular Solids 243
8.11 Twelve-connected Stretch-dominated Periodic Cellular Solids via Interference Lithography 247
8.12 Fabrication of a Simple Cubic Cellular Solid via Interference Lithography 249
8.13 Plastic Deformation of Microframes 250
Further Reading 252

9 Further Applications 255
9.1 Controlling the Spontaneous Emission of Light 256
9.2 Localization of Light: Microcavities and Waveguides 259
9.3 Simultaneous Localization of Light and Sound in Photonic–Phononic Crystals: Novel Acoustic–Optical Devices 264
9.4 Negative Refraction and Superlenses 268
9.5 Multifunctional Periodic Structures: Maximum Transport of Heat and Electricity 272
9.6 Microfluidics 273
9.7 Thermoelectric Energy 275
9.7.1 Peltier Effect 275
9.7.2 Thomson Effect 276
9.7.3 Seebeck Effect 277
Further Reading 278

Appendix A   MATLAB Program to Calculate the Optimal Electric Field Amplitude Vectors for the Interfering Light Beams   *281*

Appendix B   MATLAB Program to Calculate Reflectance versus Frequency for One-dimensional Photonic Crystals   *289*

Appendix C   MATLAB Program to Calculate Reflectance versus Frequency for One-dimensional Phononic Crystals   *297*

Index   *305*

Preface

Periodic materials have been demonstrated to have unique physical properties due to their singular interaction with waves (which are also periodic). In recent years, the discovery of an experimental technique called *interference lithography*, which can create periodic materials at very small length scales, had a strong impact on the way we think about these materials. To rationally design and fabricate periodic materials by interference lithography, it is useful to perceive a periodic material as a sum of its Fourier series components. This book studies the fascinating and strong correlation between the analytical description of periodic materials by Fourier series and the experimental realization of these materials by interference lithography. We believe this mutual relation will have a deep influence in the development of new periodic materials since the convergence of theoretical and experimental methods allows for the theoretical design of structured materials that can be experimentally realized.

The book also attempts to comprehensively study the applications of periodic materials. For example, in spite of their strong similarities, to date, photonic and phononic crystals have been studied separately. In this book, we try to integrate these two research areas by proposing photonic–phononic crystals that can combine the physical properties of these materials and may even give rise to unique acousto-optical applications. The ubiquity of periodic materials in science and technology can be demonstrated by the large number of physical processes they can control. Several of these practical applications for periodic materials are discussed in this book, which include the control of electromagnetic and elastic waves, mechanics, fluids, and heat. The broad range of applications demonstrates the multifunctional character of periodic materials and the strong impact they can have if fabricated at small length scales.

We would like to take this opportunity to thank our colleagues and friends for their valuable help and support: C. K. Ullal, J.-H. Jang, H. Koh, J. Walish, A. Urbas, S. Kooi, T. Gorishnyy, T. Choi and Profs. W. C. Carter, M. C. Boyce, and V. Tsukruk. Chapter 5 was written collaboratively by C. K. Ullal, J-H. Jang,

M. Maldovan, and E. L. Thomas. We would also like to acknowledge the financial support of our work in the area of periodic materials by the National Science Foundation – Division of Materials Research and the US Army supported Institute for Soldier Nanotechnologies at MIT.

October, 2008　　　　　　　　　　　　　　　　　　　　　　　　*Martin Maldovan*
Cambridge, Massachusetts　　　　　　　　　　　　　　　　*Edwin L. (Ned) Thomas*

Introduction

Rational Design of Materials

The control and improvement of the physical properties of materials is a central objective in many fields of science and technology. Indeed, one aspect of human evolution encompasses the ability of humans to understand, transform, and use natural materials, and our current advances in technology are a clear illustration of our deep understanding and ability to manipulate materials.

Scientists and engineers have been altering the physical properties of raw natural materials for years, from the simple combination of natural homogeneous materials to the complex fabrication of intricate structural designs that result in materials with effective properties significantly different from those corresponding to the constitutive material. Steel alloys and synthetic sponges are simple examples of the management of material properties by combination or structural design.

The history of the control of the properties of materials begins with the Stone age, the Iron age, and the Bronze age, where stone, iron, and bronze were the most sophisticated forms of materials, respectively. During the twentieth century, the control of the *mechanical* properties of materials had a significant impact on our world; buildings, highways, bridges, planes, ships, and rail-roads are just a few examples of how the understanding of the properties of materials can radically change our society. Moreover, in the last few decades, the managing of the *electrical* and *electronic* properties of materials have created an additional unpredictable revolution; radios, televisions, computers, cell-phones, digital cameras, and music players are now products that seem to have always existed.

"What is next?" ... is not a simple question to answer.

Undoubtedly, the reduction in material length scales emerges as a new frontier in science and technology since it has been demonstrated that as the intrinsic length scale is reduced, materials show different properties when compared to those exhibited at large length scales. Unfortunately, the reduction in the length scale brings difficulties in the fabrication process of these novel materials. That is,

Periodic Materials and Interference Lithography. M. Maldovan and E. Thomas
Copyright © 2009 WILEY-VCH Verlag GmbH & Co. KGaA, Weinheim
ISBN: 978-3-527-31999-2

materials at small length scales bring new physical properties as well as fabrication challenges. This is where this book comes in.

Imagine creating a material that allows us to control the propagation of light. For example, envision a material that does not allow light with certain frequencies to propagate while allowing light with other frequencies to propagate freely. Moreover, imagine a material that can make a "ray of light" turn 90° and that "ray of light" has a cross-section area in the order of microns. Or a material where a "ray of light" can pass around little objects placed within the material. Furthermore, imagine that the material can also stop the propagation of light by confining the electromagnetic wave around certain spatial regions within the material. Actually, this material exists: it is called a *photonic crystal* and it is one of the main subjects in this book.

Assume also that you want to have all of the above physical phenomena but in the case of sound. That is, imagine controlling sound in such a way that you can selectively decide whether or not a particular frequency is allowed to propagate in the material, guide sound inside of the material, or localize it within a certain region in space. This special currently emerging class of material that allows us to control sound in such innovative forms is called a *phononic crystal* and is also studied in detail in this book.

A crucial and common characteristic between photonic and phononic crystals is the fact that they are *periodic* materials. These types of materials can be understood as made of an arrangement of a specific object (e.g. sphere, cylinder, etc.) that repeats regularly in space. Importantly, the materials from which the object is made are decisive in terms of the functionality of photonic and phononic crystals.

The task to design and fabricate such periodic materials is a formidable challenge. First, there are essentially an infinite set of possible periodic geometries, and second creating such structures actually requires sculpting specific materials into the resultant structures with typical feature sizes at and below the submicron length scale.

For example, to control the propagation of visible light, which has wavelengths between 400 and 700 nm, the size of the objects and the distance between them must be on the submicron scale. That is, the fabrication of a photonic crystal that controls the propagation of visible light is analogous to the construction of a building where the periodic supports are separated by distances smaller than a micron. Needless to say, this brings an enormous challenge in terms of the techniques needed to fabricate these materials.

To be able to develop these research areas, it is essential to design periodic materials that have useful photonic and phononic properties *and* that can be actually fabricated at small length scales. This is not a trivial assignment and is one of the main topics studied in this book. We particularly concentrate on the design, fabrication, and applications of *periodic materials* at small length scales. As previously mentioned, periodic materials show striking physical effects and have surprising properties as they interact with waves. Since they do not have preferences toward a particular type of wave, periodic materials can indistinctively control the propagation of both electromagnetic and elastic waves. As a result, we have an opportunity for a radical new departure on the management of material properties, which is the control of the *optical* and *acoustical* properties of materials.

In addition, periodic materials at small length scales can present a new twist to the *mechanical* properties of materials. In the past, much attention has been paid to the fabrication of periodic structures at the millimeter scale (and above) in order to obtain stiff-strong, light-weight structures. The reduction in the length scale brings new opportunities for the applications of periodic materials by exploiting, for example, length-scale-dependent mechanical properties.

We divide the book in three sections: theory, experiments, and applications. In the theoretical section, our objective is to design useful periodic materials that can be fabricated at small length scales by a fast, low-cost experimental technique known as *interference lithography*. One of the most important advantages of periodic materials is the fact that they can be described mathematically with high precision. For example, Fourier series is a mathematical approach that can be used to analytically describe structures that repeat periodically in space. In particular, we use a scheme to design periodic structures based on the manipulation of the coefficients of the Fourier series expansions describing periodic materials. The importance of this approach is that it allows us to obtain periodic structures described by the sum of a small number of Fourier terms, which can be fabricated at small length scales by interference lithography and have exceptional optical, acoustical, and mechanical properties. In addition, we explain in detail how periodic materials can be fabricated by the use of the interference lithography technique and show that this technique is intrinsically related to the Fourier series expansions describing the periodic materials.

The experimental section of the book is intended to provide guidelines for the fabrication of periodic materials at small length scales by interference lithography. In terms of the experimental realization of submicron structured materials, joint efforts of researchers from several fields such as chemistry, engineering, and materials science are leading to good success. From the perspective of experimental fabrication of periodic materials, it is highly desirable to be able to obtain large samples while still having control over the geometry of the resultant structure. Serial writing techniques, such as two-photon lithography, three-dimensional printing, or robotic micromanipulation allow one to create arbitrary structures but are usually slow and cover only small areas. In contrast, self-assembly of colloidal spheres and block-copolymers tends to be rapid and cover large areas, but the control over the geometry of the structure is quite restricted and it is difficult to avoid random defects. Interference lithography emerges as the fabrication method that allows one to control the geometry of the periodic structure while still creating large-area, defect-free single crystals. Important experimental aspects of the interference lithography technique are described in detail in this section.

In the last section of the book, we deal with the practical applications of periodic materials. The ability to be able to precisely describe periodic structures has important benefits: the properties of periodic materials can be calculated accurately. This contributes significantly to give feedback for the design of improved periodic materials with optimized specific functionalities. We not only concentrate on explaining fundamental applications of periodic materials by employing them as photonic and phononic crystals as well as microstructures for mechanics but also demonstrate how periodic materials can be used in a large number of practical applications in several different fields.

Outline of the Book

The introductory Chapter 1 presents the fundamental concepts of periodicity. Two- and three-dimensional periodic structures are described by an arrangement of regular points known as the *point lattices*. These sets of ordered points allow us to classify periodic structures based on how they repeat in space. In this chapter, we also show how periodic materials can be described by the use of analytical functions and introduce the *Fourier series* expansions of three-dimensional periodic functions. Fourier series expansion is a mathematical technique that allows us to describe periodic structures by the use of analytical formulas, which consist of the sum of trigonometric functions.

In Chapter 2, we introduce a systematic scheme based on Fourier series expansions to obtain a large set of *simple periodic structures*. These periodic structures are *simple* in the sense that they are described by Fourier series expansions made of the sum of a small number of trigonometric terms. This is achieved by the systematic manipulation of the coefficients of the Fourier series expansions. In subsequent chapters, we demonstrate that these simple periodic structures have important applications in optics, acoustics, and mechanics, and they are appropriate for fabrication via the interference lithography technique.

Chapter 3 provides an introduction to *wave-interference* phenomena. We first study the propagation of electromagnetic plane waves in homogeneous materials. Then, we examine the superposition of monochromatic electromagnetic plane waves in the same spatial region and the physical effects associated with the interference between the waves. Drawing upon these fundamental concepts, we describe in detail the interference lithography technique, which is the fabrication method that allows us to obtain the periodic structures presented in the previous chapter.

In Chapter 4, we introduce a general scheme that allows us to obtain the electromagnetic plane waves (or interfering beams) required to create desired periodic structures by using the interference lithography technique. This scheme is based on the connection between the Fourier series expansion of a periodic structure and the spatial distribution of energy corresponding to the interference of electromagnetic plane waves. In particular, by using this general scheme, we obtain all necessary information required to fabricate the simple periodic structures presented in Chapter 2.

Chapter 5 deals with *experimental* aspects associated with the fabrication of periodic structures by interference lithography. The key concept is the ability of photosensitive materials to change their solubility characteristics in regions of either high or low light intensity, such that a subsequent chemical development leads to a material having periodic features replicating the incident light interference pattern. We examine important considerations and experimental difficulties, such as conserving the correct beams needed to create the light intensity patterns as well as drying, shrinkage, backfilling, and volume fraction control of the transfer of the pattern to a photo resist material.

In Chapter 6, we study *photonic crystals*, which are periodic structures made of two different dielectric materials. We introduce the reader to basic solid state physics concepts such as reciprocal lattice, Bloch waves, Brillouin zones, and band diagrams in order to understand the fundamental property of photonic crystals, which is the fact that electromagnetic waves having frequencies within a specific range are not allowed to propagate within the crystal. These frequency ranges are also called *photonic band gaps*. We study photonic crystals in one, two, and three dimensions and present numerical data for photonic band gaps as a function of material volume fraction for a set of periodic structures that can be fabricated via the interference lithography route.

In Chapter 7, we study the acoustic version of photonic crystals, which are *phononic crystals*. These periodic structures are made of two different elastic materials and present *phononic band gaps*, that is, frequency ranges for which mechanical waves cannot propagate within the crystal. In the first part of the chapter, we discuss the difference between the propagation of mechanical waves in fluids and solid materials. In general, when mechanical waves propagate in fluid materials they are called *acoustic* waves, whereas when they propagate in solid materials they are called *elastic* waves. As in the previous chapter, we investigate phononic crystals in one, two, and three dimensions and present numerical data for phononic band gaps as a function of material volume fraction for a set of periodic structures that can be fabricated via interference lithography.

Chapter 8 deals with the *mechanical properties* of periodic structures fabricated by interference lithography. We first present an introduction to theoretical aspects of elastic mechanical properties of materials. We explain in detail fundamental concepts in mechanics, such as Hooke's law, stress and strain tensors, elastic constants of cubic crystals, Young's modulus, Poisson's ratio, shear modulus, and bulk modulus. In the second part of the chapter, we introduce the topological design of periodic structures with application in mechanics and provide numerical calculations for effective linear elastic properties of periodic solids fabricated by interference lithography. In the last section, we briefly comment on the plastic deformation of these structures.

Finally, in Chapter 9 we examine several potential applications of periodic materials in several emerging research areas. For example, we show how periodic materials can be used to control the spontaneous emission of light, guide and localize light, manage light and sound simultaneously, focus light without diffraction limits, maximize the transport of heat and electricity, regulate the flow of fluid, and enhance thermoelectric devices. All of the above examples demonstrate the *multifunctional* character and the huge technological impact of these periodic materials.

In addition, at the end of each chapter, we give relevant references/literature for additional reading as well as background material and specialist textbooks for further study. At the end of most of the chapters, we also provide problem sets for the reader to strengthen the concepts studied in the book.

Cambridge, Massachusetts *Martin Maldovan and Edwin L. (Ned) Thomas*
October, 2008

Theory

1
Structural Periodicity

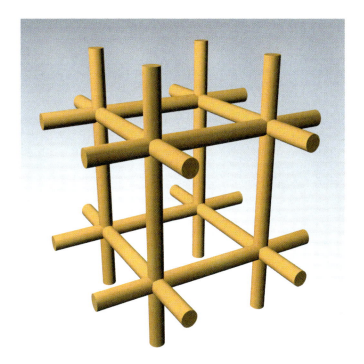

Nature is replete with fascinating examples of intricate structures. The goal of several disciplines in the physical sciences is to control and exploit the function and properties of natural and engineered structures. The first step of such an effort is to have an approach that allows us to understand, characterize, and mathematically describe any general structure. To make a start at such an ambitious effort, we focus on trying to describe one particular subset, viz. periodic structures.

1.1
Nonperiodic versus Periodic Structures

Nature attracts the attention of scientists and engineers with multiple examples of outstanding structures that fulfill specific demands. For example, wood has a huge capacity to carry sustained loads and has extraordinary mechanical properties due to a delicate action between fibers of cellulose, hemicellulose, and lignin. Another example of a fibrous natural structure that meets several specific demands is bone, which moves, supports, and protects the body of vertebrates and achieves all these functionalities by being strong and lightweight at the same time. Nature also provides wonderful examples of structures that are built by insects and animals. A supreme example of insect engineering is the honeycomb structure in bee hives, which is made of perfectly uniform repetitive cells that are efficiently organized in space to raise the young bees.

Scientists and engineers have always had curiosity for naturally occurring structures and have been *borrowing* ideas from nature since early days. Needless to say that nature has had plenty of time to develop its structural designs, whereas scientists and engineers can be considered as relative beginners in this field. Nevertheless, scientists and engineers have been successful in replicating some naturally occurring structures, but equally important, they have come up with their own structural designs that meet specific *human* demands.

We do not have to look very far to see the results of the inspiration of engineers for structures. Bridges and tall buildings that can support large loads and stand safely for years are extraordinary examples of engineering achievement and design. At smaller length scales, scientists and engineers have recently designed and fabricated structures that can interact with acoustic and electromagnetic waves in such a way that sound and light can be totally reflected by the structures. Unfortunately, the length scales of these structures do not allow for human visual perception, and so substantial magnification is needed to visualize the architectures (e.g. scanning electron microscopes with magnifications in excess of $100\,000\times$).

Irrespective of the scale at which the structure is built, natural and man-made structures can be classified into two distinct groups: *nonperiodic* and *periodic* structures. In the first case, the structure consists of elements forming a disordered arrangement, thus lacking organization or regularity in space. In the second case, however, the structure is made of a basic object that repeats at precise intervals in space and the resultant structure possesses regularity (Figure 1.1).

Illustrative examples of nonperiodic solid structures are metallic or polymeric foams. These foams are basically created by injecting a low-pressure inert gas into a melted metal or polymer. After solidification, the mixture forms a disordered structure where many gas bubbles are trapped in the solid. The bubbles are nearly randomly distributed within the structure and they may be of various sizes (e.g. different radius in the case of spheres). Some polymeric foams are widely used as packing materials, while other polymeric or metallic foams are used for sound absorption in acoustical insulators or for low thermal conductivity in thermal insulators.

1.1 Nonperiodic versus Periodic Structures

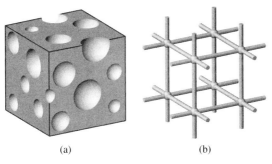

Fig. 1.1 (a) A three-dimensional *nonperiodic* structure made of various size gas bubbles randomly distributed within a solid material. (b) A portion of a three-dimensional infinite *periodic* structure created by a regular repetition of an object in space.

The theoretical prediction of the physical properties of nonperiodic structures is difficult. In fact, theoretical results for such structures are not exact and intense computational methods are required for understanding the complex relationship between structure and physical properties. For example, to properly study the physical properties of a nonperiodic structure, a large portion of the structure must be considered. If this portion is too small, the numerical results will not correspond to the actual properties of the structure. This becomes obvious in the limiting case, since, for example, the consideration of a single bubble cannot be regarded as an accurate description of a structure where many bubbles of different sizes are irregularly distributed in space. In addition to this size effect, it is also difficult to determine the number of different computational *samples* that need to be studied to obtain accurate properties. For example, one computational sample may contain numerous bubbles with large radius, whereas other samples may contain only a few. To obtain precise numerical calculations, it is therefore needed to study many computational samples consisting of randomly located bubbles with different radius and determine the physical properties statistically. This makes the theoretical prediction of the physical properties of nonperiodic structures even more difficult.

On the other hand, we have periodic structures, which are created by a regular repetition of an object in space. The frame figure provided at the beginning of this chapter, comprising a set of uniform beams having a fixed distance among them, is an example of a periodic structure. On large scales, examples of periodic structures include scaffold structures or building frameworks where the constituent objects (e.g. beams and columns) are regularly distributed in space and all have the same size. The theoretical prediction of the properties of periodic structures is less difficult than in the case of nonperiodic structures. Since the structure is formed by a repeating object, all the information needed to describe the entire structure is given by the object itself and by the manner it repeats in space. As a result, only a small portion of the structure (which includes the repeating object) needs to be considered to obtain the properties of the whole periodic structure. The application of periodic boundary conditions to this small portion of the structure allows for less complicated numerical techniques to obtain precise relationships between structure and physical properties. Moreover, in the case of periodic structures, only a single computational *sample* needs to be studied because the sample precisely describes the periodic structure, as the object that repeats in space is always of the same size.

Nonperiodic and periodic structures are both important from the point of view of practical applications. In this book, we focus on periodic structures for a number of reasons: First, periodic structures have recently shown strikingly new physical properties that permit control of elastic and electromagnetic waves. Second, an experimental technique is currently available that allows engineers to fabricate periodic structures while having certain control over the geometry of the resultant structure. Finally, there are several numerical techniques that can be used to predict the physical properties of periodic structures, which in turn allow us to design, understand, and optimize the structures prior to their experimental fabrication. All these considerations are explained throughout this book. To start studying periodic structures, we first introduce a theoretical scheme to represent and classify periodic structures.

1.2
Two-dimensional Point Lattices

In this book, we are interested in how the geometry of a periodic structure determines the resultant physical properties of the structure. To achieve this, we need a framework that allows us to classify and systematically specify the large number of possible periodic structures. In this section, we introduce an abstract model, called the *point lattice model*, that classifies periodic structures in terms of how they repeat in space.

In the text that follows, we begin with the description of two-dimensional periodic structures because the graphical representation of these structures is simpler than that of the three-dimensional case. The extension to three dimensions is then made by generalizing the concepts developed for the two-dimensional case.

A periodic structure may be defined as an assembly of an *object* that repeats regularly in space. For example, consider the two-dimensional periodic structure represented in Figure 1.2a, which is considered to extend infinitely in the plane. The structure can be divided by two sets of equally spaced parallel lines (Figure 1.2b). This division of the structure produces a set of objects, each of which is identical in size and shape. One of these constituent objects is shown in dark gray in Figure 1.2b. As a result, the two-dimensional periodic structure can be understood as formed by the regular repetition of this object in space. Note that the equally spaced parallel lines also generate a set of two-dimensional cells within which the objects are enclosed.

To understand periodic structures, it is useful, for the moment, to consider each object as an imaginary point. This consideration transforms the periodic structure into an infinite array of equivalent points (Figure 1.2c). The corresponding array

1.2 Two-dimensional Point Lattices

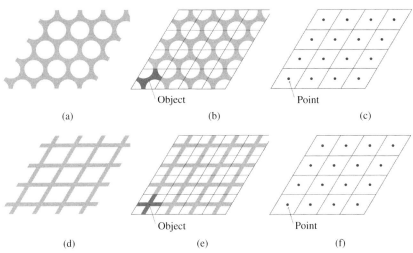

Fig. 1.2 A two-dimensional periodic structure and its point lattice. (a and d) Two distinct two-dimensional periodic structures. (b and e) The structures are divided by equally spaced parallel lines that define the object that repeats in space. (c and f) The object is represented by an imaginary point at its center and the two-dimensional periodic structure transforms into the corresponding point lattice.

of points is called the *point lattice* of the structure, and basically determines the positions of the repeating objects. Hence, the existence of spatial periodicity in the structure implies that the structure has a periodic array of lattice points associated with it. Note that the surroundings viewed from an arbitrary point in the point lattice are identical to the surroundings viewed from any other point. That is, each point in a point lattice has identical surroundings.[1]

Figure 1.2d shows a different two-dimensional periodic structure. The point lattice associated with this structure, however, is the same as the one corresponding to the structure shown in Figure 1.2a. After dividing the structure by two sets of equally spaced parallel lines (Figure 1.2e) and replacing the set of identical objects by imaginary points (Figure 1.2f), the corresponding point lattice is the same as in Figure 1.2c. That is, the periodic structures shown in Figures 1.2a and d have the same point lattice associated with them and thus have *the same spatial periodicity*. This is due to the fact that the objects corresponding to each structure repeat similarly in space.

Even though each two-dimensional periodic structure has a specific point lattice associated with it, the point lattice can be displayed in various ways. For example,

---

1) The position (relative to the underlying structure) of the two sets of equally spaced parallel lines that enclose the repeating object in Figure 1.2b is arbitrary. Therefore, the definition of the object that repeats regularly in space to form the periodic structure is also arbitrary. This means that the periodic structure shown in Figure 1.2a can alternatively be described by different choices for the repeating object. The corresponding point lattice, however, is independent on the choice of the particular repeating object.

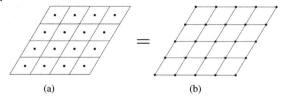

(a)   (b)

Fig. 1.3 A two-dimensional point lattice with the lattice points located (a) at the center of the cells and (b) at the corner of the cells.

the point lattice shown in Figure 1.3a has the points located at the center of the cells. However, the *same* point lattice can optionally be displayed with the points located at the corners of the cells (Figure 1.3b). In both the cases, the point lattice remains the same. That is, the overall distribution of lattice points in space is not changed. Note that additional lattice points from neighboring cells appear in Figure 1.3b since the periodic point lattice extends infinitely. In this book, we choose to represent point lattices with the points located at the corners of the cells.

Because all cells in Figure 1.3b are equivalent, we can arbitrarily choose any of those cells as a *unit cell* that repeats to form the point lattice. In this particular example, the edges of the unit cells have equal lengths. In the general case, however, the unit cell is an arbitrary parallelogram that can be described by the *primitive vectors* $\mathbf{a}_1$ and $\mathbf{a}_2$ (Figure 1.4), and when repeated by successive translations in space the unit cell forms the corresponding point lattice. The distances $a = |\mathbf{a}_1|$, $b = |\mathbf{a}_2|$, and the angle γ are called the *lattice constants*.

The primitive vectors $\mathbf{a}_1$ and $\mathbf{a}_2$ defining the unit cell also construct the entire point lattice by translations. For example, the point lattice can be viewed as an array of regularly spaced points located at the tip of the vector

$$\mathbf{R} = n_1 \mathbf{a}_1 + n_2 \mathbf{a}_2 \qquad (1.1)$$

where n_1 and n_2 are arbitrary integer numbers. Equation 1.1 indicates that for each particular set of n_1 and n_2 values there is a corresponding lattice point $\mathbf{R}$ in space. Therefore, the vector $\mathbf{R}$ fills the plane with regularly spaced points and it is said to generate the point lattice.

The number of conceivable two-dimensional periodic structures is infinite. However, when two-dimensional periodic structures are transformed into their

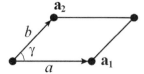

Fig. 1.4 A general two-dimensional unit cell defined by two nonorthogonal vectors, $\mathbf{a}_1$ and $\mathbf{a}_2$.

corresponding point lattices, only five distinct point lattices are found. In other words, the existence of an object that repeats regularly in space, together with the requirement that the repeating objects must have identical surroundings, determines the existence of only five types of point lattices. Therefore, we have achieved one of our goals: *we can classify the infinite set of two-dimensional periodic structures by their spatial periodicity into five categories.*

The five types of two-dimensional point lattices are known as *oblique, rectangular, square, triangular,* and *rhombus* (or centered rectangular). They are illustrated in Figure 1.5 together with the choices for the primitive vectors $\mathbf{a}_1$ and $\mathbf{a}_2$. The distinct geometrical features of the five point lattices are determined by the relative lengths a and b of the primitive vectors $\mathbf{a}_1$ and $\mathbf{a}_2$, and the value of the angle γ. These values

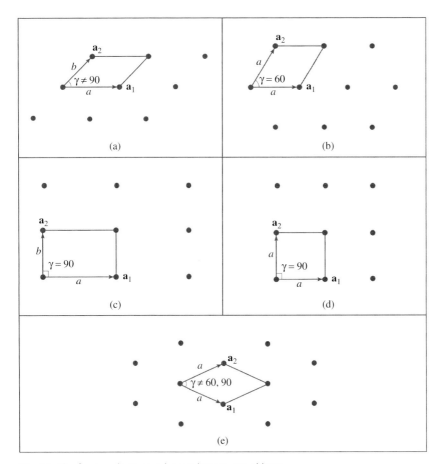

Fig. 1.5 The five two-dimensional point lattices: (a) oblique, (b) triangular, (c) rectangular, (d) square, and (e) rhombus (or centered rectangular).

are given as follows:
- oblique, $a \neq b$, $\gamma \neq 90°$
- rectangular, $a \neq b$, $\gamma = 90°$
- square, $a = b$, $\gamma = 90°$
- triangular, $a = b$, $\gamma = 60°$
- rhombus, $a = b$, $\gamma \neq 60°, 90°$

Note that rectangular, square, triangular, and rhombus lattices can all be considered as special cases of the oblique lattice.

1.3
Three-dimensional Point Lattices

In the previous section, we introduced the principles needed to classify two-dimensional periodic structures in terms of their translational spatial periodicities. In this section, we extend these concepts to the more significant three-dimensional case.

As in the previous section, to classify three-dimensional periodic structures by their translational spatial periodicity, we need to identify within the structure an object that repeats regularly in space. This repeating object is then considered as an imaginary point and the three-dimensional periodic structure transforms into a three-dimensional array of equivalent points. However, to identify and enclose a repeating object in three dimensions, we need to use three sets of equally spaced parallel planes instead of two sets of equally spaced parallel lines as those used in the two-dimensional case.

As an illustrative example, consider the three-dimensional periodic structure shown in Figure 1.6a. The structure is divided by three sets of equally spaced parallel planes (Figure 1.6b). The set of parallel planes identifies the set of objects that repeat regularly in space, each of which is identical in size and shape. The three-dimensional periodic structure is thus constructed by the regular repetition of this object (shown in dark gray in Figure 1.6b). Note that the equally spaced planes also generate a set of three-dimensional cells within which the objects are enclosed. When the repeating object is considered as an imaginary point, the three-dimensional periodic structure transforms into its corresponding point lattice (Figure 1.6c).[2]

Figures 1.6d–f show the same scheme, but applied to a slightly different three-dimensional periodic structure made of rods connected at right angles. It can be seen that the point lattice associated with this structure is the same as the

---

2) In analogy with the two-dimensional case, the location (relative to the underlying structure) of the three sets of equally spaced parallel planes that enclose the repeating object in Figure 1.6b is arbitrary. Therefore, the definition of the object that repeats regularly in space is arbitrary and the periodic structure can alternatively be described by different choices for the repeating object. However, the corresponding three-dimensional point lattice is independent on the choice of the repeating object.

3-D periodic structure 3-D point lattice

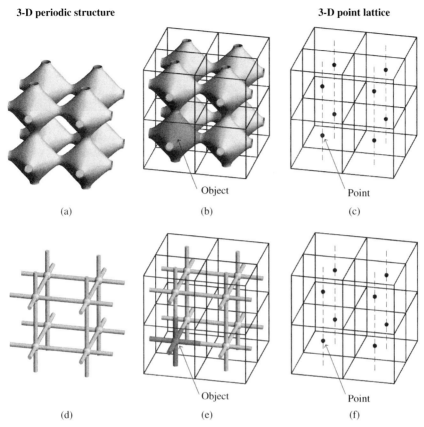

Fig. 1.6 A three-dimensional periodic structure and its point lattice. (a and d) Two different three-dimensional periodic structures. (b and e) The structures are divided by equally spaced parallel planes that define the object that is repeated. (c and f) The object is represented by an imaginary point and the three-dimensional periodic structures transform into their corresponding point lattices.

one corresponding to the previous structure. This is due to the fact that the objects corresponding to each structure repeat similarly in space.

In analogy with the two-dimensional case, three-dimensional point lattices can be displayed in various ways. For example, the point lattice shown in Figure 1.7a is displayed with the points located at the center of the cells, but it can also optionally be displayed with the points located at the corners of the cells (Figure 1.7b). In both cases, the point lattice remains the same. To be consistent with the two-dimensional case, we choose to represent this particular three-dimensional point lattice with the points located at the corners of the cells (Figure 1.7b).

Because the three-dimensional cells shown in Figure 1.7b are equivalent, we can arbitrarily choose any of these cells as a *unit cell* that repeats to form the point lattice. In this particular example, the edges of the unit cells are mutually orthogonal and have equal lengths. In the general case, however, the size and shape of the unit

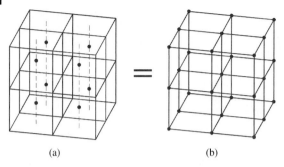

Fig. 1.7 A three-dimensional point lattice with the lattice points located (a) at the center of the cells and (b) at the corner of the cells.

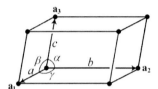

Fig. 1.8 A general three-dimensional unit cell defined by three nonorthogonal vectors a_1, a_2, and a_3.

cell are defined by three nonorthogonal *primitive vectors* a_1, a_2, and a_3 (Figure 1.8), where the lengths a, b, and c and the angles α, β, and γ are called the *lattice constants* of the three-dimensional unit cell.

As mentioned in Section 1.2, the primitive vectors can be used to construct the point lattice by translation. For example, a three-dimensional point lattice can be described by an array of regularly spaced points located at the tip of the vector

$$\mathbf{R} = n_1 \mathbf{a}_1 + n_2 \mathbf{a}_2 + n_3 \mathbf{a}_3 \tag{1.2}$$

where n_1, n_2, and n_3 are arbitrary integer numbers. By choosing different values for n_1, n_2, and n_3, the vector **R** fills the space with regularly spaced points and generates the corresponding three-dimensional point lattice.

In 1848, the French physicist Bravais demonstrated that there are only 14 possible arrangements of regularly spaced points in space, which satisfy the requirement that each point has identical surroundings. These point lattices were named *Bravais lattices* in his honor. This means that, although the number of conceivable three-dimensional periodic structures is infinite, when three-dimensional periodic structures are transformed into their corresponding point lattices only 14 distinct point lattices are found. Figure 1.9 displays the 14 Bravais lattices using unit cells that are adopted by convention. These *conventional unit cells* easily build up the Bravais lattices by repeating themselves in space, but conventional unit cells do not necessarily have points only at the corners (Figure 1.9). The connection between

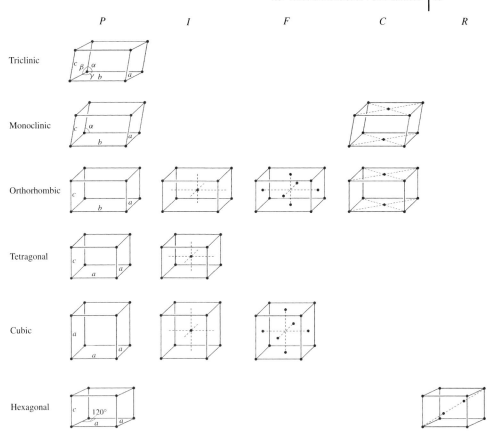

Fig. 1.9 The 14 Bravais lattices. The lattices are grouped in rows according to the shape of the conventional unit cell: triclinic, monoclinic, orthorhombic, tetragonal, cubic, and hexagonal. The six distinct shapes of conventional unit cells define the so-called *six crystal systems*. The lattices are also grouped in columns according to the location of the points within the conventional unit cells.

conventional unit cells and primitive unit cells (which have points only at the corners) is explained in Section 1.3.1.

The 14 Bravais lattices are organized into six *crystal systems*: triclinic, monoclinic, orthorhombic, tetragonal, cubic, and hexagonal. The size and shape of the conventional unit cells used to define the six crystal systems are determined by the relative lengths a, b, and c of the edges of the unit cell and the interaxial angles α, β, and γ (Figure 1.9). These values are given as follows:

- triclinic, $\quad a \neq b \neq c, \quad \alpha \neq \beta \neq \gamma$
- monoclinic, $\quad a \neq b \neq c, \quad \alpha \neq 90° = \beta = \gamma$
- orthorhombic, $\quad a \neq b \neq c, \quad \alpha = \beta = \gamma = 90°$
- tetragonal, $\quad a = b \neq c, \quad \alpha = \beta = \gamma = 90°$

- cubic, $a = b = c$, $\alpha = \beta = \gamma = 90°$
- hexagonal, $a = b \neq c$, $\alpha = \beta = 90°, \gamma = 120°$

Note: for simplicity sake, interaxial angles which are equal to 90° are not marked in Figure 1.9.

We can observe different types of conventional unit cells within each crystal system. For example, in the first column the conventional unit cells have points only at the corners of the unit cells. These cells are called *primitive unit cells* and they are designated by the symbol P. In the second column, the conventional unit cells have an additional point at the center of the unit cell. In this case, the cells are called *body-centered unit cells* and they are designated by the symbol I. In the third column, the cells have additional points at the centers of each face and they are called *face-centered unit cells* and designated by the symbol F. In the fourth column, the cells have additional points at the center of two faces. They are called *side-centered unit cells* and are designated by the symbol C. Finally, in the last column, the cell contains two additional points located at equal distances along a diagonal of the conventional unit cell. This cell is called a *rhombohedral unit cell* and is designated by the symbol R.

We now consider the number of points contained within each type of conventional unit cell. For instance, primitive (P) unit cells have points located only at the corners of the unit cells. Since the fractional parts of the points at the corners of the cell always sum to one point within the cell, primitive unit cells have one point per unit cell. Body-centered (I) unit cells have an additional point at the center. Hence, this type of unit cell has two points per unit cell. Face-centered (F) unit cells have additional points located at the faces, each of which contributes in half to the total number of points. Because there are six faces, the total number of points within the face-centered cells is three (from the faces) + one (from the corners) = four. By following the same reasoning, side-centered (C) unit cells have two points per unit cell, whereas rhombohedral (R) unit cells have three points per unit cell. Note that the number of points per unit cell is equivalent to the number of objects that are included in the conventional unit cells.

It is important to mention that *any* three-dimensional periodic structure will have one of the 14 Bravais lattices associated with it. To find the Bravais lattice that corresponds to a particular structure, it is only necessary to replace the repeating object in the structure by an imaginary point. Therefore, as we did in the case of two-dimensional periodic structures, *we can classify the infinite set of three-dimensional periodic structures by their spatial periodicity into 14 categories.* On the basis of how the object in the structure repeats in space, the 14 Bravais lattices provide a powerful framework to classify three-dimensional periodic structures.

1.3.1
Primitive and Nonprimitive Unit Cells

We conclude the description of the Bravais lattices by observing the arbitrariness in the choice of the conventional unit cells shown in Figure 1.9. It is important to remark that the arrangement of the regularly spaced points in each of the 14 Bravais lattices is fixed and unique. That is, each Bravais lattice describes a distinct distribution of points in space. However, the unit cell that can be used to graphically represent a Bravais lattice is absolutely arbitrary. We would like to stress the point that the unit cells used to represent the Bravais lattices, shown in Figure 1.9, are chosen by convention to help visualize the symmetric distribution of points (and this explains why they are called *conventional unit cells*).

By definition, the unit cell must construct the Bravais lattice by repeating itself in space. However, nothing is said about the number of points that must be contained within the unit cell. If the unit cell precisely contains a single lattice point, it is called a *primitive unit cell*. For example, unit cells with lattice points only at the corners of the cell are primitive unit cells (first column in Figure 1.9). On the other hand, if the unit cell contains more than one lattice point within the cell, it is called a *nonprimitive unit cell*. For example, the body-centered (*I*), face-centered (*F*), side-centered (*C*), and rhombohedral (*R*) unit cells (Figure 1.9) are nonprimitive unit cells because they contain two, four, two, and three points per unit cell, respectively. Therefore, the conventional unit cells used to represent the 14 Bravais lattices, in Figure 1.9, are primitive in some cases and nonprimitive in others.

It is important to mention, however, that for *any* Bravais lattice we can always find a primitive unit cell to represent the lattice. As an example, we examine in Figure 1.10 the three cubic Bravais lattices. In the first case, the conventional unit cell is a primitive cell, whereas in the case of the face-centered-cubic and body-centered-cubic Bravais lattices, the conventional unit cells are nonprimitive cells. In spite of this, primitive unit cells that describe the corresponding lattices for these two latter cases can be found. Figure 1.10 shows both the conventional nonprimitive and the primitive unit cell for the face-centered-cubic and body-centered-cubic Bravais lattices. The large cubes in solid lines define the conventional nonprimitive unit cells (which contain more than one lattice point), whereas the gray rhombs with six parallelogram faces define the primitive unit cells (which contain a single lattice point). When repeated in space, both the unit cells (nonprimitive and primitive) build up exactly the same lattice. The reason to use the conventional nonprimitive unit cell instead of the primitive unit cell is the fact that the former helps visualize the symmetric distribution of the points in the lattice.

Figure 1.10 also shows the primitive vectors that define the primitive unit cells and generate the corresponding cubic Bravais lattices by the use of Equation 1.2.

To summarize, in the last two sections, we have developed a powerful framework that allows us to classify *any* periodic structure based on its periodicity in space. In the case of two-dimensional structures, in whatever shape or geometry the structure is made, it can be classified into one of the five two-dimensional point lattices (Figure 1.5). On the other hand, in the three-dimensional case, independent of shape or geometry, a three-dimensional structure can always be classified into one of the 14 Bravais lattices (Figure 1.9).

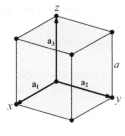

Simple cubic lattice

$$\mathbf{a}_1 = a\,\hat{\mathbf{x}}$$

$$\mathbf{a}_2 = a\,\hat{\mathbf{y}}$$

$$\mathbf{a}_3 = a\,\hat{\mathbf{z}}$$

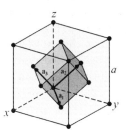

Face-centered-cubic lattice

$$\mathbf{a}_1 = \frac{a}{2}\hat{\mathbf{x}} + \frac{a}{2}\hat{\mathbf{y}}$$

$$\mathbf{a}_2 = \frac{a}{2}\hat{\mathbf{y}} + \frac{a}{2}\hat{\mathbf{z}}$$

$$\mathbf{a}_3 = \frac{a}{2}\hat{\mathbf{z}} + \frac{a}{2}\hat{\mathbf{x}}$$

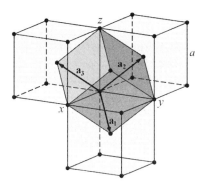

Body-centered-cubic lattice

$$\mathbf{a}_1 = +\frac{a}{2}\hat{\mathbf{x}} + \frac{a}{2}\hat{\mathbf{y}} - \frac{a}{2}\hat{\mathbf{z}}$$

$$\mathbf{a}_2 = -\frac{a}{2}\hat{\mathbf{x}} + \frac{a}{2}\hat{\mathbf{y}} + \frac{a}{2}\hat{\mathbf{z}}$$

$$\mathbf{a}_3 = +\frac{a}{2}\hat{\mathbf{x}} - \frac{a}{2}\hat{\mathbf{y}} + \frac{a}{2}\hat{\mathbf{z}}$$

Fig. 1.10 Simple cubic, face-centered-cubic, and body-centered-cubic Bravais lattices. The face-centered-cubic and body-centered-cubic Bravais lattices can be equivalently represented by a conventional nonprimitive unit cell (large cube) or a primitive unit cell (gray rhomb). The primitive vectors, which define the primitive unit cells, are in the right column.

1.4
Mathematical Description of Periodic Structures

Now that we have a graphic picture of translationally periodic structures we next focus on a mathematical grounding. In the next sections, we describe periodic structures by using analytical formulas. As we show in subsequent chapters, this analytical approach allows us to design periodic structures that can be fabricated by a low-cost experimental technique known as *interference lithography*.

1.4 Mathematical Description of Periodic Structures

We have shown that the point lattices determine the positions of the object that repeats in space to form two- and three-dimensional periodic structures. In fact, the object is considered to be a point in the abstract lattice model. In this section, we consider the repeating object as a full component having a definite shape. That is, *to classify periodic structures in terms of translational spatial periodicities, the repeating object is considered as an imaginary point, whereas to describe analytically periodic structures the whole object is considered.*

We are interested in periodic structures made of two different constituent materials. These types of structures are called *binary composite structures*. The constituent materials A and B can be *any* type of material (including solids and/or fluids). For example, the composite structure can be made of two different solid materials or of a combination of solid and fluid materials. The aim of this section is to present a method that permits us to quantitatively describe the distribution of materials A and B in a periodic structure by using an analytical function.

Consider a real function $f(x, y)$ defined within a two-dimensional unit cell and assume that the function $f(x, y)$ is positive in some regions and negative in others. By considering that the regions for which $f(x, y)$ is greater than zero are made of material A and that the regions for which $f(x, y)$ is less than zero are made of material B, we have a simple scheme to define the distribution of materials A and B within the unit cell. In fact, the contour line $f(x, y) = 0$ divides the two-dimensional unit cell in regions filled with materials A and B, respectively. Since the two-dimensional unit cell repeats indefinitely in space to form the infinite periodic structure, we are defining the distribution of materials A and B for the whole periodic structure. Analogously, in the case of three-dimensional periodic structures, the scheme can be applied by considering a real function $f(x, y, z)$ and assuming that if $f(x, y, z)$ is greater than zero, the space is filled with material A, whereas if $f(x, y, z)$ is less than zero, the space is filled with material B. In this case, the three-dimensional surface $f(x, y, z) = 0$ divides the space into regions filled with materials A and B.

Therefore, the distribution of materials A and B within the unit cell of a periodic structure (and consequently within the whole structure) is defined as follows:

Two-dimensional structure

If $\quad f(x, y) > 0 \Rightarrow$ Material A $\quad\quad\quad\quad$ (1.3a)

If $\quad f(x, y) < 0 \Rightarrow$ Material B $\quad\quad\quad\quad$ (1.3b)

Three-dimensional structure

If $\quad f(x, y, z) > 0 \Rightarrow$ Material A $\quad\quad\quad\quad$ (1.4a)

If $\quad f(x, y, z) < 0 \Rightarrow$ Material B $\quad\quad\quad\quad$ (1.4b)

Equations 1.3 and 1.4 determine that the analytical description of a periodic structure is now dependent on the definition of the real function $f(x, y, z)$ within the unit cell of structure. (Note: in the rest of this section we focus on the three-dimensional case.)

There exist two inverse situations. On the one hand, we can have some specific periodic structure that is to be described analytically and we want to find the real function $f(x, y, z)$ that represents the structure by the use of Equations 1.4. On the other hand, we can have a real function $f(x, y, z)$ defined within a unit cell and we want to find the geometry of the corresponding periodic structure, which is determined by Equations 1.4. These two inverse situations are as follows:

Periodic structure $\longrightarrow$ corresponding real function $f(x, y, z)$
Real function $f(x, y, z) \longrightarrow$ corresponding periodic structure

To better understand this important concept, we next consider one example of each case. We first show the case in which we want to find the function $f(x, y, z)$ that describes a particular periodic structure. For example, consider a periodic structure made of solid spheres in air as in Figure 1.11a. This structure is shown in Figure 1.11b together with the planes that define the repeating object (the sphere) and the unit cells. The corresponding simple cubic Bravais lattice is shown in Figure 1.11c.

As previously mentioned, we need to describe the structure only within a single unit cell because the unit cell repeats itself in space to form the periodic structure. Figure 1.12a shows the unit cell of this structure in a Cartesian coordinate system. The center of the sphere is located at $(x_0, y_0, z_0) = (a/2, a/2, a/2)$, where a is the lattice constant of the cubic unit cell. Because the spheres in Figure 1.11a are in contact, the radius of the sphere in Figure 1.12a is $R = a/2$.

We consider that the sphere is made of material B, whereas the background is made of material A. Therefore, the set of points (x, y, z) in the unit cell for which the distance d to the center of the sphere (x_0, y_0, z_0) is less than the radius R form

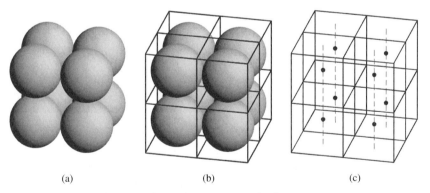

(a) (b) (c)

Fig. 1.11 (a) A three-dimensional periodic structure made of solid spheres in air. (b) The structure is divided by equally spaced parallel planes that define the basic object (which in this case is a sphere) and the unit cells. (c) The corresponding simple cubic Bravais lattice. Note that there is one lattice point per unit cell, but our choice here is nonconventional.

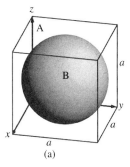 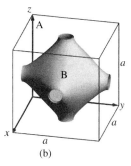

Fig. 1.12 Analytical representation of periodic structures. The distribution of materials A and B within the unit cell of the structure is determined by a function $f(x, y, z)$ together with the inequalities $f(x, y, z) > 0 \Rightarrow$ material A; $f(x, y, z) < 0 \Rightarrow$ material B. (a) $f(x, y, z) = \sqrt{(x - x_0)^2 + (y - y_0)^2 + (z - z_0)^2} - R$, where $(x_0, y_0, z_0) = (a/2, a/2, a/2)$ and $R = a/2$. (b) $f(x, y, z) = \cos(2\pi x/a) + \cos(2\pi y/a) + \cos(2\pi z/a) + 0.85$.

the region filled with material B. On the other hand, the set of points for which the distance d is larger than the radius R form the region filled with material A. This means that the distribution of materials A and B within the unit cell of the structure can be described analytically as follows:

$$\text{If} \quad d = \sqrt{(x - x_0)^2 + (y - y_0)^2 + (z - z_0)^2} > R \Rightarrow \text{Material A} \tag{1.5a}$$

$$\text{If} \quad d = \sqrt{(x - x_0)^2 + (y - y_0)^2 + (z - z_0)^2} < R \Rightarrow \text{Material B} \tag{1.5b}$$

By defining the function

$$f(x, y, z) = \sqrt{(x - x_0)^2 + (y - y_0)^2 + (z - z_0)^2} - R \tag{1.6}$$

Equations 1.5 transform into

$$\text{If} \quad f(x, y, z) > 0 \Rightarrow \text{Material A} \tag{1.7a}$$

$$\text{If} \quad f(x, y, z) < 0 \Rightarrow \text{Material B} \tag{1.7b}$$

Therefore, we have succeeded in finding an analytical function $f(x, y, z)$ that describes the periodic structure within the unit cell. We note that this illustrative example is simple due to the fact that the periodic structure is made of spheres. Periodic structures having more complex geometries will certainly require more complicated analytical formulas.

We now consider the inverse case, which is the case frequently found in this book. We have an analytical function $f(x, y, z)$ defined within a unit cell and we want to find the geometry of the periodic structure, which is defined by Equations 1.4. As an illustrative example, consider a function $f(x, y, z)$ made of the sum of three cosine terms plus a constant defined within a cubic unit cell of lattice constant a

$$f(x, y, z) = \cos\left(\frac{2\pi x}{a}\right) + \cos\left(\frac{2\pi y}{a}\right) + \cos\left(\frac{2\pi z}{a}\right) + 0.85 \qquad (1.8)$$

In general, to display the geometry of the periodic structure, we need to use some mathematical software such as Maple, Mathematica, or Matlab. In Figure 1.12b, we show the distribution of materials A and B within the unit cell of the periodic structure defined by Equations 1.4 and the analytical function given in Equation 1.8. In this inverse case, we were thus able to graphically find the geometry of a periodic structure given its analytical formula.

In general, *any* real function $f(x, y, z)$ that changes sign within a unit cell will have associated with it a particular three-dimensional periodic structure defined by Equations 1.4. Owing to the fact that the set of possible analytical functions $f(x, y, z)$ is infinite, the number of periodic structures that can be created by using this scheme is also infinite. In spite of this, in Chapter 2 our main goal is to establish a scheme to *systematically create analytical functions $f(x, y, z)$ that represent periodic structures* with the further objective of fabricating these structures by interference lithography. To do this, however, we first need to introduce the Fourier series expansion of a periodic function $f(x, y, z)$, which is a powerful analytical method that is extensively used in Chapter 2 to represent periodic structures.

1.5
Fourier Series

Fourier series is basically a powerful, yet simple mathematical technique that allows us to represent an arbitrary periodic function as a weighted sum of cosine and sine functions. By using this technique, each periodic function has an associated distinctive set of weighted coefficients (or Fourier coefficients) that univocally represent the periodic function. In the next section, we define the Fourier coefficients and show how to calculate them for arbitrary two- and three-dimensional periodic functions.

1.5.1
Fourier Series for Two-dimensional Periodic Functions

Suppose we have a two-dimensional real function $f(x, y)$ defined within a rectangular unit cell of sides a and b. Since the unit cell repeats itself in space to form the periodic structure (Figure 1.13), the function $f(x, y)$ also repeats itself over and over along the x and y directions at intervals of a and b, respectively. This means that

$$f(x + a, y) = f(x, y) \qquad (1.9a)$$
$$f(x, y + b) = f(x, y) \qquad (1.9b)$$

and the function $f(x, y)$ is said to be *periodic* in two dimensions.

1.5 Fourier Series

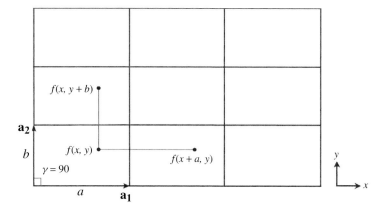

Fig. 1.13 Two-dimensional periodic functions. A function $f(x, y)$ defined within a rectangular unit cell repeats at intervals a and b along the x- and y-axis coordinates, that is, $f(x + a, y) = f(x, y + b) = f(x, y)$. As a result, the function $f(x, y)$ is periodic and it can be represented by its Fourier series expansion.

In general, a two-dimensional *periodic* function $f(x, y)$ can be written as an infinite sum of cosine and sine functions as

$$f(x,y) = a_{00} + \sum_n \sum_m \left\{ a_{nm} \cos\left[2\pi \left(\frac{nx}{a} + \frac{my}{b}\right)\right] \right.$$
$$\left. + b_{nm} \sin\left[2\pi \left(\frac{nx}{a} + \frac{my}{b}\right)\right] \right\} \tag{1.10}$$

where n and m are integer numbers, a_{00}, a_{nm}, and b_{nm} are real constants, and a and b are the lengths of the sides of the rectangular unit cell. Equation 1.10 is called the *Fourier series expansion* of the periodic function $f(x, y)$, and the constants a_{00}, a_{nm}, and b_{nm} are called the *Fourier coefficients*.

Equation 1.10 establishes that each periodic function $f(x, y)$ has an associated set of Fourier coefficients $\{a_{00}, a_{nm}, b_{nm}\}$. If the formula of the periodic function $f(x, y)$ is known, these coefficients can be obtained by calculating the following integrals.

1.5.1.1 Calculation of a_{00}

By integrating both sides of Equation 1.10 over the area of the rectangular unit cell, we have

$$\int_0^a \int_0^b f(x,y)\,dx\,dy = \int_0^a \int_0^b \left\{ a_{00} + \sum_n \sum_m \left\{ a_{nm} \cos\left[2\pi \left(\frac{nx}{a} + \frac{my}{b}\right)\right] \right.\right.$$
$$\left.\left. + b_{nm} \sin\left[2\pi \left(\frac{nx}{a} + \frac{my}{b}\right)\right] \right\} \right\} dx\,dy$$
$$\tag{1.11a}$$

or equivalently

$$\int_0^a \int_0^b f(x,y)\,dx\,dy = a_{00} \int_0^a \int_0^b dx\,dy + \sum_n \sum_m \left\{ a_{nm} \int_0^a \int_0^b \right.$$
$$\left. \times \cos\left[2\pi \left(\frac{nx}{a} + \frac{my}{b}\right)\right] dx\,dy + b_{nm} \int_0^a \int_0^b \sin\left[2\pi \left(\frac{nx}{a} + \frac{my}{b}\right)\right] dx\,dy \right\}$$

(1.11b)

Because the integrals that contain either cosine or sine functions are equal to zero, independent of the values of n and m, Equation 1.11b determines that

$$a_{00} = \frac{1}{ab} \int_0^a \int_0^b f(x,y)\,dx\,dy \qquad (1.11c)$$

1.5.1.2 Calculation of a_{nm}

Multiplying Equation 1.10 by $\cos\left[2\pi\left(\frac{n'x}{a} + \frac{m'y}{b}\right)\right]$ and integrating over the area of the rectangular unit cell, we have

$$\int_0^a \int_0^b f(x,y) \cos\left[2\pi\left(\frac{n'x}{a} + \frac{m'y}{b}\right)\right] dx\,dy$$
$$= \int_0^a \int_0^b \left\{ a_{00} + \sum_n \sum_m \left\{ a_{nm} \cos\left[2\pi\left(\frac{nx}{a} + \frac{my}{b}\right)\right] \right.\right.$$
$$\left.\left. + b_{nm} \sin\left[2\pi\left(\frac{nx}{a} + \frac{my}{b}\right)\right] \right\} \right\} \cos\left[2\pi\left(\frac{n'x}{a} + \frac{m'y}{b}\right)\right] dx\,dy \quad (1.12a)$$

The integrals on the right-hand side of Equation 1.12a are given by

$$\int_0^a \int_0^b \cos\left[2\pi\left(\frac{n'x}{a} + \frac{m'y}{b}\right)\right] dx\,dy = 0 \quad \text{for all } n', m' \qquad (1.12b)$$

$$\int_0^a \int_0^b \cos\left[2\pi\left(\frac{nx}{a} + \frac{my}{b}\right)\right] \cos\left[2\pi\left(\frac{n'x}{a} + \frac{m'y}{b}\right)\right] dx\,dy = \frac{ab}{2} \delta_{n,n'} \delta_{m,m'}$$

(1.12c)

$$\int_0^a \int_0^b \sin\left[2\pi\left(\frac{nx}{a} + \frac{my}{b}\right)\right] \cos\left[2\pi\left(\frac{n'x}{a} + \frac{m'y}{b}\right)\right] dx\,dy = 0$$
for all n, n', m, m' (1.12d)

where $\delta_{n,n'}$ is the delta function with properties $\delta_{n,n'} = 1$ if $n = n'$ and $\delta_{n,n'} = 0$ if $n \neq n'$.

By replacing Equations 1.12b–d in Equation 1.12a, the coefficients a_{nm} can be calculated by the integral

$$a_{nm} = \frac{2}{ab} \int_0^a \int_0^b f(x,y) \cos\left[2\pi\left(\frac{nx}{a} + \frac{my}{b}\right)\right] dx\,dy \qquad (1.12e)$$

1.5.1.3 Calculation of b_{nm}

Analogously, multiplying Equation 1.10 by $\sin\left[2\pi\left(\frac{n'x}{a} + \frac{m'y}{b}\right)\right]$ and integrating over the area of the rectangular unit cell, we have

$$\int_0^a \int_0^b f(x,y) \sin\left[2\pi\left(\frac{n'x}{a} + \frac{m'y}{b}\right)\right] dx\,dy$$

$$= \int_0^a \int_0^b \left\{ a_{00} + \sum_n \sum_m \left\{ a_{nm} \cos\left[2\pi\left(\frac{nx}{a} + \frac{my}{b}\right)\right]\right.\right.$$

$$\left.\left. + b_{nm} \sin\left[2\pi\left(\frac{nx}{a} + \frac{my}{b}\right)\right]\right\}\right\} \sin\left[2\pi\left(\frac{n'x}{a} + \frac{m'y}{b}\right)\right] dx\,dy \quad (1.13a)$$

In this case, the integrals on the right-hand side of Equation 1.13a are given by

$$\int_0^a \int_0^b \sin\left[2\pi\left(\frac{n'x}{a} + \frac{m'y}{b}\right)\right] dx\,dy = 0 \quad \text{for all } n', m' \quad (1.13b)$$

$$\int_0^a \int_0^b \cos\left[2\pi\left(\frac{nx}{a} + \frac{my}{b}\right)\right] \sin\left[2\pi\left(\frac{n'x}{a} + \frac{m'y}{b}\right)\right] dx\,dy = 0$$

$$\text{for all } n, n', m, m' \quad (1.13c)$$

$$\int_0^a \int_0^b \sin\left[2\pi\left(\frac{nx}{a} + \frac{my}{b}\right)\right] \sin\left[2\pi\left(\frac{n'x}{a} + \frac{m'y}{b}\right)\right] dx\,dy = \frac{ab}{2}\delta_{n,n'}\delta_{m,m'} \quad (1.13d)$$

By replacing Equations 1.13b–d in Equation 1.13a, the coefficients b_{nm} can be calculated by the integral

$$b_{nm} = \frac{2}{ab}\int_0^a \int_0^b f(x,y) \sin\left[2\pi\left(\frac{nx}{a} + \frac{my}{b}\right)\right] dx\,dy \quad (1.13e)$$

In summary, Equations 1.11c, 1.12e, and 1.13e determine the set of Fourier coefficients $\{a_{00}, a_{nm}, b_{nm}\}$ associated with the periodic function $f(x,y)$. This set of coefficients univocally define the periodic function $f(x,y)$.

1.5.2
Fourier Series for Three-dimensional Periodic Functions

In complete analogy with the two-dimensional case, consider a three-dimensional real function $f(x,y,z)$ defined within an orthorhombic unit cell of sides a, b, and c. Since the unit cell repeats itself in space to form the periodic structure (Figure 1.14), the function $f(x,y,z)$ also repeats itself over and over along the x, y, and z directions and we have

$$f(x+a, y, z) = f(x, y, z) \quad (1.14a)$$
$$f(x, y+b, z) = f(x, y, z) \quad (1.14b)$$
$$f(x, y, z+c) = f(x, y, z) \quad (1.14c)$$

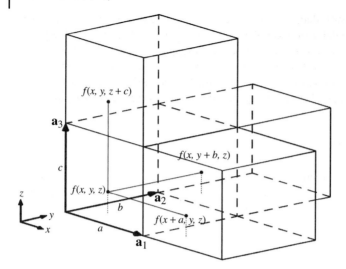

Fig. 1.14 Three-dimensional periodic functions. A function $f(x, y, z)$ defined within an orthorhombic unit cell repeats at intervals a, b, and c along the x-, y-, and z-axis coordinates, that is, $f(x + a, y, z) = f(x, y + b, z) = f(x, y, z + c) = f(x, y, z)$. As a result, the function is periodic and it can be represented by its Fourier series expansion.

Because the function $f(x, y, z)$ is periodic in three dimensions, it can be written as an infinite sum of cosine and sine functions as

$$f(x, y, z) = a_{000} + \sum_n \sum_m \sum_p \left\{ a_{nmp} \cos\left[2\pi \left(\frac{nx}{a} + \frac{my}{b} + \frac{pz}{c}\right)\right] \right.$$
$$\left. + b_{nmp} \sin\left[2\pi \left(\frac{nx}{a} + \frac{my}{b} + \frac{pz}{c}\right)\right] \right\} \quad (1.15)$$

where n, m, and p are integer numbers, a_{000}, a_{nmp}, and b_{nmp} are the Fourier coefficients, and a, b, and c are the lengths of the edges of the orthorhombic unit cell.

In analogy with the two-dimensional case, the Fourier coefficients a_{000}, a_{nmp}, and b_{nmp} can be calculated by the integrals

$$a_{000} = \frac{1}{abc} \int_0^a \int_0^b \int_0^c f(x, y, z)\, dx\, dy\, dz \quad (1.16a)$$

$$a_{nmp} = \frac{2}{abc} \int_0^a \int_0^b \int_0^c f(x, y, z) \cos\left[2\pi \left(\frac{nx}{a} + \frac{my}{b} + \frac{pz}{c}\right)\right] dx\, dy\, dz \quad (1.16b)$$

$$b_{nmp} = \frac{2}{abc} \int_0^a \int_0^b \int_0^c f(x, y, z) \sin\left[2\pi \left(\frac{nx}{a} + \frac{my}{b} + \frac{pz}{c}\right)\right] dx\, dy\, dz \quad (1.16c)$$

For the sake of simplicity, in the last two sections, we considered two- and three-dimensional periodic functions defined within unit cells with mutually orthogonal edges. This means that, in the two-dimensional case, the periodic functions must be defined within rectangular or square unit cells (Figure 1.5), whereas in the three-dimensional case, the functions must be defined within orthorhombic, tetragonal, or cubic unit cells (Figure 1.9). In the next section, we briefly discuss the Fourier series expansion of a periodic function defined within an arbitrary unit cell, where the angles between the primitive vectors can have arbitrary values.

1.5.3
Arbitrary Unit Cells

Consider a three-dimensional real function $f(\mathbf{r})$ defined within an arbitrary unit cell given by the primitive vectors $\mathbf{a}_1$, $\mathbf{a}_2$, and $\mathbf{a}_3$ (Figure 1.8). Here, $\mathbf{r}$ is the position vector $\mathbf{r} = (x, y, z)$. Because the unit cell repeats indefinitely in space, the function $f(\mathbf{r})$ is periodic and satisfies

$$f(\mathbf{r}) = f(\mathbf{r} + \mathbf{R}) \qquad (1.17)$$

where $\mathbf{R} = n_1\mathbf{a}_1 + n_2\mathbf{a}_2 + n_3\mathbf{a}_3$, and n_1, n_2, and n_3 are arbitrary integer numbers.

A periodic function $f(\mathbf{r})$ defined within an arbitrary unit cell can be written as an infinite sum of cosine and sine functions as

$$f(\mathbf{r}) = a_{000} + \sum_n \sum_m \sum_p \left\{ a_{nmp} \cos\left[\mathbf{G}_{nmp} \cdot \mathbf{r}\right] + b_{nmp} \sin\left[\mathbf{G}_{nmp} \cdot \mathbf{r}\right] \right\} \qquad (1.18)$$

where n, m, and p are integer numbers, a_{000}, a_{nmp}, and b_{nmp} are the Fourier coefficients, and $\mathbf{G}_{nmp}$ are called the *reciprocal lattice vectors*. The $\mathbf{G}_{nmp}$ vectors are, in fact, determined by the primitive vectors $\mathbf{a}_1$, $\mathbf{a}_2$, and $\mathbf{a}_3$ that define the primitive unit cell. The calculation of the reciprocal lattice vectors $\mathbf{G}_{nmp}$ is introduced in Chapter 6, where general formulas are provided to obtain these vectors from arbitrary unit cells. Equation 1.18 thus represents the Fourier series expansion of a periodic function $f(\mathbf{r})$ defined within an arbitrary unit cell.

To summarize, in this chapter, we introduced the reader to the concept of periodic structures by examining different theoretical aspects. We first examined the notion of spatial periodicity and introduced the point lattice model, which allows us to classify periodic structures depending on how the object that forms the structure repeats itself in space. It turns out that two-dimensional periodic structures can be classified into five two-dimensional point lattices, whereas three-dimensional periodic structures can be classified into 14 three-dimensional Bravais lattices. Then, we examined how periodic structures can be represented by analytical functions. We showed that these analytical functions can in fact be written as a sum of trigonometric terms known as the *Fourier series expansion*. The Fourier series expansion of periodic functions will be the basis to systematically create analytical functions representing periodic structures. We elaborate the procedure in Chapter 2, where we also graphically display the corresponding periodic structures obtained by using this scheme. From this formulation, we arrive at a set of periodic structures that can be fabricated by a currently available experimental technique known as *interference lithography* and that have applications on a variety of fields that include optics, acoustics, and mechanics.

Further Reading

1. Buerger, M. (1978) *Elementary Crystallography: An Introduction to the Fundamental Geometric Features of Crystals*, MIT Press, Massachusetts.
2. Kittel, C. (1996) *Introduction to Solid State Physics*, John Wiley & Sons, Ltd, New York.
3. Ashcroft, N.W. and Mermin, N.D. (1976) *Solid State Physics*, Saunders College Publishing, New York.
4. Cullity, B.D. (1978) *Elements of X-Ray Diffraction*, Addison-Wesley, Massachusetts, Chapter 2.
5. Allen, S.M. and Thomas, E.L. (1999) *The Structure of Materials*, John Wiley & Sons, Ltd, New York.

Problems

1.1 Rectangular lattice

(a) Determine the two-dimensional primitive vectors a_1 and a_2 for a rectangular lattice of sides $a = 1.5$ cm and $b = 1$ cm.

(b) By using Equation 1.1, determine the (x, y) positions of the eight nearest points to the origin $(0, 0)$.

(c) Draw the lattice.

1.2 Triangular lattice

(a) Determine the two-dimensional primitive vectors a_1 and a_2 for a triangular lattice of side $a = 1$ cm.

(b) By using Equation 1.1, determine the (x, y) positions of the six nearest points to the origin $(0, 0)$.

(c) Draw the lattice.

1.3 Why is the following two-dimensional periodic arrangement of points *not* a point lattice?

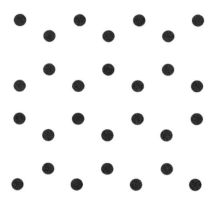

1.4 Honeycomb structure

(a) On the diagram below, draw two sets of parallel lines to identify and enclose a repeating object that forms the periodic structure.

(b) Replace the repeating object by a point and form the corresponding point lattice. What is the lattice associated with the honeycomb structure?

1.5 Nearest-neighbor points are the points in the Bravais lattice that are closest to a particular point.

(a) Show that the number of nearest-neighbors in the simple cubic, body-centered-cubic, and face-centered-cubic Bravais lattices is 6, 8, and 12, respectively.

(b) Demonstrate that the distance between nearest-neighbors for the simple cubic, body-centered-cubic, and face-centered-cubic Bravais lattices is a, $\sqrt{3}a/2$, and $\sqrt{2}a/2$, respectively, where a is the lattice constant of the cubic unit cell.

1.6 A diamond structure

(a) In the structure shown below, identify an object that repeats regularly in space to form the three-dimensional diamond structure.

(b) Replace the object by an imaginary point and show that the Bravais lattice associated with the diamond structure is the face-centered-cubic lattice.

Note: the cubic unit cell of the structure repeats infinitely in space.

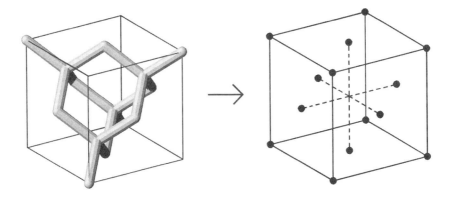

1.7 Find the function $f(x, y, z)$ that describes the three-dimensional periodic structure shown in Figure 1.6e.

1.8 Sketch the two-dimensional periodic structure described by the function $f(x, y) = \cos(2\pi x) + \cos(2\pi y)$.

1.9 Demonstrate the values of the Integrals 1.12b–d and 1.13b–d. Hint: use the following trigonometric identities

$$\cos\alpha \cos\beta = \tfrac{1}{2}[\cos(\alpha + \beta) + \cos(\alpha - \beta)]$$
$$\sin\alpha \sin\beta = \tfrac{1}{2}[\cos(\alpha - \beta) - \cos(\alpha + \beta)]$$
$$\sin\alpha \cos\beta = \tfrac{1}{2}[\sin(\alpha + \beta) + \sin(\alpha - \beta)]$$

1.10 Calculate the Fourier coefficients $\{a_{00}, a_{nm}, b_{nm}\}$ for a periodic function $f(x, y)$ defined on a square lattice of side a, where

$$f(x, y) = 1 \quad 0.25a < x < 0.75a \text{ and } 0.25a < y < 0.75a$$
$$f(x, y) = 0 \quad \text{otherwise}$$

2
Periodic Functions and Structures

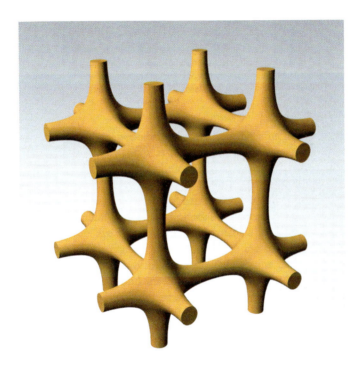

A periodic function can be decomposed into a weighted sum of cosine and sine functions known as the Fourier series expansion of the function. In this chapter, we use Fourier series expansions to systematically create periodic functions that represent two- and three-dimensional periodic structures. By making a few assumptions on the relative weights of the cosine and sine functions, we develop a method that allows us to create and classify periodic functions based on their Fourier series expansions.

Periodic Materials and Interference Lithography. M. Maldovan and E. Thomas
Copyright © 2009 WILEY-VCH Verlag GmbH & Co. KGaA, Weinheim
ISBN: 978-3-527-31999-2

2.1 Introduction

In Chapter 1, we showed that two- and three-dimensional periodic functions, defined in rectangular or orthorhombic unit cells respectively, can be written as an infinite sum of cosine and sine functions as

$$f(x,y) = a_{00} + \sum_n \sum_m \left\{ a_{nm} \cos\left[2\pi\left(\frac{nx}{a} + \frac{my}{b}\right)\right] \right.$$
$$\left. + b_{nm} \sin\left[2\pi\left(\frac{nx}{a} + \frac{my}{b}\right)\right] \right\} \quad (2.1a)$$

$$f(x,y,z) = a_{000} + \sum_n \sum_m \sum_p \left\{ a_{nmp} \cos\left[2\pi\left(\frac{nx}{a} + \frac{my}{b} + \frac{pz}{c}\right)\right] \right.$$
$$\left. + b_{nmp} \sin\left[2\pi\left(\frac{nx}{a} + \frac{my}{b} + \frac{pz}{c}\right)\right] \right\} \quad (2.1b)$$

where n, m, and p are integer numbers, a, b, and c are the lengths of the edges of the unit cells, and $\{a_{00}, a_{nm}, b_{nm}\}$ and $\{a_{000}, a_{nmp}, b_{nmp}\}$ are the set of Fourier coefficients corresponding to two- and three-dimensional periodic functions, respectively.

As mentioned earlier, Equations 2.1 determine that each periodic function has an associated unique set of Fourier coefficients. That is, given an arbitrary periodic function, we can find the associated Fourier coefficients by calculating the integrals introduced in Section 1.5. We therefore have

$$f(x,y) \rightarrow \{a_{00}, a_{nm}, b_{nm}\} \quad (2.2a)$$
$$f(x,y,z) \rightarrow \{a_{000}, a_{nmp}, b_{nmp}\} \quad (2.2b)$$

In this chapter, however, we are interested in systematically creating periodic functions (which represent periodic structures) by using Fourier series expansions and we therefore consider the inverse case. That is, given a specific set of Fourier coefficients we are interested in finding the corresponding periodic function.

$$\{a_{00}, a_{nm}, b_{nm}\} \rightarrow f(x,y) \quad (2.3a)$$
$$\{a_{000}, a_{nmp}, b_{nmp}\} \rightarrow f(x,y,z) \quad (2.3b)$$

If the Fourier coefficients are known, this can easily be done by replacing the set of Fourier coefficients in Equations 2.1, which gives us the formula of the periodic function in the form of a Fourier series expansion. We discuss this method in the subsequent sections. However, because the number of possible sets of Fourier coefficients is infinite, we need an organized scheme to systematically create periodic functions by using this method. In fact, our main goal in this chapter is to create *simple* periodic functions. By simple, we mean periodic functions for which the infinite sum in expressions 2.1 is reduced to the sum of only a few trigonometric terms. *It will turn out that many of the resultant periodic structures represented by these simple periodic functions can be fabricated by using interference lithography (Chapters 3–5) and have useful optical, acoustical, and mechanical properties (Chapters 6–8).*

In the subsequent sections, we show how to obtain these simple periodic functions and graphically display the corresponding periodic structures. We begin with the description of the method in the case of two-dimensional periodic functions and subsequently extend the description to the three-dimensional case.

2.2 Creating Simple Periodic Functions in Two Dimensions

2.2.1 The Square Lattice

In this section, we are interested in creating two-dimensional simple periodic functions $f(x, y)$ defined within square unit cells of side a (Figure 1.5d). By replacing $a = b$ in Equation 2.1a, the Fourier series expansion for these periodic functions is given as follows:

$$f(x, y) = a_{00} + \sum_{n} \sum_{m} \left\{ a_{nm} \cos\left[\frac{2\pi}{a}(nx + my)\right] + b_{nm} \sin\left[\frac{2\pi}{a}(nx + my)\right] \right\} \quad (2.4)$$

A reasonable option to systematically create simple periodic functions by the use of the Fourier series expansion (Equation 2.4) is to consider periodic functions $f(x, y)$ for which the sum in Equation 2.4 consists of cosine and sine functions *with the same spatial period*.

For example, the distance (or spatial period) λ at which the cosine or sine functions in Equation 2.4 repeat in space is given by $\lambda = a/\sqrt{n^2 + m^2}$. Therefore, if we define the parameter $d = n^2 + m^2$, the cosine and sine functions in Equation 2.4 with equal values of d have the same spatial period λ. However, it is important to mention that depending on the specific values of n and m, the spatial periodicity of the cosine and sine functions will be oriented along different directions in the two-dimensional space.

In Table 2.1, we sort the cosine and sine functions in Equation 2.4 by decreasing values of their period λ, which is equivalent to sorting the integer numbers n and m by increasing values of d.

Table 2.1 The integer numbers n and m corresponding to the sum in Equation 2.4 are sorted by increasing values of d, where $d = n^2 + m^2$

λ	d	n, m			
$a/\sqrt{1}$	1	1, 0	0, 1		
$a/\sqrt{2}$	2	1, 1	1, −1		
$a/\sqrt{4}$	4	2, 0	0, 2		
$a/\sqrt{5}$	5	1, 2	1, −2	2, 1	2, −1
⋮	⋮	⋮	⋮	⋮	⋮

In this manner, the infinite sum in the Fourier series expansion (Equation 2.4) can be organized into groups, which are labeled by the parameter d.

$$f(x, y) = a_{00} + \sum_{n,m}^{d=1} \left\{ a_{nm} \cos\left[\frac{2\pi}{a}(nx + my)\right] + b_{nm} \sin\left[\frac{2\pi}{a}(nx + my)\right] \right\}$$
$$+ \sum_{n,m}^{d=2} \left\{ a_{nm} \cos\left[\frac{2\pi}{a}(nx + my)\right] + b_{nm} \sin\left[\frac{2\pi}{a}(nx + my)\right] \right\}$$
$$+ \sum_{n,m}^{d=4} \left\{ a_{nm} \cos\left[\frac{2\pi}{a}(nx + my)\right] + b_{nm} \sin\left[\frac{2\pi}{a}(nx + my)\right] \right\} + \cdots \quad (2.5)$$

Note that each d group in Equation 2.5 has cosine and sine functions with the same spatial period (which is determined by the value of the parameter d). Also note that expression 2.5 is just another way to write the Fourier series expansion of a periodic function $f(x, y)$ in which cosine and sine functions with the same spatial period are grouped together.

As mentioned earlier, a route to systematically create simple periodic functions is to consider each group in Equation 2.5 separately. We denote the set of simple periodic functions created by each group as $f_i(x, y)$, where the subindex i indicates the corresponding value of the parameter d. For example,

$$f_1(x, y) = \sum_{n,m}^{d=1} \left\{ a_{nm} \cos\left[\frac{2\pi}{a}(nx + my)\right] + b_{nm} \sin\left[\frac{2\pi}{a}(nx + my)\right] \right\} \quad (2.6)$$

$$f_2(x, y) = \sum_{n,m}^{d=2} \left\{ a_{nm} \cos\left[\frac{2\pi}{a}(nx + my)\right] + b_{nm} \sin\left[\frac{2\pi}{a}(nx + my)\right] \right\} \quad (2.7)$$

$$f_4(x, y) = \sum_{n,m}^{d=4} \left\{ a_{nm} \cos\left[\frac{2\pi}{a}(nx + my)\right] + b_{nm} \sin\left[\frac{2\pi}{a}(nx + my)\right] \right\} \quad (2.8)$$

2.2 Creating Simple Periodic Functions in Two Dimensions

Because the coefficients a_{nm} and b_{nm} can have arbitrary values, each d group in Equations 2.6–2.8 creates an infinite set of simple periodic functions. As a means to obtain a *finite* set of simple periodic functions for each group, we make an assumption about the relative weights of the coefficients a_{nm} and b_{nm}. We consider that each cosine and sine function is equally relevant to the final expression of the simple periodic function. That is, we assume that the coefficients a_{nm} and b_{nm} have the same magnitude. Because an overall factor in the expressions of the periodic functions (Equations 2.6–2.8) does not alter the periodic structure they represent (Equations 1.3), we can arbitrarily choose

$$|a_{nm}| = |b_{nm}| = 1 \tag{2.9}$$

By substituting Equation 2.9 in Equations 2.6–2.8, we obtain

$$f_1(x,y) = \sum_{n,m}^{d=1} \left\{ \pm \cos\left[\frac{2\pi}{a}(nx+my)\right] \pm \sin\left[\frac{2\pi}{a}(nx+my)\right] \right\} \tag{2.10}$$

$$f_2(x,y) = \sum_{n,m}^{d=2} \left\{ \pm \cos\left[\frac{2\pi}{a}(nx+my)\right] \pm \sin\left[\frac{2\pi}{a}(nx+my)\right] \right\} \tag{2.11}$$

$$f_4(x,y) = \sum_{n,m}^{d=4} \left\{ \pm \cos\left[\frac{2\pi}{a}(nx+my)\right] \pm \sin\left[\frac{2\pi}{a}(nx+my)\right] \right\} \tag{2.12}$$

and, in general, we have

$$f_i(x,y) = \sum_{n,m}^{d=i} \left\{ \pm \cos\left[\frac{2\pi}{a}(nx+my)\right] \pm \sin\left[\frac{2\pi}{a}(nx+my)\right] \right\} \tag{2.13}$$

Equations 2.10–2.13 are the starting point to systematically create two-dimensional simple periodic functions that represent periodic structures in square unit cells. We next illustrate the proposed scheme by finding the set of simple periodic functions corresponding to the groups $d = 1, 2,$ and 4.

2.2.1.1 $f_1(x, y)$ Functions

By using the $d = 1$ terms given in Table 2.1, we expand expression 2.10 to obtain the formula that generates the simple periodic functions $f_1(x, y)$

$$\begin{aligned} f_1(x,y) = &\pm \cos\left[\frac{2\pi}{a}(1x+0y)\right] \pm \sin\left[\frac{2\pi}{a}(1x+0y)\right] \\ &\pm \cos\left[\frac{2\pi}{a}(0x+1y)\right] \pm \sin\left[\frac{2\pi}{a}(0x+1y)\right] \end{aligned} \tag{2.14}$$

For simplicity, we separately consider simple periodic functions $f_1(x, y)$ consisting of either cosine or sine functions. By examining all possible combinations of plus

Table 2.2 The set of simple periodic functions $f_1(x, y)$

n, m	1, 0	0, 1
$f_1(x, y) =$	$+\cos\left(\dfrac{2\pi x}{a}\right)$	$+\cos\left(\dfrac{2\pi y}{a}\right)$
$f_1(x, y) =$	$+\cos\left(\dfrac{2\pi x}{a}\right)$	$-\cos\left(\dfrac{2\pi y}{a}\right)$

n, m	1, 0	0, 1
$f_1(x, y) =$	$+\sin\left(\dfrac{2\pi x}{a}\right)$	$+\sin\left(\dfrac{2\pi y}{a}\right)$
$f_1(x, y) =$	$+\sin\left(\dfrac{2\pi x}{a}\right)$	$-\sin\left(\dfrac{2\pi y}{a}\right)$

and minus signs in Equation 2.14, we obtain the set of four simple periodic functions $f_1(x, y)$ shown in Table 2.2, in which each simple periodic function is made of the sum of two cosine or sine terms.

Note that we consider only functions $f_1(x, y)$ and not their corresponding negative versions $-f_1(x, y)$. This is because the functions $f_1(x, y)$ and $-f_1(x, y)$ represent two periodic structures where materials A and B interchange their position in space (Equations 1.3). These two inverse structures are taken into account by considering only the functions given by $f_1(x, y)$ and defining "direct" and "inverse" structures

$$
\begin{array}{ll}
\text{Direct structure} & \text{Inverse structure} \\
\text{If } f_1(x, y) > 0 \Rightarrow \text{Material A} & \text{If } f_1(x, y) > 0 \Rightarrow \text{Material B} \\
\text{If } f_1(x, y) < 0 \Rightarrow \text{Material B} & \text{If } f_1(x, y) < 0 \Rightarrow \text{Material A}
\end{array}
\quad (2.15)
$$

A careful analysis of the set of four simple periodic functions $f_1(x, y)$ in Table 2.2 shows that the functions are indeed spatial translations of the same function; for example, a spatial translation along the y axis, where $x' = x$, $y' = y - 0.5a$, transforms the second function into the first function. Similarly, a translation $x' = x - 0.25a$, $y' = y - 0.25a$ transforms the third function into the first function. Because spatial translations of a periodic function do not change the geometry of the periodic structure that the function represents, all the simple periodic functions $f_1(x, y)$ in Table 2.2 represent the *same* two-dimensional periodic structure. In fact, spatial translations of a function only move the periodic structure represented by the function to different locations in space. Therefore, we can arbitrarily choose any of the four simple periodic functions in Table 2.2 as representative of the whole set of simple periodic functions $f_1(x, y)$.

2.2 Creating Simple Periodic Functions in Two Dimensions

We choose to represent the simple periodic functions $f_1(x, y)$ by the formula

$$f_1(x, y) = +\cos\left(\frac{2\pi x}{a}\right) + \cos\left(\frac{2\pi y}{a}\right) \tag{2.16}$$

Therefore, by examining the group $d = 1$, we obtained only one independent simple periodic function $f_1(x, y)$. In particular, this simple periodic function is made of the sum of only two trigonometric terms.

We are now interested in graphically displaying the periodic structure that corresponds to the simple periodic function (Equation 2.16). Note that if we use Equations 2.15 and 2.16, the amount of material within the structure is fixed. To control the relative amount of materials A and B within the structure, we add to the formula of the periodic function a parameter t. That is, periodic structures with different amounts of materials A and B are described by the formulas

$$\begin{array}{ll} \textit{Direct structure} & \textit{Inverse structure} \\ \text{If } f_1(x, y) + t > 0 \Rightarrow \text{Material A} & \text{If } f_1(x, y) + t > 0 \Rightarrow \text{Material B} \\ \text{If } f_1(x, y) + t < 0 \Rightarrow \text{Material B} & \text{If } f_1(x, y) + t < 0 \Rightarrow \text{Material A} \end{array} \tag{2.17}$$

where the parameter t determines the precise amounts of materials A and B in the structure.

Figure 2.1a shows the direct periodic structure corresponding to the formula $f_1(x, y) - 0.5$, where materials A and B are illustrated in white and gray, respectively. The unit cell of the structure is shown on the left, whereas the periodic structure is shown on the right. In this case, the amount of material A within the unit cell is 31% of the total volume of the unit cell. This means that the *volume fraction* of material A is $f_A = 31\%$. Because materials A and B completely fill the unit cell of the structure, the volume fraction of the complementary material B is therefore $f_B = 100\% - f_A = 69\%$.

Figures 2.1b and c show the direct periodic structures corresponding to the formulas $f_1(x, y) + 0.0$ and $f_1(x, y) + 0.5$, respectively. In these cases, the volume fractions of materials A and B are 50–50% and 69–31%, respectively. We can see that the volume fraction of material A increases as the value of the parameter t increases. Therefore, the addition of a constant to the simple periodic function $f_1(x, y)$ allows us to control the volume fraction of material A (and consequently material B), as different constants determine structures with different amounts of materials A and B. In the case of the simple periodic function $f_1(x, y)$, if $t > 2$, the corresponding periodic structure is completely made of material A, whereas if $t < -2$, the structure is completely made of material B. This means that to have a periodic structure made of materials A and B, the parameter t should be in the range $-2 < t < 2$.

Note that if we interchange materials A and B in the periodic structure shown in Figure 2.1a, we obtain the structure shown in Figure 2.1c. That is, in this particular case of the simple periodic function $f_1(x, y)$, no additional structures are obtained by considering inverse structures.

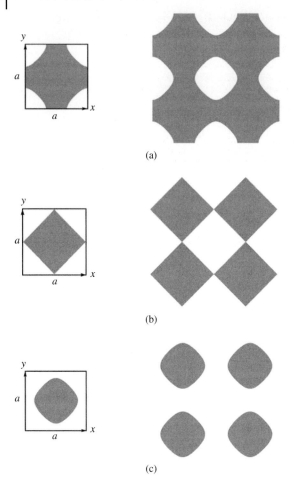

Fig. 2.1 Two-dimensional direct periodic structures that correspond to the simple periodic function $f_1(x, y) = \cos(2\pi x/a) + \cos(2\pi y/a)$. The unit cell of the structure is shown on the left and the periodic structure on the right. The contour line $f_1(x, y) + t = 0$ divides the two-dimensional space in regions filled with materials A (white) and B (gray). If $f_1(x, y) + t > 0 \rightarrow$ Material A, and if $f_1(x, y) + t < 0 \rightarrow$ Material B. (a) $t = -0.5$, $f_A = 31\%$; (b) $t = 0.0$, $f_A = 50\%$; and (c) $t = +0.5$, $f_A = 69\%$.

2.2.1.2 $f_2(x, y)$ and $f_4(x, y)$ Functions

Analogously, in this section we obtain the formulas for the simple periodic functions $f_2(x, y)$ and $f_4(x, y)$ and graphically display the corresponding periodic structures.

By using the $d = 2$ and $d = 4$ terms given in Table 2.1, we expand expressions 2.11 and 2.12 and obtain the set of simple periodic functions shown in Tables 2.3 and 2.4, respectively, where each simple periodic function is made of the sum of two cosine or sine terms.

Table 2.3 The set of simple periodic functions $f_2(x, y)$

n, m	1, 1	1, −1
$f_2(x, y) =$	$+\cos\left[\dfrac{2\pi}{a}(x+y)\right]$	$+\cos\left[\dfrac{2\pi}{a}(x-y)\right]$
$f_2(x, y) =$	$+\cos\left[\dfrac{2\pi}{a}(x+y)\right]$	$-\cos\left[\dfrac{2\pi}{a}(x-y)\right]$

n, m	1, 1	1, −1
$f_2(x, y) =$	$+\sin\left[\dfrac{2\pi}{a}(x+y)\right]$	$+\sin\left[\dfrac{2\pi}{a}(x-y)\right]$
$f_2(x, y) =$	$+\sin\left[\dfrac{2\pi}{a}(x+y)\right]$	$-\sin\left[\dfrac{2\pi}{a}(x-y)\right]$

Table 2.4 The set of simple periodic functions $f_4(x, y)$

n, m	2, 0	0, 2
$f_4(x, y) =$	$+\cos\left(\dfrac{4\pi x}{a}\right)$	$+\cos\left(\dfrac{4\pi y}{a}\right)$
$f_4(x, y) =$	$+\cos\left(\dfrac{4\pi x}{a}\right)$	$-\cos\left(\dfrac{4\pi y}{a}\right)$

n, m	2, 0	0, 2
$f_4(x, y) =$	$+\sin\left(\dfrac{4\pi x}{a}\right)$	$+\sin\left(\dfrac{4\pi y}{a}\right)$
$f_4(x, y) =$	$+\sin\left(\dfrac{4\pi x}{a}\right)$	$-\sin\left(\dfrac{4\pi y}{a}\right)$

Similar to the $f_1(x, y)$ case, spatial translations reduce the number of independent functions in Tables 2.3 and 2.4 into a single function for each case. Therefore, the sets of simple periodic functions $f_2(x, y)$ and $f_4(x, y)$ can be represented by the formulas

$$f_2(x, y) = +\cos\left[\frac{2\pi}{a}(x+y)\right] + \cos\left[\frac{2\pi}{a}(x-y)\right] \tag{2.18}$$

$$f_4(x, y) = +\cos\left(\frac{4\pi x}{a}\right) + \cos\left(\frac{4\pi y}{a}\right) \tag{2.19}$$

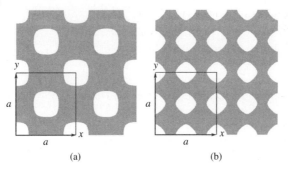

Fig. 2.2 The two-dimensional direct periodic structures that correspond to the simple periodic functions $f_2(x, y)$ and $f_4(x, y)$. Regions with materials A (white) and B (gray) are divided by contour lines $f_i(x, y) + t = 0$ ($i =$ 2, 4). If $f_i(x, y) + t > 0 \rightarrow$ Material A, and if $f_i(x, y) + t < 0 \rightarrow$ Material B. The periodic structures are given by (a) $f_2(x, y) - 0.5$ and (b) $f_4(x, y) - 0.5$.

To graphically display the corresponding two-dimensional periodic structures, we use Equations 2.17, but employ the $f_2(x, y)$ and $f_4(x, y)$ functions (Equations 2.18 and 2.19). Figure 2.2 shows the direct periodic structures corresponding to the formulas $f_2(x, y) - 0.5$ and $f_4(x, y) - 0.5$, where materials A and B are illustrated in white and gray, respectively. As in the $f_1(x, y)$ case, no additional structures are obtained by considering inverse structures.

It is important to note that the periodic structure in Figure 2.2a is equivalent to the periodic structure in Figure 2.1a. That is, if we rotate the structure in Figure 2.2a by 45° and increase its intrinsic length, we obtain the structure in Figure 2.1a. As a result, the simple periodic functions $f_1(x, y)$ and $f_2(x, y)$ represent the same two-dimensional periodic structure. Also note that the periodic structure in Figure 2.2b is equivalent to the periodic structure in Figure 2.1a. In fact, the periodic structures given by the $f_4(x, y)$ functions are the same as the structures given by the $f_1(x, y)$ functions, but compressed by a factor of 2. That is, in the case of the square lattice, by considering the groups $d = 2$ and $d = 4$, we obtain additional simple periodic functions, but the corresponding periodic structures are the same as the $f_1(x, y)$ structures except for a scaling factor.

To obtain periodic structures in square unit cells with different geometries from those generated by the simple periodic function $f_1(x, y)$, we need to consider simple periodic functions with higher d values. The calculation of the set of simple periodic functions corresponding to the group $d = 5$ and the displaying of the corresponding periodic structures are left as an exercise for the reader (Problem 2.1).

2.2.2
The Triangular Lattice

We now focus in creating two-dimensional simple periodic functions $f(x, y)$ defined within triangular unit cells of side a (Figure 1.5b). The Fourier series expansion of such periodic functions is given by the formula

2.2 Creating Simple Periodic Functions in Two Dimensions

$$f(x, y) = a_{00} + \sum_n \sum_m \left\{ a_{nm} \cos\left[\frac{2\pi}{a}\left(-nx + \left(\frac{1}{\sqrt{3}}n + \frac{2}{\sqrt{3}}m\right)y\right)\right] \right.$$
$$\left. + b_{nm} \sin\left[\frac{2\pi}{a}\left(-nx + \left(\frac{1}{\sqrt{3}}n + \frac{2}{\sqrt{3}}m\right)y\right)\right] \right\} \quad (2.20)$$

where Equation 2.20 is obtained by replacing the reciprocal lattice vectors for the triangular lattice

$$G_{nm} = \frac{2\pi}{a}\left(-n, \frac{1}{\sqrt{3}}n + \frac{2}{\sqrt{3}}m\right) \quad (2.21)$$

into the Fourier series expansion (Equation 1.18) for periodic functions in arbitrary unit cells. (Note: general expressions for reciprocal lattice vectors are given in Chapter 6).

In complete analogy with the square case, to systematically create simple periodic functions in the triangular lattice by using the Fourier series expansion (Equation 2.20), we group together in Equation 2.20 cosine and sine functions with the same spatial period λ. In the triangular case, however, the period λ at which the cosine or sine functions repeat in space is given as follows:

$$\lambda = a \Big/ \sqrt{(-n)^2 + \left(\frac{1}{\sqrt{3}}n + \frac{2}{\sqrt{3}}m\right)^2} \quad (2.22)$$

This means that by defining the parameter

$$d = (-n)^2 + \left(\frac{1}{\sqrt{3}}n + \frac{2}{\sqrt{3}}m\right)^2 \quad (2.23)$$

we can group in Equation 2.20 cosine and sine functions with the same spatial period. All we need to do is to sort the integer numbers n and m in Equation 2.20 by increasing values of d (Table 2.5).

In the development that follows, we use the same scheme introduced for the square lattice case. That is, we systematically create two-dimensional simple periodic functions by separately considering groups of cosine and sine functions, in Equation 2.20, with the same d values.

Table 2.5 The integer numbers n and m corresponding to the sum in Equation 2.20 are sorted by increasing values of d

λ	d	n, m		
$a/\sqrt{4/3}$	4/3	1, 0	0, 1	1, −1
$a/\sqrt{4}$	4	1, 1	1, −2	2, −1
$a/\sqrt{16/3}$	16/3	2, 0	0, 2	2, −2
⋮	⋮	⋮	⋮	⋮

2.2.2.1 $f_{4/3}(x, y)$ Functions

By substituting the $d = 4/3$ terms given in Table 2.5 in Equation 2.20, we obtain the formula that generates the simple periodic functions $f_{4/3}(x, y)$

$$\begin{aligned} f_{4/3}(x, y) = &\pm \cos\left[\frac{2\pi}{a}\left(-x + \frac{1}{\sqrt{3}}y\right)\right] \pm \sin\left[\frac{2\pi}{a}\left(-x + \frac{1}{\sqrt{3}}y\right)\right] \\ &\pm \cos\left[\frac{2\pi}{a}\left(+\frac{2}{\sqrt{3}}y\right)\right] \pm \sin\left[\frac{2\pi}{a}\left(+\frac{2}{\sqrt{3}}y\right)\right] \\ &\pm \cos\left[\frac{2\pi}{a}\left(-x - \frac{1}{\sqrt{3}}y\right)\right] \pm \sin\left[\frac{2\pi}{a}\left(-x - \frac{1}{\sqrt{3}}y\right)\right] \end{aligned} \quad (2.24)$$

By separately considering $f_{4/3}(x, y)$ functions consisting of either cosine or sine functions and examining all possible combinations of plus and minus signs in Equation 2.24, we obtain the set of eight simple periodic functions $f_{4/3}(x, y)$ shown in Table 2.6, where each simple periodic function is made of the sum of three cosine or sine terms.

As in the previous cases, spatial translations reduce the number of independent functions in Table 2.6 to two. Therefore, the group $d = 4/3$ generates two independent simple periodic functions labeled $f^I_{4/3}(x, y)$ and $f^{II}_{4/3}(x, y)$, and they are given by the formulas

$$\begin{aligned} f^I_{4/3}(x, y) = &\cos\left[\frac{2\pi}{a}\left(-x + \frac{1}{\sqrt{3}}y\right)\right] + \cos\left[\frac{2\pi}{a}\left(\frac{2}{\sqrt{3}}y\right)\right] \\ &+ \cos\left[\frac{2\pi}{a}\left(-x - \frac{1}{\sqrt{3}}y\right)\right] \end{aligned} \quad (2.25)$$

$$\begin{aligned} f^{II}_{4/3}(x, y) = &\sin\left[\frac{2\pi}{a}\left(-x + \frac{1}{\sqrt{3}}y\right)\right] + \sin\left[\frac{2\pi}{a}\left(\frac{2}{\sqrt{3}}y\right)\right] \\ &+ \sin\left[\frac{2\pi}{a}\left(-x - \frac{1}{\sqrt{3}}y\right)\right] \end{aligned} \quad (2.26)$$

To graphically display the corresponding two-dimensional periodic structures, we use Equations 2.17, but employ the $f^I_{4/3}(x, y)$ and $f^{II}_{4/3}(x, y)$ functions. Figure 2.3 shows the direct and inverse periodic structures represented by the simple periodic functions (Equations 2.25 and 2.26), where materials A and B are, as usual, illustrated in white and gray, respectively. The first row in Figure 2.3 shows the direct structures given by the formulas $f^I_{4/3}(x, y) - 1.2$, $f^I_{4/3}(x, y)$ and $f^I_{4/3}(x, y) + 1.2$, respectively, whereas the second row shows the inverse structures, which are obtained by interchanging the materials A and B in the direct structures. The third row shows the direct structures given by the formulas $f^{II}_{4/3}(x, y) - 1.2$, $f^{II}_{4/3}(x, y)$ and $f^{II}_{4/3}(x, y) + 1.2$. Note that in this latest case, no additional structures are obtained by considering the inverse structures.

We would like to mention that additional simple periodic functions (and corresponding periodic structures) can be obtained in the triangular lattice by considering higher d groups (Table 2.5). These functions, however, are left as exercises for the reader (Problem 2.2).

Table 2.6 The set of simple periodic functions $f_{4/3}(x, y)$

n, m	1, 0	0, 1	1, −1
$f_{4/3}(x,y) = +\cos\left[\frac{2\pi}{a}\left(-x+\frac{1}{\sqrt{3}}y\right)\right]$		$+\cos\left[\frac{2\pi}{a}\left(+\frac{2}{\sqrt{3}}y\right)\right]$	$+\cos\left[\frac{2\pi}{a}\left(-x-\frac{1}{\sqrt{3}}y\right)\right]$
$f_{4/3}(x,y) = +\cos\left[\frac{2\pi}{a}\left(-x+\frac{1}{\sqrt{3}}y\right)\right]$		$+\cos\left[\frac{2\pi}{a}\left(+\frac{2}{\sqrt{3}}y\right)\right]$	$-\cos\left[\frac{2\pi}{a}\left(-x-\frac{1}{\sqrt{3}}y\right)\right]$
$f_{4/3}(x,y) = +\cos\left[\frac{2\pi}{a}\left(-x+\frac{1}{\sqrt{3}}y\right)\right]$		$-\cos\left[\frac{2\pi}{a}\left(+\frac{2}{\sqrt{3}}y\right)\right]$	$+\cos\left[\frac{2\pi}{a}\left(-x-\frac{1}{\sqrt{3}}y\right)\right]$
$f_{4/3}(x,y) = +\cos\left[\frac{2\pi}{a}\left(-x+\frac{1}{\sqrt{3}}y\right)\right]$		$-\cos\left[\frac{2\pi}{a}\left(+\frac{2}{\sqrt{3}}y\right)\right]$	$-\cos\left[\frac{2\pi}{a}\left(-x-\frac{1}{\sqrt{3}}y\right)\right]$

n, m	1, 0	0, 1	1, −1
$f_{4/3}(x,y) = +\sin\left[\frac{2\pi}{a}\left(-x+\frac{1}{\sqrt{3}}y\right)\right]$		$+\sin\left[\frac{2\pi}{a}\left(+\frac{2}{\sqrt{3}}y\right)\right]$	$+\sin\left[\frac{2\pi}{a}\left(-x-\frac{1}{\sqrt{3}}y\right)\right]$
$f_{4/3}(x,y) = +\sin\left[\frac{2\pi}{a}\left(-x+\frac{1}{\sqrt{3}}y\right)\right]$		$+\sin\left[\frac{2\pi}{a}\left(+\frac{2}{\sqrt{3}}y\right)\right]$	$-\sin\left[\frac{2\pi}{a}\left(-x-\frac{1}{\sqrt{3}}y\right)\right]$
$f_{4/3}(x,y) = +\sin\left[\frac{2\pi}{a}\left(-x+\frac{1}{\sqrt{3}}y\right)\right]$		$-\sin\left[\frac{2\pi}{a}\left(+\frac{2}{\sqrt{3}}y\right)\right]$	$+\sin\left[\frac{2\pi}{a}\left(-x-\frac{1}{\sqrt{3}}y\right)\right]$
$f_{4/3}(x,y) = +\sin\left[\frac{2\pi}{a}\left(-x+\frac{1}{\sqrt{3}}y\right)\right]$		$-\sin\left[\frac{2\pi}{a}\left(+\frac{2}{\sqrt{3}}y\right)\right]$	$-\sin\left[\frac{2\pi}{a}\left(-x-\frac{1}{\sqrt{3}}y\right)\right]$

In summary, in the last two sections we showed how to create simple periodic functions in two dimensions. The formulas of the simple periodic functions are written as the sum of a small number of cosine and sine functions (or Fourier terms). In particular, we studied simple periodic functions that represent periodic structures in the square and triangular lattices.[3] As mentioned earlier, periodic structures represented by simple periodic functions are important since they can be fabricated by interference lithography and offer useful physical properties. In Section 2.3, we apply the above scheme to the important three-dimensional case.

2.3
Creating Simple Periodic Functions in Three Dimensions

In complete analogy with the two-dimensional case, in this section we want to systematically create simple periodic functions in three dimensions by using Fourier series expansions. In particular, we consider three-dimensional periodic functions $f(x, y, z)$ defined within cubic unit cells of side a (Figure 1.9). By replacing

---

[3] In the case that the lattice is not square or triangular, a similar procedure can be applied. However, some of the resultant simple periodic functions may have trigonometric terms with different spatial periodicities.

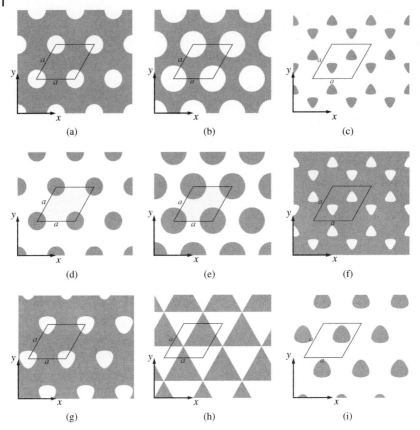

Fig. 2.3 Two-dimensional periodic structures in the triangular lattice. (a–c) Direct structures where regions filled with materials A (white) and B (gray) are defined by the contour lines $f^I_{4/3}(x, y) + t = 0$, where $t = -1.2$, 0.0, and 1.2, respectively. (d–f) The inverse structures that correspond to the previous structures. (g–i) Direct structures where the regions filled with materials A and B are divided by the contour lines $f^{II}_{4/3}(x, y) + t = 0$, where $t = -1.2$, 0.0, and 1.2, respectively.

$a = b = c$ in Equation 2.1b, the Fourier series expansion for these periodic functions is given as follows:

$$f(x, y, z) = a_{000} + \sum_n \sum_m \sum_p \left\{ a_{nmp} \cos\left[\frac{2\pi}{a}(nx + my + pz)\right] \right.$$
$$\left. + b_{nmp} \sin\left[\frac{2\pi}{a}(nx + my + pz)\right] \right\} \quad (2.27)$$

The distance (or period) λ at which the cosine or sine functions in Equation 2.27 repeat in space is given by $\lambda = a/\sqrt{n^2 + m^2 + p^2}$. Therefore, to group cosine

2.3 Creating Simple Periodic Functions in Three Dimensions

Table 2.7 The integer numbers n, m, and p corresponding to the sum in Equation 2.27 are arranged by increasing values of d, where $d = n^2 + m^2 + p^2$

d	n, m, p
1	1, 0, 0 0, 1, 0 0, 0, 1
2	1, 1, 0 1, −1, 0 1, 0, 1 1, 0, −1 0, 1, 1 0, 1, −1
3	1, 1, 1 1, 1, −1 1, −1, 1 1, −1, −1
4	2, 0, 0 0, 2, 0 0, 0, 2
5	1, 2, 0 1, −2, 0 2, 1, 0 2, −1, 0 1, 0, 2 1, 0, −2 2, 0, 1 2, 0, −1 0, 1, 2 0, 1, −2 0, 2, 1 0, 2, −1
⋮	⋮

and sine functions with the same spatial period λ, we define the parameter $d = n^2 + m^2 + p^2$ and sort the integer numbers n, m, and p in Equation 2.27 by increasing values of d (Table 2.7).

As in the previous sections, we assume that $|a_{nm}| = |b_{nm}| = 1$ in Equation 2.27 and separately consider groups of cosine and sine functions with the same d values. As a result, we obtain the following analytical formulas for different d groups of simple periodic functions in three dimensions

$$f_1(x, y, z) = \sum_{n,m,p}^{d=1} \left\{ \pm \cos\left[\frac{2\pi}{a}(nx + my + pz)\right] \pm \sin\left[\frac{2\pi}{a}(nx + my + pz)\right] \right\} \quad (2.28)$$

$$f_2(x, y, z) = \sum_{n,m,p}^{d=2} \left\{ \pm \cos\left[\frac{2\pi}{a}(nx + my + pz)\right] \pm \sin\left[\frac{2\pi}{a}(nx + my + pz)\right] \right\} \quad (2.29)$$

$$f_3(x, y, z) = \sum_{n,m,p}^{d=3} \left\{ \pm \cos\left[\frac{2\pi}{a}(nx + my + pz)\right] \pm \sin\left[\frac{2\pi}{a}(nx + my + pz)\right] \right\} \quad (2.30)$$

And, in general, we have

$$f_i(x, y, z) = \sum_{n,m,p}^{d=i} \left\{ \pm \cos\left[\frac{2\pi}{a}(nx + my + pz)\right] \pm \sin\left[\frac{2\pi}{a}(nx + my + pz)\right] \right\} \quad (2.31)$$

We next obtain explicit formulas for three-dimensional simple periodic functions defined in cubic lattices by separately examining the groups $d = 1$, 2, and 3, and graphically display the corresponding periodic structures.

2.3.1
The Simple Cubic Lattice

By using the $d = 1$ terms given in Table 2.7, we expand Equation 2.28 to obtain the formula for the simple periodic functions $f_1(x, y, z)$

$$f_1(x, y, z) = \pm \cos\left[\frac{2\pi}{a}(1x + 0y + 0z)\right] \pm \sin\left[\frac{2\pi}{a}(1x + 0y + 0z)\right]$$
$$\pm \cos\left[\frac{2\pi}{a}(0x + 1y + 0z)\right] \pm \sin\left[\frac{2\pi}{a}(0x + 1y + 0z)\right]$$
$$\pm \cos\left[\frac{2\pi}{a}(0x + 0y + 1z)\right] \pm \sin\left[\frac{2\pi}{a}(0x + 0y + 1z)\right] \quad (2.32)$$

By separately considering $f_1(x, y, z)$ functions consisting of either cosine or sine functions, and examining all possible combinations of plus and minus signs in Equation 2.32, we obtain the set of eight simple periodic functions $f_1(x, y, z)$ shown in Table 2.8, where each simple periodic function is made of the sum of three cosine or sine terms.

A careful analysis of the simple periodic functions $f_1(x, y, z)$ in Table 2.8 shows that all eight functions are indeed spatial translations of a single function. For

Table 2.8 The set of simple periodic functions $f_1(x, y, z)$

n, m, p	1, 0, 0	0, 1, 0	0, 0, 1
$f_1(x, y, z) =$	$+\cos\left(\frac{2\pi x}{a}\right)$	$+\cos\left(\frac{2\pi y}{a}\right)$	$+\cos\left(\frac{2\pi z}{a}\right)$
$f_1(x, y, z) =$	$+\cos\left(\frac{2\pi x}{a}\right)$	$+\cos\left(\frac{2\pi y}{a}\right)$	$-\cos\left(\frac{2\pi z}{a}\right)$
$f_1(x, y, z) =$	$+\cos\left(\frac{2\pi x}{a}\right)$	$-\cos\left(\frac{2\pi y}{a}\right)$	$+\cos\left(\frac{2\pi z}{a}\right)$
$f_1(x, y, z) =$	$+\cos\left(\frac{2\pi x}{a}\right)$	$-\cos\left(\frac{2\pi y}{a}\right)$	$-\cos\left(\frac{2\pi z}{a}\right)$

n, m, p	1, 0, 0	0, 1, 0	0, 0, 1
$f_1(x, y, z) =$	$+\sin\left(\frac{2\pi x}{a}\right)$	$+\sin\left(\frac{2\pi y}{a}\right)$	$+\sin\left(\frac{2\pi z}{a}\right)$
$f_1(x, y, z) =$	$+\sin\left(\frac{2\pi x}{a}\right)$	$+\sin\left(\frac{2\pi y}{a}\right)$	$-\sin\left(\frac{2\pi z}{a}\right)$
$f_1(x, y, z) =$	$+\sin\left(\frac{2\pi x}{a}\right)$	$-\sin\left(\frac{2\pi y}{a}\right)$	$+\sin\left(\frac{2\pi z}{a}\right)$
$f_1(x, y, z) =$	$+\sin\left(\frac{2\pi x}{a}\right)$	$-\sin\left(\frac{2\pi y}{a}\right)$	$-\sin\left(\frac{2\pi z}{a}\right)$

2.3 Creating Simple Periodic Functions in Three Dimensions

example, a spatial translation along the z axis, where $x' = x$, $y' = y$, and $z' = z - 0.5a$, transforms the second function into the first function. Similarly, a translation $x' = x - 0.25a$, $y' = y - 0.25a$, and $z' = z - 0.25a$ transforms the fifth function into the first function. As mentioned earlier, spatial translations of a periodic function do not change the geometry of the corresponding periodic structure. As a result, all simple periodic functions $f_1(x, y, z)$ in Table 2.8 are equivalent and they can arbitrarily be represented by the formula

$$f_1(x, y, z) = +\cos\left(\frac{2\pi x}{a}\right) + \cos\left(\frac{2\pi y}{a}\right) + \cos\left(\frac{2\pi z}{a}\right) \tag{2.33}$$

We are now interested in graphically displaying the three-dimensional periodic structure that corresponds to the simple periodic function (Equation 2.33). As in the two-dimensional case, to control the relative amount of materials A and B within the structure, we add to the formula of the periodic function a parameter t. Three-dimensional periodic structures with different amounts of materials A and B are therefore described by the formulas

Direct structure | Inverse structure

If $f_1(x, y, z) + t > 0 \Rightarrow$ Material A   If $f_1(x, y, z) + t > 0 \Rightarrow$ Material B
If $f_1(x, y, z) + t < 0 \Rightarrow$ Material B   If $f_1(x, y, z) + t < 0 \Rightarrow$ Material A   (2.34)

where the parameter t determines the precise amounts of materials A and B in the three-dimensional periodic structure.

Figure 2.4 is a graphic illustration of the periodic structure that corresponds to the simple periodic function $f_1(x, y, z)$ given by Equation 2.33. In the three-dimensional case, the structures are illustrated by displaying the three-dimensional surface that divides the regions between materials A and B. For example, in Figure 2.4a, the structure is represented by the level-set surface $f_1(x, y, z) + 0 = 0$. If we consider direct structures, the inner region of the surface (which includes the center of the unit cell) is filled with material B, whereas the outer region (which includes the origin point) is filled with material A. On the other hand, if we consider inverse structures, the distribution of materials A and B is just the inverse.

In particular, the three-dimensional surface shown in Figure 2.4a divides the space into two regions of equal volume fraction. This means that the volume fractions of materials A and B are each 50% in this case. In Figure 2.4b, the structure is represented by the three-dimensional surface $f_1(x, y, z) + 0.8 = 0$. We can see that the addition of the constant $t = 0.8$ does not change the geometry of the surface, but it modifies the volume fractions of the inner and outer regions, which are now 27 and 73%, respectively.

In the case of the simple periodic function $f_1(x, y, z)$, if $t > 3$, the corresponding direct structure is entirely made of material A, whereas if $t < -3$, the direct structure is entirely made of material B. This means that to obtain a periodic structure made of materials A and B, the parameter t should be in the range $-3 < t < 3$.

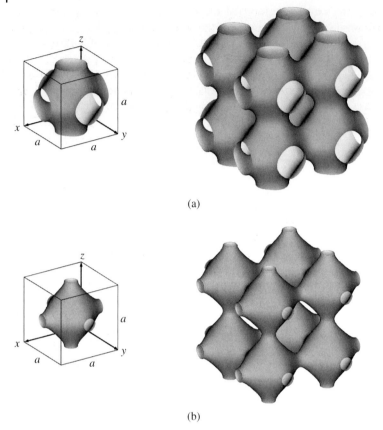

Fig. 2.4 The three-dimensional periodic structure that corresponds to the simple periodic function $f_1(x, y, z)$. The unit cell of the structure is shown on the left, whereas a $2 \times 2 \times 2$ set of unit cells from the periodic structure is shown on the right. (a) The structure is represented by the surface $f_1(x, y, z) + 0 = 0$. The volume fractions of the inner and outer regions of the surface are equal to 50%. (b) The structure is represented by the surface $f_1(x, y, z) + 0.8 = 0$. In this case, the volume fractions of the inner and outer regions are 27 and 73%, respectively.

In brief, despite the large number of simple periodic functions $f_1(x, y, z)$ in Table 2.8, only a single independent function is obtained for the $d = 1$ group. This simple periodic function is given by Equation 2.33. A family of $f_1(x, y, z)$ structures with different volume fractions is obtained by adding an arbitrary constant t to the analytical formula of the structure (Figure 2.4). Note that the resultant three-dimensional periodic structures are made of objects located at the sites of the simple cubic Bravais lattice (Figure 1.9).[4]

---

4) This is because the initial formula for the simple periodic functions $f_1(x, y, z)$ given by Equation 2.32 is invariant under the spatial translations $[x' = x - a]$, $[y' = y - a]$, and $[z' = z - a]$, which are characteristics of the simple cubic Bravais lattice.

2.3.2
The Face-centered-cubic Lattice

We next obtain explicit formulas for simple periodic functions corresponding to higher d values (Table 2.7). Owing to the large number of simple periodic functions that can be obtained for the group $d = 2$, we first consider the group $d = 3$. Therefore, we expand Equation 2.30 by using the $d = 3$ terms given in Table 2.7 and obtain the formula for the simple periodic functions $f_3(x, y, z)$

$$
\begin{aligned}
f_3(x, y, z) = &\pm \cos\left[\frac{2\pi}{a}(1x + 1y + 1z)\right] \pm \sin\left[\frac{2\pi}{a}(1x + 1y + 1z)\right] \\
&\pm \cos\left[\frac{2\pi}{a}(1x + 1y - 1z)\right] \pm \sin\left[\frac{2\pi}{a}(1x + 1y - 1z)\right] \\
&\pm \cos\left[\frac{2\pi}{a}(1x - 1y + 1z)\right] \pm \sin\left[\frac{2\pi}{a}(1x - 1y + 1z)\right] \\
&\pm \cos\left[\frac{2\pi}{a}(1x - 1y - 1z)\right] \pm \sin\left[\frac{2\pi}{a}(1x - 1y - 1z)\right] \quad (2.35)
\end{aligned}
$$

As in the previous sections, we consider simple periodic functions consisting of either cosine or sine functions. By examining all possible combinations of plus and minus signs in Equation 2.35, we obtain the 16 simple periodic functions shown in Table 2.9, which are made of the sum of four cosine or sine terms.

Although 16 functions are obtained in Table 2.9, spatial translations reduce the number of independent functions to two. These functions are given by the formulas

$$
\begin{aligned}
f_3^{4,\mathrm{I}}(x, y, z) = &+\cos\left[\frac{2\pi}{a}(x + y + z)\right] + \cos\left[\frac{2\pi}{a}(x + y - z)\right] \\
&+ \cos\left[\frac{2\pi}{a}(x - y + z)\right] + \cos\left[\frac{2\pi}{a}(x - y - z)\right] \quad (2.36)
\end{aligned}
$$

$$
\begin{aligned}
f_3^{4,\mathrm{II}}(x, y, z) = &+\cos\left[\frac{2\pi}{a}(x + y + z)\right] + \cos\left[\frac{2\pi}{a}(x + y - z)\right] \\
&+ \cos\left[\frac{2\pi}{a}(x - y + z)\right] - \cos\left[\frac{2\pi}{a}(x - y - z)\right] \quad (2.37)
\end{aligned}
$$

where the subscript 3 denotes that the functions belong to the third group ($d = 3$), the superscript 4 denotes that they consist of four trigonometric functions, and the Roman numerals I and II indicate that there exist two independent functions. The periodic structures corresponding to these simple periodic functions are shown in Figures 2.5 and 2.6. The unit cell of the structures are shown on the left, whereas portions of the structures made of $2 \times 2 \times 2$ number of unit cells along the x, y, and z directions are shown on the right.

We previously mentioned that one of the motivations to obtain simple periodic functions, which are made of the sum of a small number of trigonometric terms, is the fact that the corresponding periodic structures can be fabricated by interference

2 Periodic Functions and Structures

Table 2.9 The set of simple periodic functions $f_3^4(x, y, z)$

n, m, p	1, 1, 1	1, 1, −1	1, −1, 1	1, −1, −1
	$\cos\left[\frac{2\pi}{a}(x+y+z)\right]$	$\cos\left[\frac{2\pi}{a}(x+y-z)\right]$	$\cos\left[\frac{2\pi}{a}(x-y+z)\right]$	$\cos\left[\frac{2\pi}{a}(x-y-z)\right]$
$f_3^{4,\mathrm{I}}(x,y,z) =$	+	+	+	+
$f_3^{4,\mathrm{II}}(x,y,z) =$	+	+	+	−
$f_3^{4,\mathrm{II}}(x,y,z) =$	+	+	−	+
$f_3^{4,\mathrm{I}}(x,y,z) =$	+	+	−	−
$f_3^{4,\mathrm{II}}(x,y,z) =$	+	−	+	+
$f_3^{4,\mathrm{I}}(x,y,z) =$	+	−	+	−
$f_3^{4,\mathrm{I}}(x,y,z) =$	+	−	−	+
$f_3^{4,\mathrm{II}}(x,y,z) =$	+	−	−	−

n, m, p	1, 1, 1	1, 1, −1	1, −1, 1	1, −1, −1
	$\sin\left[\frac{2\pi}{a}(x+y+z)\right]$	$\sin\left[\frac{2\pi}{a}(x+y-z)\right]$	$\sin\left[\frac{2\pi}{a}(x-y+z)\right]$	$\sin\left[\frac{2\pi}{a}(x-y-z)\right]$
$f_3^{4,\mathrm{I}}(x,y,z) =$	+	+	+	+
$f_3^{4,\mathrm{II}}(x,y,z) =$	+	+	+	−
$f_3^{4,\mathrm{II}}(x,y,z) =$	+	+	−	+
$f_3^{4,\mathrm{I}}(x,y,z) =$	+	+	−	−
$f_3^{4,\mathrm{II}}(x,y,z) =$	+	−	+	+
$f_3^{4,\mathrm{I}}(x,y,z) =$	+	−	+	−
$f_3^{4,\mathrm{I}}(x,y,z) =$	+	−	−	+
$f_3^{4,\mathrm{II}}(x,y,z) =$	+	−	−	−

lithography. Therefore, in our aim to obtain simple periodic functions, we can reduce the number of trigonometric terms that form the simple periodic functions given in Table 2.9. By noting that the minimum number of trigonometric terms required to create three-dimensional periodic structures is three, we successively drop one of the four terms in the formulas of the simple periodic functions given by Table 2.9. Therefore, the following additional simple periodic functions made of the sum of three cosine terms (Table 2.10) and a similar table (not shown) where the simple periodic functions consist of a sum of three sine terms are obtained.

Although 32 functions are obtained in this case, translations and rotations reduce the number of simple periodic functions to a single independent function. That is, all simple periodic functions shown in Table 2.10 represent the same periodic structure (either at different locations in space or rotated around some axis). Therefore, the set of simple periodic functions $f_3^3(x, y, z)$ is represented by the formula

Table 2.10 The set of simple periodic functions $f_3^3(x, y, z)$

n, m, p	1, 1, 1	1, 1, −1	1, −1, 1	1, −1, −1
	$\cos\left[\dfrac{2\pi}{a}(x+y+z)\right]$	$\cos\left[\dfrac{2\pi}{a}(x+y-z)\right]$	$\cos\left[\dfrac{2\pi}{a}(x-y+z)\right]$	$\cos\left[\dfrac{2\pi}{a}(x-y-z)\right]$
$f_3^3(x, y, z) =$		+	+	+
$f_3^3(x, y, z) =$		+	+	−
$f_3^3(x, y, z) =$		+	−	+
$f_3^3(x, y, z) =$		+	−	−
$f_3^3(x, y, z) =$	+		+	+
$f_3^3(x, y, z) =$	+		+	−
$f_3^3(x, y, z) =$	+		−	+
$f_3^3(x, y, z) =$	+		−	−
$f_3^3(x, y, z) =$	+	+		+
$f_3^3(x, y, z) =$	+	+		−
$f_3^3(x, y, z) =$	+	−		+
$f_3^3(x, y, z) =$	+	−		−
$f_3^3(x, y, z) =$	+	+	+	
$f_3^3(x, y, z) =$	+	+	−	
$f_3^3(x, y, z) =$	+	−	+	
$f_3^3(x, y, z) =$	+	−	−	

$$f_3^3(x, y, z) = +\cos\left[\frac{2\pi}{a}(x+y-z)\right] + \cos\left[\frac{2\pi}{a}(x-y+z)\right]$$
$$+ \cos\left[\frac{2\pi}{a}(x-y-z)\right] \qquad (2.38)$$

The periodic structure corresponding to this simple periodic function is shown in Figure 2.7.

In brief, by considering the group $d = 3$, we obtained three independent simple periodic functions. These functions are labeled as $f_3^{4,\,\mathrm{I}}(x, y, z)$, $f_3^{4,\,\mathrm{II}}(x, y, z)$, and $f_3^3(x, y, z)$ and their formulas are given by Equations 2.36–2.38, respectively. The three-dimensional periodic structures represented by these simple periodic functions are shown in Figures 2.5–2.7. In this case, the structures are made of objects located at the sites of the face-centered-cubic Bravais lattice (Figure 1.9).[5]

Note: The three-dimensional periodic structures shown in Figures 2.6 and 2.7, which correspond to the simple periodic functions $f_3^{4,\,\mathrm{II}}(x, y, z) - 1.5$ and $f_3^3(x, y, z) - 0.8$, are bicontinuous. That is, both inner and outer regions of the three-dimensional surface, which are made of materials A and B, are self-connected (or continuous). This is not the case with the periodic structure shown in Figure 2.5 because the flattened spheres in this structure are disconnected. It is important to remark that the

---

5) This is because the initial formula for the simple periodic functions $f_3(x, y, z)$ given by Equation 2.35 is invariant under the spatial translations $[x' = x, y' = y - (a/2), z' = z - (a/2)]$, $[x' = x - (a/2), y' = y, z' = z - (a/2)]$, and $[x' = x - (a/2), y' = y - (a/2), z' = z]$, which are characteristics of the face-centered-cubic Bravais lattice.

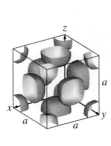
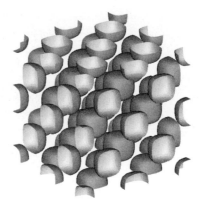

Fig. 2.5 The three-dimensional periodic structure that corresponds to the simple periodic function $f_3^{4,I}(x, y, z)$. The structure is represented by the surface $f_3^{4,I}(x, y, z) - 0.8 = 0$ and the volume fraction of the inner region of the surface is $f = 23.5\%$. Note that the structure is made of nonconnected objects (flattened spheres) located at the sites of the face-centered-cubic Bravais lattice (Figure 1.9).

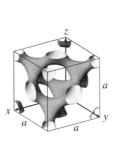
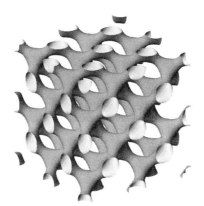

Fig. 2.6 The three-dimensional periodic structure that corresponds to the simple periodic function $f_3^{4,II}(x, y, z)$. To show that the structure is a face-centered-cubic diamond structure, we translate the corresponding function and display the structure given by the surface $f_3^{4,II}(x - 0.125, y + 0.125, z + 0.125) - 1.5 = 0$. The volume fraction of the inner region of the surface is $f = 19\%$. Note that the structure is made of objects located at the sites of the face-centered-cubic Bravais lattice (Figure 1.9).

presence of bicontinuity in the three-dimensional periodic structure depends on the particular simple periodic function and also on the specific value of the parameter t. For example, the periodic structure shown in Figure 2.6 is no longer bicontinuous when the parameter $t = -2.1$. As discussed in Chapter 3, bicontinuous structures are very convenient in terms of their experimental fabrication by interference lithography.

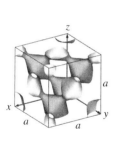
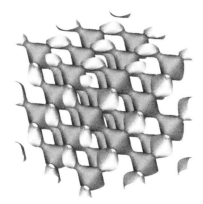

Fig. 2.7 The three-dimensional periodic structure that corresponds to the simple periodic function $f_3^3(x, y, z)$. The structure is represented by the surface $f_3^3(x - 0.125, y - 0.125, z - 0.125) - 0.8 = 0$, and the volume fraction of the inner region of the surface is $f = 27\%$. The structure is made of objects located at the sites of the face-centered-cubic Bravais lattice (Figure 1.9).

2.3.3
The Body-centered-cubic Lattice

We now consider the group $d = 2$ (Table 2.7) and expand the sum in Equation 2.29 to obtain the formula for the simple periodic functions $f_2(x, y, z)$

$$\begin{aligned}
f_2(x, y, z) = &\pm \cos\left[\frac{2\pi}{a}(1x + 1y + 0z)\right] \pm \sin\left[\frac{2\pi}{a}(1x + 1y + 0z)\right] \\
&\pm \cos\left[\frac{2\pi}{a}(1x - 1y + 0z)\right] \pm \sin\left[\frac{2\pi}{a}(1x - 1y + 0z)\right] \\
&\pm \cos\left[\frac{2\pi}{a}(1x + 0y + 1z)\right] \pm \sin\left[\frac{2\pi}{a}(1x + 0y + 1z)\right] \\
&\pm \cos\left[\frac{2\pi}{a}(1x + 0y - 1z)\right] \pm \sin\left[\frac{2\pi}{a}(1x + 0y - 1z)\right] \\
&\pm \cos\left[\frac{2\pi}{a}(0x + 1y + 1z)\right] \pm \sin\left[\frac{2\pi}{a}(0x + 1y + 1z)\right] \\
&\pm \cos\left[\frac{2\pi}{a}(0x + 1y - 1z)\right] \pm \sin\left[\frac{2\pi}{a}(0x + 1y - 1z)\right]
\end{aligned} \quad (2.39)$$

By considering simple periodic functions consisting of either cosine or sine functions and examining all possible combinations of plus and minus signs in Equation 2.39, we obtain the 64 simple periodic functions shown in Table 2.11, which are made of the sum of six cosine or sine terms.

2 Periodic Functions and Structures

Although 64 simple periodic functions are obtained in this case, translations and rotations reduce the number of independent functions in Table 2.11 to four. These four simple periodic functions are given by the formulas

$$f_2^{6,\text{I}}(x,y,z) = +\cos\left[\frac{2\pi}{a}(x+y)\right] + \cos\left[\frac{2\pi}{a}(x-y)\right] + \cos\left[\frac{2\pi}{a}(x+z)\right]$$
$$+ \cos\left[\frac{2\pi}{a}(x-z)\right] + \cos\left[\frac{2\pi}{a}(y+z)\right] + \cos\left[\frac{2\pi}{a}(y-z)\right]$$
(2.40)

$$f_2^{6,\text{II}}(x,y,z) = +\cos\left[\frac{2\pi}{a}(x+y)\right] + \cos\left[\frac{2\pi}{a}(x-y)\right] + \cos\left[\frac{2\pi}{a}(x+z)\right]$$
$$+ \cos\left[\frac{2\pi}{a}(x-z)\right] + \cos\left[\frac{2\pi}{a}(y+z)\right] - \cos\left[\frac{2\pi}{a}(y-z)\right]$$
(2.41)

$$f_2^{6,\text{III}}(x,y,z) = +\sin\left[\frac{2\pi}{a}(x+y)\right] + \sin\left[\frac{2\pi}{a}(x-y)\right] + \sin\left[\frac{2\pi}{a}(x+z)\right]$$
$$+ \sin\left[\frac{2\pi}{a}(x-z)\right] + \sin\left[\frac{2\pi}{a}(y+z)\right] + \sin\left[\frac{2\pi}{a}(y-z)\right]$$
(2.42)

$$f_2^{6,\text{IV}}(x,y,z) = +\sin\left[\frac{2\pi}{a}(x+y)\right] + \sin\left[\frac{2\pi}{a}(x-y)\right] + \sin\left[\frac{2\pi}{a}(x+z)\right]$$
$$- \sin\left[\frac{2\pi}{a}(x-z)\right] + \sin\left[\frac{2\pi}{a}(y+z)\right] + \sin\left[\frac{2\pi}{a}(y-z)\right]$$
(2.43)

As in the previous section, to create additional simple periodic functions $f_2(x,y,z)$, we can reduce the number of trigonometric terms in Table 2.11. For example, if we reduce the number of terms from six to five, 192 simple periodic functions are obtained. However, only four of these functions are independent. These four simple periodic functions are given by the formulas

$$f_2^{5,\text{I}}(x,y,z) = \cos\left[\frac{2\pi}{a}(x-y)\right] + \cos\left[\frac{2\pi}{a}(x+z)\right] + \cos\left[\frac{2\pi}{a}(x-z)\right]$$
$$+ \cos\left[\frac{2\pi}{a}(y+z)\right] + \cos\left[\frac{2\pi}{a}(y-z)\right]$$
(2.44)

$$f_2^{5,\text{II}}(x,y,z) = \cos\left[\frac{2\pi}{a}(x-y)\right] + \cos\left[\frac{2\pi}{a}(x+z)\right] + \cos\left[\frac{2\pi}{a}(x-z)\right]$$
$$+ \cos\left[\frac{2\pi}{a}(y+z)\right] - \cos\left[\frac{2\pi}{a}(y-z)\right]$$
(2.45)

2.3 Creating Simple Periodic Functions in Three Dimensions

$$f_2^{5,\mathrm{III}}(x,y,z) = \sin\left[\frac{2\pi}{a}(x-y)\right] + \sin\left[\frac{2\pi}{a}(x+z)\right] + \sin\left[\frac{2\pi}{a}(x-z)\right]$$

$$+ \sin\left[\frac{2\pi}{a}(y+z)\right] + \sin\left[\frac{2\pi}{a}(y-z)\right] \quad (2.46)$$

$$f_2^{5,\mathrm{IV}}(x,y,z) = \sin\left[\frac{2\pi}{a}(x-y)\right] + \sin\left[\frac{2\pi}{a}(x+z)\right] + \sin\left[\frac{2\pi}{a}(x-z)\right]$$

$$+ \sin\left[\frac{2\pi}{a}(y+z)\right] - \sin\left[\frac{2\pi}{a}(y-z)\right] \quad (2.47)$$

Table 2.11 The set of simple periodic functions $f_2^6(x,y,z)$

n, m, p	1, 1, 0	1, −1, 0	1, 0, 1	1, 0, −1	0, 1, 1	0, 1, −1
	$\cos\left[\frac{2\pi}{a}(x+y)\right]$	$\cos\left[\frac{2\pi}{a}(x-y)\right]$	$\cos\left[\frac{2\pi}{a}(x+z)\right]$	$\cos\left[\frac{2\pi}{a}(x-z)\right]$	$\cos\left[\frac{2\pi}{a}(y+z)\right]$	$\cos\left[\frac{2\pi}{a}(y-z)\right]$
$f_2^{6,\mathrm{I}}(x,y,z) =$	+	+	+	+	+	+
$f_2^{6,\mathrm{II}}(x,y,z) =$	+	+	+	+	+	−
$f_2^{6,\mathrm{II}}(x,y,z) =$	+	+	+	+	−	+
$f_2^{6,\mathrm{I}}(x,y,z) =$	+	+	+	+	−	−
$f_2^{6,\mathrm{II}}(x,y,z) =$	+	+	+	−	+	+
$f_2^{6,\mathrm{II}}(x,y,z) =$	+	+	+	−	+	−
$f_2^{6,\mathrm{II}}(x,y,z) =$	+	+	+	−	−	+
$f_2^{6,\mathrm{II}}(x,y,z) =$	+	+	+	−	−	−
$f_2^{6,\mathrm{II}}(x,y,z) =$	+	+	−	+	+	+
$f_2^{6,\mathrm{II}}(x,y,z) =$	+	+	−	+	+	−
$f_2^{6,\mathrm{II}}(x,y,z) =$	+	+	−	+	−	+
$f_2^{6,\mathrm{II}}(x,y,z) =$	+	+	−	+	−	−
$f_2^{6,\mathrm{I}}(x,y,z) =$	+	+	−	−	+	+
$f_2^{6,\mathrm{II}}(x,y,z) =$	+	+	−	−	+	−
$f_2^{6,\mathrm{II}}(x,y,z) =$	+	+	−	−	−	+
$f_2^{6,\mathrm{I}}(x,y,z) =$	+	+	−	−	−	−
$f_2^{6,\mathrm{II}}(x,y,z) =$	+	−	+	+	+	+
$f_2^{6,\mathrm{II}}(x,y,z) =$	+	−	+	+	+	−
$f_2^{6,\mathrm{II}}(x,y,z) =$	+	−	+	+	−	+
$f_2^{6,\mathrm{II}}(x,y,z) =$	+	−	+	+	−	−
$f_2^{6,\mathrm{II}}(x,y,z) =$	+	−	+	−	+	+
$f_2^{6,\mathrm{I}}(x,y,z) =$	+	−	+	−	+	−
$f_2^{6,\mathrm{I}}(x,y,z) =$	+	−	+	−	−	+
$f_2^{6,\mathrm{II}}(x,y,z) =$	+	−	+	−	−	−
$f_2^{6,\mathrm{II}}(x,y,z) =$	+	−	−	+	+	+
$f_2^{6,\mathrm{I}}(x,y,z) =$	+	−	−	+	+	−
$f_2^{6,\mathrm{II}}(x,y,z) =$	+	−	−	+	−	+
$f_2^{6,\mathrm{II}}(x,y,z) =$	+	−	−	+	−	−
$f_2^{6,\mathrm{II}}(x,y,z) =$	+	−	−	−	+	+
$f_2^{6,\mathrm{II}}(x,y,z) =$	+	−	−	−	+	−
$f_2^{6,\mathrm{II}}(x,y,z) =$	+	−	−	−	−	+
$f_2^{6,\mathrm{II}}(x,y,z) =$	+	−	−	−	−	−

2 Periodic Functions and Structures

Table 2.11 (continued)

n, m, p	1, 1, 0	1, −1, 0	1, 0, 1	1, 0, −1	0, 1, 1	0, 1, −1
	$\sin\left[\dfrac{2\pi}{a}(x+y)\right]$	$\sin\left[\dfrac{2\pi}{a}(x-y)\right]$	$\sin\left[\dfrac{2\pi}{a}(x+z)\right]$	$\sin\left[\dfrac{2\pi}{a}(x-z)\right]$	$\sin\left[\dfrac{2\pi}{a}(y+z)\right]$	$\sin\left[\dfrac{2\pi}{a}(y-z)\right]$
$f_2^{6,III}(x, y, z) =$	+	+	+	+	+	+
$f_2^{6,III}(x, y, z) =$	+	+	+	+	+	−
$f_2^{6,III}(x, y, z) =$	+	+	+	+	−	+
$f_2^{6,III}(x, y, z) =$	+	+	+	+	−	−
$f_2^{6,IV}(x, y, z) =$	+	+	+	−	+	+
$f_2^{6,III}(x, y, z) =$	+	+	+	−	+	−
$f_2^{6,III}(x, y, z) =$	+	+	+	−	−	+
$f_2^{6,IV}(x, y, z) =$	+	+	+	−	−	−
$f_2^{6,IV}(x, y, z) =$	+	+	−	+	+	+
$f_2^{6,III}(x, y, z) =$	+	+	−	+	+	−
$f_2^{6,III}(x, y, z) =$	+	+	−	+	−	+
$f_2^{6,IV}(x, y, z) =$	+	+	−	+	−	−
$f_2^{6,III}(x, y, z) =$	+	+	−	−	+	+
$f_2^{6,III}(x, y, z) =$	+	+	−	−	+	−
$f_2^{6,III}(x, y, z) =$	+	+	−	−	−	+
$f_2^{6,III}(x, y, z) =$	+	+	−	−	−	−
$f_2^{6,III}(x, y, z) =$	+	−	+	+	+	+
$f_2^{6,IV}(x, y, z) =$	+	−	+	+	+	−
$f_2^{6,IV}(x, y, z) =$	+	−	+	+	−	+
$f_2^{6,III}(x, y, z) =$	+	−	+	+	−	−
$f_2^{6,III}(x, y, z) =$	+	−	+	−	+	+
$f_2^{6,III}(x, y, z) =$	+	−	+	−	+	−
$f_2^{6,III}(x, y, z) =$	+	−	+	−	−	+
$f_2^{6,III}(x, y, z) =$	+	−	−	+	+	+
$f_2^{6,III}(x, y, z) =$	+	−	−	+	+	−
$f_2^{6,III}(x, y, z) =$	+	−	−	+	−	+
$f_2^{6,III}(x, y, z) =$	+	−	−	−	−	−
$f_2^{6,III}(x, y, z) =$	+	−	−	−	+	+
$f_2^{6,IV}(x, y, z) =$	+	−	−	−	+	−
$f_2^{6,IV}(x, y, z) =$	+	−	−	−	−	+
$f_2^{6,III}(x, y, z) =$	+	−	−	−	−	−

Similarly, by reducing the number of trigonometric terms to four and three, respectively, we obtain the simple periodic functions

$$f_2^{4,I}(x, y, z) = \cos\left[\frac{2\pi}{a}(x+y)\right] + \cos\left[\frac{2\pi}{a}(x-y)\right] + \cos\left[\frac{2\pi}{a}(x+z)\right]$$
$$+ \cos\left[\frac{2\pi}{a}(x-z)\right] \quad (2.48)$$

$$f_2^{4,II}(x, y, z) = \cos\left[\frac{2\pi}{a}(x+y)\right] + \cos\left[\frac{2\pi}{a}(x-y)\right] + \cos\left[\frac{2\pi}{a}(x+z)\right]$$
$$- \cos\left[\frac{2\pi}{a}(x-z)\right] \quad (2.49)$$

2.3 Creating Simple Periodic Functions in Three Dimensions

$$f_2^{4,\text{III}}(x,y,z) = \cos\left[\frac{2\pi}{a}(x+y)\right] + \cos\left[\frac{2\pi}{a}(x-y)\right] + \cos\left[\frac{2\pi}{a}(x+z)\right]$$

$$+ \cos\left[\frac{2\pi}{a}(y+z)\right] \qquad (2.50)$$

$$f_2^{4,\text{IV}}(x,y,z) = \sin\left[\frac{2\pi}{a}(x+y)\right] + \sin\left[\frac{2\pi}{a}(x-y)\right] + \sin\left[\frac{2\pi}{a}(x+z)\right]$$

$$+ \sin\left[\frac{2\pi}{a}(y+z)\right] \qquad (2.51)$$

and

$$f_2^{3,\text{I}}(x,y,z) = \cos\left[\frac{2\pi}{a}(x+y)\right] + \cos\left[\frac{2\pi}{a}(x-y)\right] + \cos\left[\frac{2\pi}{a}(x+z)\right] \qquad (2.52)$$

$$f_2^{3,\text{II}}(x,y,z) = \cos\left[\frac{2\pi}{a}(x+y)\right] + \cos\left[\frac{2\pi}{a}(x+z)\right] + \cos\left[\frac{2\pi}{a}(y+z)\right] \qquad (2.53)$$

In brief, the group $d = 2$ generates 14 independent simple periodic functions. The corresponding three-dimensional periodic structures are shown in Figure 2.8, where different constants t are added to the formulas of the simple periodic functions to have a better view of the structures. In the case of the group $d = 2$, the three-dimensional periodic structures are made of objects located at the sites of the body-centered-cubic Bravais lattice (Figure 1.9).[6]

The scheme to create three-dimensional simple periodic functions defined in cubic lattices can certainly be extended to higher groups (e.g. $d > 3$). However, it is important to mention that as we increase the value of d (and consequently the values of n, m, p), more geometrically complex structures are obtained due to the fact that the periods λ of the cosine and sine functions in the Fourier series expansion are smaller.

In summary, in the last three sections we showed how to systematically create simple periodic functions by the use of Fourier series in three dimensions. In particular, we studied simple periodic functions that represent three-dimensional periodic structures in the cubic Bravais lattices.[7] As we show in the following chapters, these periodic structures are suitable for fabrication by interference lithography.

---

[6] This is because the initial formula for the simple periodic functions $f_2(x, y, z)$ given by Equation 2.39 is invariant under the spatial translation $[x' = x - (a/2), y' = y - (a/2), z' = z - (a/2)]$, which is characteristic of the body-centered-cubic Bravais lattice.

[7] In the case that the lattice is not cubic, a similar procedure can be applied. However, some of the resultant simple periodic functions may have trigonometric terms with different spatial periodicities.

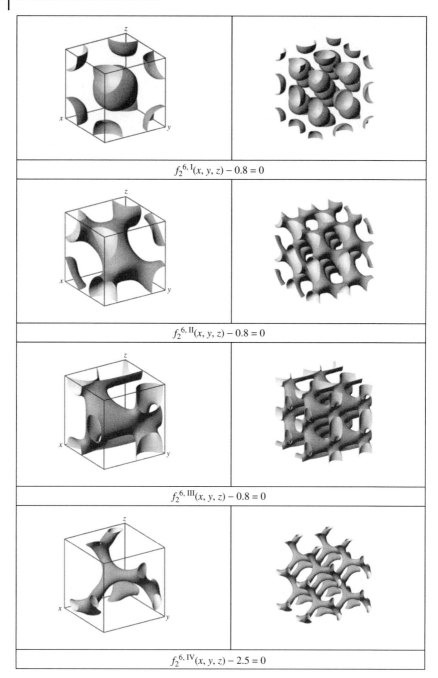

Fig. 2.8 The three-dimensional periodic structures that correspond to the simple periodic functions $f_2(x, y, z)$. The structures are made of objects located at the sites of the body-centered-cubic Bravais lattice (Figure 1.9).

2.3 Creating Simple Periodic Functions in Three Dimensions

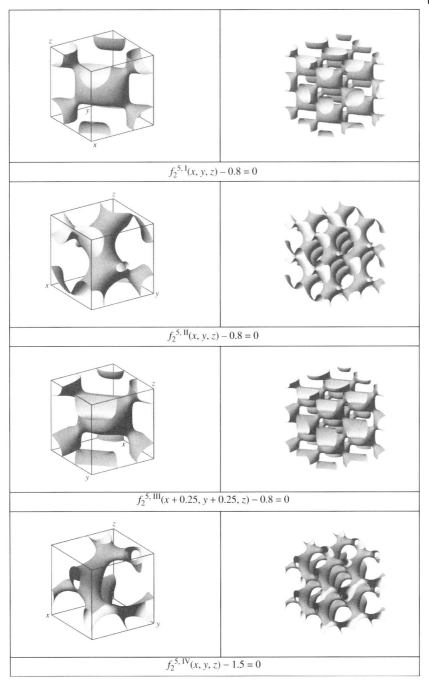

$f_2^{5,\text{I}}(x, y, z) - 0.8 = 0$

$f_2^{5,\text{II}}(x, y, z) - 0.8 = 0$

$f_2^{5,\text{III}}(x + 0.25, y + 0.25, z) - 0.8 = 0$

$f_2^{5,\text{IV}}(x, y, z) - 1.5 = 0$

Fig. 2.8 *(continued)*

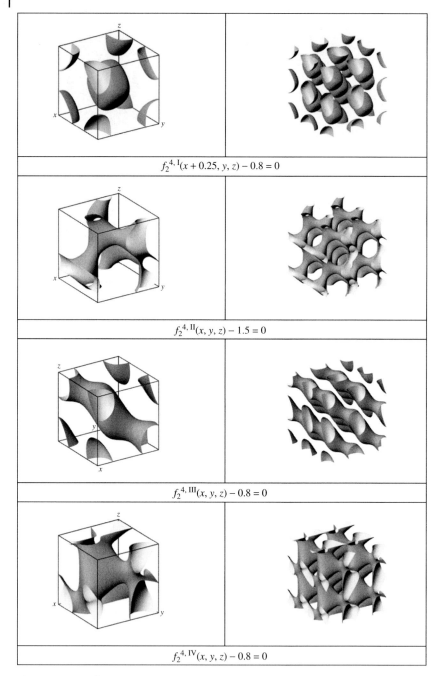

Fig. 2.8 *(continued)*

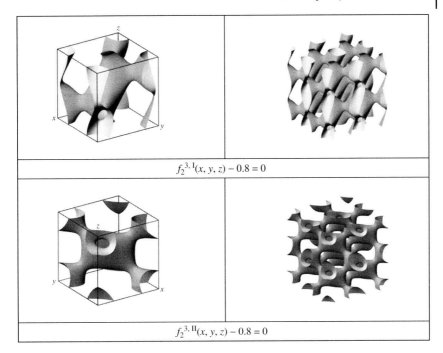

$f_2^{3,\,\mathrm{I}}(x, y, z) - 0.8 = 0$

$f_2^{3,\,\mathrm{II}}(x, y, z) - 0.8 = 0$

Fig. 2.8 (continued)

2.4 Combination of Simple Periodic Functions

We want to mention that additional periodic functions can be created by combining simple periodic functions corresponding to different d groups. For example, we can combine two simple periodic functions $f_i(x, y, z)$ and $f_j(x, y, z)$ from groups i and j, respectively, and have

$$f(x, y, z) = s f_i(x, y, z) + (1 - s) f_j(x, y, z) \qquad (2.54)$$

where $0 < s < 1$ is a real parameter.

As an example, we consider the three-dimensional function

$$f(x, y, z) = 0.6 f_3^{4,\,1}(x, y, z) + 0.4 f_8^{6,\,1}(x, y, z) \qquad (2.55)$$

which is obtained by combining simple periodic functions defined in cubic lattices belonging to the groups $d = 3$ and $d = 8$.

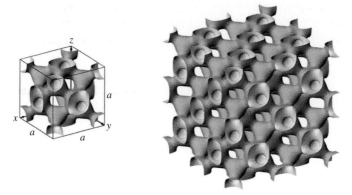

Fig. 2.9 The three-dimensional periodic structure that corresponds to the function $f(x, y, z) = 0.6\, f_3^{4,I}(x, y, z) + 0.4\, f_8^{6,I}(x, y, z) - 0.16 = 0$. The structure is made of objects located at the sites of the face-centered-cubic Bravais lattice (Figure 1.9).

After replacing the corresponding formulas for the functions $f_3^{4,I}$ and $f_8^{6,I}$, we have

$$\begin{aligned} f(x,y,z) = +0.6 &\left\{ \cos\left[\frac{2\pi}{a}(x+y+z)\right] + \cos\left[\frac{2\pi}{a}(x+y-z)\right] \right. \\ &\left. + \cos\left[\frac{2\pi}{a}(x-y+z)\right] + \cos\left[\frac{2\pi}{a}(x-y-z)\right] \right\} \\ +0.4 &\left\{ \cos\left[\frac{2\pi}{a}(2x+2y)\right] + \cos\left[\frac{2\pi}{a}(2x-2y)\right] \right. \\ &+ \cos\left[\frac{2\pi}{a}(2x+2z)\right] + \cos\left[\frac{2\pi}{a}(2x-2z)\right] \\ &\left. + \cos\left[\frac{2\pi}{a}(2y+2z)\right] + \cos\left[\frac{2\pi}{a}(2y-2z)\right] \right\} \end{aligned} \qquad (2.56)$$

This particular example of a combined simple periodic function is of importance in Chapter 6, when we study some optical properties of periodic structures. Figure 2.9 shows the corresponding three-dimensional periodic structure.

To summarize, we presented in this chapter a systematic method to create simple periodic functions by the use of the Fourier series expansion. We examined simple periodic functions defined in the two-dimensional square and triangular lattices as well as in the three-dimensional cubic Bravais lattices. The main characteristic of these simple periodic functions is that their formulas are given by the sum of a small number of trigonometric terms. This determines that the periodic structures they represent can be fabricated by interference lithography, which is theoretically explained in Chapter 3. Because the fabrication of periodic structures by interference lithography requires the superposition of electromagnetic waves, in the following chapter, we first present an introduction to the propagation of electromagnetic waves in homogeneous materials. Subsequently, we explain how periodic structures represented by simple periodic functions can be fabricated by the use of the interference lithography technique.

Problems

2.1 Square lattice

(a) Find the 16 simple periodic functions that correspond to the group $d = 5$ (Table 2.1).

(b) Show that the following two-dimensional periodic structures respectively correspond to the formulas

$$f_5(x,y) = +\cos\left[\frac{2\pi}{a}(x+2y)\right] + \cos\left[\frac{2\pi}{a}(x-2y)\right] + \cos\left[\frac{2\pi}{a}(2x+y)\right]$$
$$+ \cos\left[\frac{2\pi}{a}(2x-y)\right] - 1.5$$

$$f_5(x,y) = +\cos\left[\frac{2\pi}{a}(x+2y)\right] - \cos\left[\frac{2\pi}{a}(x-2y)\right] + \cos\left[\frac{2\pi}{a}(2x+y)\right]$$
$$- \cos\left[\frac{2\pi}{a}(2x-y)\right] - 0.5$$

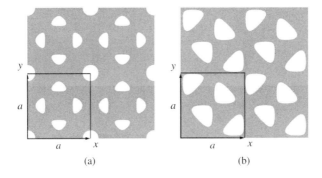

(a)   (b)

2.2 Triangular lattice

(a) Find the n and m values for the fourth group of simple periodic functions (Table 2.5).

(b) Find the simple periodic functions corresponding to the group $d = 4$ and display the corresponding periodic structures.

2.3 Simple cubic lattice

Demonstrate that the simple periodic functions in Table 2.8 are all spatial translations of the function

$$f_1(x,y,z) = +\cos\left(\frac{2\pi x}{a}\right) + \cos\left(\frac{2\pi y}{a}\right) + \cos\left(\frac{2\pi z}{a}\right)$$

2.4 Show that the three-dimensional periodic structure given by the simple periodic function $f_1(x, y, z)$ is bicontinuous only for values of the parameter t in the range $-1 < t < 1$. Hint: in a bicontinuous structure both the inner and outer regions to the surface $f_1(x, y, z) + t = 0$ are self-connected.

2.5 Show that if a translation or rotation transforms $f(x, y)$ into $-f(x, y)$, then direct and inverse structures given by the formula $f(x, y) + t$ are equivalent.

3

Interference of Waves and Interference Lithography

The interference of electromagnetic waves is the basic physical phenomenon that allows us to fabricate periodic structures such as those presented in the previous sections. In this chapter, we present a theoretical introduction to wave interference. We examine the propagation of electromagnetic waves in homogeneous materials, demonstrate the transverse character of these waves, describe different states of polarizations, and examine the transport of energy associated with the propagation of electromagnetic waves. Then we study the interference of electromagnetic waves and show how this physical phenomenon can be used to create periodic structures.

3.1
Electromagnetic Waves

An electromagnetic wave is basically a progressive disturbance generated by time-varying electric and magnetic fields, which propagates from point to point while transporting electromagnetic energy. Common examples of electromagnetic waves include visible light from a candle or lamp, radio waves from antennas of radio stations, and X-rays from metallic tubes bombarded by high-energy electrons. All the above are examples of electromagnetic waves. These waves can be described by a simultaneous propagation of an electric field and a magnetic field wave (Figure 3.1). At any given point in space, the electric field vector **E** and the magnetic field vector **H** of the waves oscillate at right angles to each other and to the direction of propagation of the wave given by the wave vector **k**, and at the same time the wave profile moves forward along the propagation direction. An important characteristic of electromagnetic waves is the distance, measured in the direction of propagation of the wave, between two successive points where the character of the wave is the same. This distance is called the *wavelength* λ.

The number of cycles per unit time, at which the electric or magnetic field vectors oscillate at a given point in space, is called the *frequency f* of the wave. After each of these cycles is completed, the wave has moved forward by a distance λ. Therefore, the *velocity v* of the wave is given by the product of the wavelength and the frequency of the wave

$$v = \lambda f \tag{3.1}$$

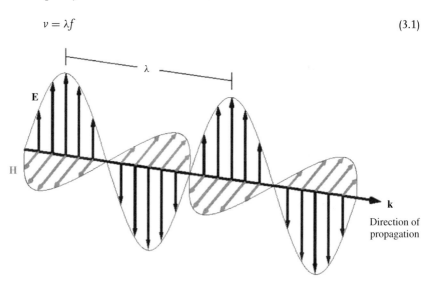

Fig. 3.1 An electromagnetic wave with its characteristic wavelength λ indicated. The electric field vector **E** and the magnetic field vector **H** oscillate at right angles to each other and to the direction of propagation of the wave, which is given by the wave vector **k**.

Table 3.1 The electromagnetic spectrum classified according to the wavelengths and frequencies of the electromagnetic waves (http://imagine.gsfc.nasa.gov).

	Wavelength (m)	Frequency (s^{-1})
Radio waves	$>1 \times 10^{-1}$	$<3 \times 10^9$
Microwaves	$1 \times 10^{-3} - 1 \times 10^{-1}$	$3 \times 10^9 - 3 \times 10^{11}$
Infrared light	$7 \times 10^{-7} - 1 \times 10^{-3}$	$3 \times 10^{11} - 4 \times 10^{14}$
Visible light	$4 \times 10^{-7} - 7 \times 10^{-7}$	$4 \times 10^{14} - 7.5 \times 10^{14}$
Ultraviolet light	$1 \times 10^{-8} - 4 \times 10^{-7}$	$7.5 \times 10^{14} - 3 \times 10^{16}$
X rays	$1 \times 10^{-11} - 1 \times 10^{-8}$	$3 \times 10^{16} - 3 \times 10^{19}$
γ rays	$<1 \times 10^{-11}$	$>3 \times 10^{19}$

For electromagnetic waves propagating in vacuum, the velocity v is constant and equal to 3×10^8 m s^{-1}. As a result, there is a one-to-one correspondence between the wavelength and the frequency of the wave, which is given by

$$f = \frac{3 \times 10^8 \text{ m s}^{-1}}{\lambda} \quad \text{in vacuum} \tag{3.2}$$

Electromagnetic waves can be classified according to their wavelengths, from largest to smallest, into radio waves, microwaves, infrared light, visible light, ultraviolet light, X rays, and γ rays. All these types of electromagnetic waves form the so-called electromagnetic spectrum. Table 3.1 shows the approximate wavelength (and frequency) ranges for the electromagnetic spectrum.

3.2
The Wave Equation

All physical phenomena associated with the propagation of electromagnetic waves in different materials are mathematically described by a set of differential equations known as *Maxwell's equations*.

In the most general case, the materials in which the waves propagate can be charged or have electrical currents flowing through them. However, in many cases (including those of interest in this book), it can be considered that the electromagnetic waves propagate within materials in which no free electric charges or currents exist. In this latter case, Maxwell's equations can be written as

$$\nabla \cdot \mathbf{D} = 0, \quad \nabla \times \mathbf{E} = -\frac{\partial \mathbf{B}}{\partial t} \tag{3.3}$$

$$\nabla \cdot \mathbf{B} = 0, \quad \nabla \times \mathbf{H} = \frac{\partial \mathbf{D}}{\partial t} \tag{3.4}$$

where **E** and **H** are the *electric* and *magnetic* field vectors respectively, **D** and **B** are the *electric displacement* and *magnetic induction* field vectors, ∇ represents the gradient operator, which in Cartesian coordinates is given by $\nabla = (\partial/\partial x, \partial/\partial y, \partial/\partial z)$, and $\partial/\partial t$ is the partial derivative with respect to time.

The two equations on the left are known as *Maxwell's divergence equations*, whereas the two equations on the right are known as *Maxwell's curl equations*.

The vectors **E**, **H**, **D**, and **B** are all real functions of the position vector **r** and the time t. In addition, the field vectors **D** and **B** are related to the electric and magnetic field vectors **E** and **H** by the *constitutive parameters* that characterize the material in which the waves propagate. That is,

$$\mathbf{D} = \varepsilon \mathbf{E} \tag{3.5}$$

$$\mathbf{B} = \mu \mathbf{H} \tag{3.6}$$

where ε and μ are, respectively, the *dielectric constant* (also called *electric permittivity*) and the *magnetic permeability* of the material.

In general, the parameters ε and μ could depend on a large set of variables that include the direction of propagation of the wave within the material, the intensity of the electric and magnetic fields, the frequency of the wave, and the particular location within the material. The dependence of ε and μ on all these variables is determined by the type of material in which the waves propagate. In this chapter, we assume that the underlying material is as follows:

- *Isotropic.* As a result, ε and μ do not depend on the direction of the propagation of the wave and they are scalar parameters (and not second-rank tensors as in the general case).
- *Linear.* This means that ε and μ do not depend on the intensity of the electric and magnetic field vectors. That is, $\varepsilon \neq \varepsilon(\mathbf{E}, \mathbf{H})$ and $\mu \neq \mu(\mathbf{E}, \mathbf{H})$. This consideration is valid when the intensities of the electric and magnetic fields are sufficiently small.
- *Nondispersive.* In this case, ε and μ do not depend on the frequency f of the wave that propagates in the material. That is, $\varepsilon \neq \varepsilon(f)$ and $\mu \neq \mu(f)$.
- *Nonconducting.* This means that the electromagnetic energy is conserved and ε and μ are purely real parameters. This consideration is valid when the conductivity of the material is sufficiently small.
- *Homogeneous.* This means that ε and μ do not depend on the particular location within the material and have the same values at every point in space. That is, $\varepsilon \neq \varepsilon(\mathbf{r})$ and $\mu \neq \mu(\mathbf{r})$, where **r** is the position vector.

Because here we assume that the waves propagate in isotropic, linear, nondispersive, nonconducting, and homogeneous materials, the dielectric constant ε and the magnetic permeability μ are *real* and *constant* numbers. In this case, the replacement of Equations 3.5 and 3.6 into Equations 3.3 and 3.4 transforms Maxwell's equations into

$$\nabla \cdot \mathbf{E} = 0, \qquad \nabla \times \mathbf{E} = -\mu \frac{\partial \mathbf{H}}{\partial t} \tag{3.7}$$

$$\nabla \cdot \mathbf{H} = 0, \qquad \nabla \times \mathbf{H} = \varepsilon \frac{\partial \mathbf{E}}{\partial t} \tag{3.8}$$

By combining Maxwell's equations (3.7 and 3.8), we next obtain simple differential equations that contain either the electric field **E** or the magnetic field **H**. These differential equations are known as the *wave equations* and they describe the propagation of electromagnetic waves in materials characterized by the parameters ε and μ. Note that we focus only on the electric and magnetic fields **E** and **H** owing to the fact that **D** and **B** can be obtained by using Equations 3.5 and 3.6 once **E** and **H** are determined.

By taking the curl ($\nabla \times$) of Maxwell's curl equation (3.7), we have

$$\nabla \times (\nabla \times \mathbf{E}) = -\nabla \times \left(\mu \frac{\partial \mathbf{H}}{\partial t} \right) = -\mu \frac{\partial (\nabla \times \mathbf{H})}{\partial t} \tag{3.9}$$

After replacing Maxwell's curl equation (3.8), the above equation transforms into

$$\nabla \times (\nabla \times \mathbf{E}) = -\mu\varepsilon \frac{\partial^2 \mathbf{E}}{\partial t^2} \tag{3.10}$$

By applying the vector identity $\nabla \times (\nabla \times \mathbf{E}) = -\nabla^2 \mathbf{E} + \nabla(\nabla \cdot \mathbf{E})$, where ∇^2 is the Laplacian operator $\nabla^2 = \partial^2/\partial x^2 + \partial^2/\partial y^2 + \partial^2/\partial z^2$, and noting that $\nabla \cdot \mathbf{E} = 0$, we have

$$\nabla^2 \mathbf{E} = \frac{1}{v^2} \frac{\partial^2 \mathbf{E}}{\partial t^2} \tag{3.11}$$

where

$$v = \frac{1}{\sqrt{\varepsilon\mu}} \tag{3.12}$$

Analogously, by taking the curl of Maxwell's curl equation (3.8), we find an identical differential equation for the magnetic field **H**, which is given by

$$\nabla^2 \mathbf{H} = \frac{1}{v^2} \frac{\partial^2 \mathbf{H}}{\partial t^2} \tag{3.13}$$

Equations 3.11 and 3.13 are known as the *wave equations* for the electric field **E** and the magnetic field **H**, respectively. As we show next, these equations are fundamental because they govern the propagation of electromagnetic waves.

Importantly, the parameter v in Equations 3.11 and 3.13 is the *velocity* of the wave. Therefore, Equation 3.12 shows that the velocity of the wave is defined by the dielectric constant ε and the magnetic permeability μ of the material in which the wave propagates. For example, in the case of an electromagnetic wave propagating in vacuum, we have $\varepsilon = \varepsilon_0 = 8.85 \times 10^{-12}\,\mathrm{s^2\,C^2\,m^{-3}\,kg^{-1}}$ and $\mu = \mu_0 = 4\pi \times 10^{-7}\,\mathrm{m\,kg\,C^{-2}}$. By considering Equation 3.12, the velocity of the wave is thus equal to $v = 1/\sqrt{\varepsilon_0\mu_0} = 3 \times 10^8\,\mathrm{m\,s^{-1}}$ (which corresponds to the speed of light c in vacuum). On the other hand, in the case of an electromagnetic wave propagating within a material with $\varepsilon = 3\varepsilon_0$ and $\mu = 2\mu_0$, Equation 3.12 gives the velocity of the wave as $v = 1/\sqrt{3\varepsilon_0 2\mu_0} = c/\sqrt{6}$.

3.3
Electromagnetic Plane Waves

One of the main advantages of transforming Maxwell's equations (3.7 and 3.8) for isotropic, linear, nondispersive, nonconducting, and homogeneous materials into the wave equations (3.11 and 3.13) is the fact that the solutions of the wave equations are well known. For example, Equations 3.11 and 3.13 have analytical solutions for the electric and magnetic fields **E** and **H** in the form of *plane waves*:

$$\mathbf{E}(\mathbf{r}, t) = \mathrm{Re}(\mathbf{E}_0\, e^{i(\mathbf{k}\cdot\mathbf{r}-\omega t)}) \tag{3.14}$$

$$\mathbf{H}(\mathbf{r}, t) = \mathrm{Re}(\mathbf{H}_0\, e^{i(\mathbf{k}\cdot\mathbf{r}-\omega t)}) \tag{3.15}$$

where Re is an abbreviation for 'real part of'; **E** and **H** are the instantaneous electric and magnetic field vectors whose Cartesian components are real functions of the position vector **r** and time t; **E**$_0$, **H**$_0$ are the complex-constant electric and magnetic field *amplitude* vectors whose Cartesian components are complex numbers; **k** is a real vector called the *wave vector* whose direction determines the direction of propagation of the wave; **r** is the position vector; ω is the *angular frequency* of the wave, that is, $\omega = 2\pi f$; and t is the time.

The complex amplitude vectors **E**$_0$ and **H**$_0$, whose Cartesian components are complex numbers *constant* in space and time, determine the direction, amplitude, and polarization of the instantaneous electric and magnetic fields **E** and **H** (Section 3.5).

The direction of the wave vector **k** determines the direction of propagation of the electromagnetic plane wave, whereas its magnitude $k = \sqrt{\mathbf{k}\cdot\mathbf{k}}$ determines the wavelength λ. This can be demonstrated by replacing the plane-wave expression for **E**(**r**, t) given by Equation 3.14 into the wave equation (3.11), which gives

$$\nabla^2[\mathrm{Re}(\mathbf{E}_0\, e^{i(\mathbf{k}\cdot\mathbf{r}-\omega t)})] = \frac{1}{v^2}\frac{\partial^2}{\partial t^2}[\mathrm{Re}(\mathbf{E}_0\, e^{i(\mathbf{k}\cdot\mathbf{r}-\omega t)})] \tag{3.16}$$

$$\mathrm{Re}[\nabla^2(\mathbf{E}_0\, e^{i(\mathbf{k}\cdot\mathbf{r}-\omega t)})] = \frac{1}{v^2}\mathrm{Re}\left[\frac{\partial^2}{\partial t^2}(\mathbf{E}_0\, e^{i(\mathbf{k}\cdot\mathbf{r}-\omega t)})\right] \tag{3.17}$$

$$-(\mathbf{k}\cdot\mathbf{k})\mathrm{Re}(\mathbf{E}_0\, e^{i(\mathbf{k}\cdot\mathbf{r}-\omega t)}) = -\frac{\omega^2}{v^2}\mathrm{Re}(\mathbf{E}_0\, e^{i(\mathbf{k}\cdot\mathbf{r}-\omega t)}) \tag{3.18}$$

$$\mathbf{k}\cdot\mathbf{k} = k^2 = \frac{\omega^2}{v^2} \tag{3.19}$$

Or, equivalently,

$$k = \frac{\omega}{v} = \frac{2\pi f}{v} = \frac{2\pi}{\lambda} \qquad (3.20)$$

Note that we used the following rule, which is valid for operators Ξ such as ∇^2, ∇, $\partial^2/\partial t^2$, or $\partial/\partial t$

$$\Xi[\text{Re}(z)] = \text{Re}[\Xi(z)] \qquad (3.21)$$

where $z = z(\mathbf{r}, t)$ is an arbitrary complex function of $\mathbf{r}$ and t.

Equation 3.20 establishes that there is only one independent variable in the set of variables $\{k, \omega, \lambda\}$. This means that if one of them is known, the other two can be calculated by using Equation 3.20.

Finally, we would like to mention that the plane waves described by Equations 3.14 and 3.15 are called *monochromatic* waves because they vary in time with a single angular frequency ω, or equivalently a single λ, and therefore are single color.

■ **Example**

Consider an electromagnetic plane wave with frequency ω propagating along the z direction with a complex amplitude vector $\mathbf{E}_0 = 2\hat{\mathbf{x}} + 3i\hat{\mathbf{y}}$. Because the wave propagates along the z direction, the wave vector $\mathbf{k}$ is given by $\mathbf{k} = k\hat{\mathbf{z}}$. By replacing these quantities in Equation 3.14, the instantaneous electric field vector $\mathbf{E}(\mathbf{r}, t)$ is given by

$$\mathbf{E}(\mathbf{r}, t) = \text{Re}[(2\hat{\mathbf{x}} + 3i\hat{\mathbf{y}})\, e^{i(kz - \omega t)}]$$
$$= \text{Re}[(2\hat{\mathbf{x}} + 3i\hat{\mathbf{y}})(\cos(kz - \omega t) + i\sin(kz - \omega t))]$$
$$= 2\cos(kz - \omega t)\hat{\mathbf{x}} - 3\sin(kz - \omega t)\hat{\mathbf{y}}$$

3.4
The Transverse Character of Electromagnetic Plane Waves

We mentioned that the electric and magnetic field vectors oscillate at right angles to each other and to the direction of propagation of the wave. We next demonstrate this unique characteristic of electromagnetic waves by using Maxwell's equations and obtain some physically interesting insight into the relationship of $\mathbf{k}$, $\mathbf{E}$, and $\mathbf{H}$. The fact that the electric field $\mathbf{E}$ and the magnetic field $\mathbf{H}$ are perpendicular to the direction of propagation of the wave determines that electromagnetic waves are *transverse* waves.

Because Maxwell's equations govern all electromagnetic phenomena, they must be satisfied without exception. As a result, the plane waves 3.14 and 3.15, which satisfy the wave equations (3.11 and 3.13), must also satisfy Maxwell's equations (3.7 and 3.8).

By substituting Equations 3.14 and 3.15 into Maxwell's curl equation (3.7), we have

$$\nabla \times \mathbf{E} + \mu \frac{\partial \mathbf{H}}{\partial t} = 0 \tag{3.22}$$

$$\nabla \times [\text{Re}(\mathbf{E}_0\, e^{i(\mathbf{k}\cdot\mathbf{r}-\omega t)})] + \mu \frac{\partial}{\partial t}[\text{Re}(\mathbf{H}_0\, e^{i(\mathbf{k}\cdot\mathbf{r}-\omega t)})] = 0 \tag{3.23}$$

$$\text{Re}[\nabla \times (\mathbf{E}_0\, e^{i(\mathbf{k}\cdot\mathbf{r}-\omega t)})] + \mu \text{Re}\left[\frac{\partial}{\partial t}(\mathbf{H}_0\, e^{i(\mathbf{k}\cdot\mathbf{r}-\omega t)})\right] = 0 \tag{3.24}$$

$$\text{Re}[i\mathbf{k} \times (\mathbf{E}_0\, e^{i(\mathbf{k}\cdot\mathbf{r}-\omega t)})] + \mu \text{Re}[-i\omega \mathbf{H}_0\, e^{i(\mathbf{k}\cdot\mathbf{r}-\omega t)}] = 0 \tag{3.25}$$

$$\text{Re}[i(\mathbf{k} \times \mathbf{E}_0 - \omega\mu\mathbf{H}_0)\, e^{i(\mathbf{k}\cdot\mathbf{r}-\omega t)}] = 0 \tag{3.26}$$

Equation 3.26 can be rewritten as

$$\text{Re}[Z\, e^{-i\omega t}] = 0 \tag{3.27}$$

where

$$Z = i(\mathbf{k} \times \mathbf{E}_0 - \omega\mu\mathbf{H}_0)\, e^{i\mathbf{k}\cdot\mathbf{r}} \tag{3.28}$$

Since Equation 3.27 is valid for all times t, the complex quantity $Z = \text{Re}(Z) + i\text{Im}(Z)$ must be equal to zero. For example, at $\omega t = 0$, Equation 3.27 indicates that $\text{Re}(Z) = 0$, whereas at $\omega t = \pi/2$ the equation indicates that $\text{Im}(Z) = 0$. As a result, Z is equal to zero and we have

$$\mathbf{k} \times \mathbf{E}_0 = \omega\mu\mathbf{H}_0 \tag{3.29}$$

Analogously, by inserting Equations 3.14 and 3.15 into Maxwell's curl equation (3.8), we have

$$\mathbf{k} \times \mathbf{H}_0 = -\omega\varepsilon\mathbf{E}_0 \tag{3.30}$$

On the other hand, by inserting the electric field (Equation 3.14) into Maxwell's divergence equation (3.7), we obtain

$$\nabla \cdot \mathbf{E} = 0 \tag{3.31}$$

$$\nabla \cdot [\text{Re}(\mathbf{E}_0\, e^{i(\mathbf{k}\cdot\mathbf{r}-\omega t)})] = 0 \tag{3.32}$$

$$\text{Re}[\nabla \cdot (\mathbf{E}_0\, e^{i(\mathbf{k}\cdot\mathbf{r}-\omega t)})] = 0 \tag{3.33}$$

$$\text{Re}[i\mathbf{k} \cdot \mathbf{E}_0\, e^{i(\mathbf{k}\cdot\mathbf{r}-\omega t)}] = 0 \tag{3.34}$$

$$\mathbf{k} \cdot \mathbf{E}_0 = 0 \tag{3.35}$$

Analogously, by inserting the magnetic field (Equation 3.15) into Maxwell's divergence equation (3.8), we have

$$\mathbf{k} \cdot \mathbf{H}_0 = 0 \tag{3.36}$$

3.4 The Transverse Character of Electromagnetic Plane Waves

In summary, the requirement that the plane waves (Equations 3.14 and 3.15) satisfy Maxwell's equations (3.7 and 3.8) translates into

$$\mathbf{k} \cdot \mathbf{E}_0 = 0, \qquad \mathbf{k} \times \mathbf{E}_0 = \omega\mu\mathbf{H}_0 \tag{3.37}$$

$$\mathbf{k} \cdot \mathbf{H}_0 = 0, \qquad \mathbf{k} \times \mathbf{H}_0 = -\omega\varepsilon\mathbf{E}_0 \tag{3.38}$$

After multiplying Equations 3.37 and 3.38 by $e^{i(\mathbf{k}\cdot\mathbf{r}-\omega t)}$ and taking the real part, we find that the wave vector $\mathbf{k}$, the instantaneous electric field $\mathbf{E}$, and the instantaneous magnetic field $\mathbf{H}$ are related by

$$\mathbf{k} \cdot \mathbf{E} = 0, \qquad \mathbf{k} \times \mathbf{E} = \omega\mu\mathbf{H} \tag{3.39}$$

$$\mathbf{k} \cdot \mathbf{H} = 0, \qquad \mathbf{k} \times \mathbf{H} = -\omega\varepsilon\mathbf{E} \tag{3.40}$$

Finally, by using Equations 3.12 and 3.20, the previous equations transform into

$$\hat{\mathbf{k}} \cdot \mathbf{E} = 0, \qquad \hat{\mathbf{k}} \times \mathbf{E} = \sqrt{\frac{\mu}{\varepsilon}}\mathbf{H} \tag{3.41}$$

$$\hat{\mathbf{k}} \cdot \mathbf{H} = 0, \qquad \hat{\mathbf{k}} \times \mathbf{H} = -\sqrt{\frac{\varepsilon}{\mu}}\mathbf{E} \tag{3.42}$$

where $\hat{\mathbf{k}} = \mathbf{k}/k$ is a unit vector along the direction of propagation of the wave.

We next examine the physical implications of the above equations. The dot products in Equations 3.41 and 3.42 indicate that $\mathbf{E}$ and $\mathbf{H}$ are perpendicular to the wave vector $\mathbf{k}$ and hence to the direction of propagation of the wave. In addition, the cross-products in Equations 3.41 and 3.42 show that $\mathbf{H}$ is always perpendicular to $\mathbf{E}$. As a result, Equations 3.41 and 3.42 determine that the vectors $\mathbf{E}$, $\mathbf{H}$, and $\mathbf{k}$ are *always mutually orthogonal* as in Figure 3.2.

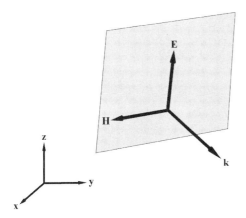

Fig. 3.2 Electromagnetic waves are transverse waves where the instantaneous electric field $\mathbf{E}$, the instantaneous magnetic field $\mathbf{H}$, and the wave vector $\mathbf{k}$ are mutually orthogonal.

In addition, since the constitutive parameters ε and μ are real numbers, the cross-product equations (3.41 and 3.42) establish that the oscillating electric and magnetic fields **E** and **H** are *in phase*. That is, a maximum strength in the electric field is accompanied by a maximum strength in the magnetic field. These equations also determine that the ratio between the magnitudes of the electric and magnetic field vectors is constant. For example, by taking the absolute value on both sides of the cross-product equations (3.41 and 3.42), we obtain

$$\frac{|\mathbf{E}|}{|\mathbf{H}|} = \sqrt{\frac{\mu}{\varepsilon}} \tag{3.43}$$

This means that the constant ratio between the magnitudes of the electric and magnetic fields is determined by the constitutive parameters ε and μ of the material in which the wave propagates.

3.5
Polarization

Electromagnetic plane waves can be classified according to how the electric and magnetic field vectors **E** and **H** evolve in time in the plane perpendicular to the direction of propagation of the wave (Figure 3.2). In the general case, the direction of the vectors **E** and **H** may not follow a specific pattern, and the electromagnetic wave is said to be *nonpolarized*. The directions of **E** and **H**, however, are constrained by the fact that the electric and magnetic field vectors **E** and **H** must always be perpendicular to each other and perpendicular to the direction of propagation of the wave.

In particular cases, however, the directions of **E** and **H** as a function of time follow a well-defined pattern, and the electromagnetic wave is said to be *polarized*. For example, the electric and magnetic field vectors **E** and **H** can oscillate along fixed directions in space (*linearly polarized* waves) or they can rotate maintaining their magnitudes constant (*circularly polarized* waves). Because the magnetic field **H** is always perpendicular and proportional to the electric field **E**, the time evolution of the magnetic field **H** is completely determined by the time evolution of the electric field **E**. This means that the states of polarization can be studied by considering only the time evolution of the electric field **E**.

Consider the electric field vector **E** of an electromagnetic plane wave given by Equation 3.14. The electric field **E** can be arbitrarily separated into two perpendicular components along the directions $\hat{\mathbf{n}}_1$ and $\hat{\mathbf{n}}_2$ as (Figure 3.3)

$$\mathbf{E}(\mathbf{r}, t) = \text{Re}[(\hat{\mathbf{n}}_1 E_{01} + \hat{\mathbf{n}}_2 E_{02}) e^{i(\mathbf{k}\cdot\mathbf{r}-\omega t)}] \tag{3.44}$$

where $\hat{\mathbf{n}}_1$ and $\hat{\mathbf{n}}_2$ are real and perpendicular unit vectors, constant in space and time, and E_{01} and E_{02} are independent complex numbers. Because the electric field **E** must be perpendicular to the direction of propagation **k**, the unit vectors $\hat{\mathbf{n}}_1$ and $\hat{\mathbf{n}}_2$ and the wave vector **k** are mutually orthogonal (Figure 3.3).

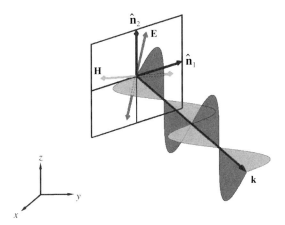

Fig. 3.3 A linearly polarized electromagnetic plane wave.

We show next that the direction of the electric field vector **E** as a function of time, and consequently the state of polarization of the electromagnetic plane wave, is indeed determined by the relative magnitudes and phases of the complex amplitudes E_{01} and E_{02} in Equation 3.44.

3.5.1
Linearly Polarized Electromagnetic Plane Waves

When the complex amplitudes E_{01} and E_{02} of the component waves have the same phase φ, we have

$$E_{01} = |E_{01}| e^{i\varphi} \tag{3.45}$$

$$E_{02} = |E_{02}| e^{i\varphi} \tag{3.46}$$

and the electric field (Equation 3.44) transforms into

$$\mathbf{E}(\mathbf{r}, t) = \text{Re}[(\hat{\mathbf{n}}_1 |E_{01}| + \hat{\mathbf{n}}_2 |E_{02}|)\, e^{i(\mathbf{k}\cdot\mathbf{r} - \omega t + \varphi)}] \tag{3.47}$$

Equation 3.47 indicates that **E** oscillates along a fixed direction, which is given by the constant real vector $\hat{\mathbf{n}}_1 |E_{01}| + \hat{\mathbf{n}}_2 |E_{02}|$. This direction does not change in time and forms an angle θ with respect to $\hat{\mathbf{n}}_1$ given by $\theta = \tan^{-1}(|E_{02}|/|E_{01}|)$. In this case, the wave is said to be *linearly polarized*.

By taking the real part in Equation 3.47, the components of the electric field **E** along the $\hat{\mathbf{n}}_1$ and $\hat{\mathbf{n}}_2$ directions are given by

$$E_1(\mathbf{r}, t) = |E_{01}| \cos(\mathbf{k}\cdot\mathbf{r} - \omega t + \varphi) \tag{3.48}$$

$$E_2(\mathbf{r}, t) = |E_{02}| \cos(\mathbf{k}\cdot\mathbf{r} - \omega t + \varphi) \tag{3.49}$$

Figure 3.3 shows both the electric and magnetic fields **E** and **H** as a function of time for a linearly polarized electromagnetic plane wave.

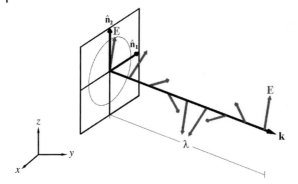

Fig. 3.4 A circularly polarized electromagnetic plane wave.

3.5.2
Circularly Polarized Electromagnetic Plane Waves

When the complex amplitudes E_{01} and E_{02} of the component waves have the same magnitude but differ in phase by $\pi/2$, we have

$$E_{01} = |E_{01}| e^{i\varphi} \tag{3.50}$$

$$E_{02} = |E_{01}| e^{i(\varphi \pm \pi/2)} \tag{3.51}$$

and the electric field (Equation 3.44) transforms into

$$\mathbf{E}(\mathbf{r}, t) = \mathrm{Re}[(\hat{\mathbf{n}}_1 |E_{01}| + \hat{\mathbf{n}}_2 |E_{01}| e^{\pm i\pi/2}) e^{i(\mathbf{k}\cdot\mathbf{r} - \omega t + \varphi)}] \tag{3.52}$$

In this case, the components of the electric field $\mathbf{E}$ along the $\hat{\mathbf{n}}_1$ and $\hat{\mathbf{n}}_2$ directions are

$$E_1(\mathbf{r}, t) = |E_{01}| \cos(\mathbf{k}\cdot\mathbf{r} - \omega t + \varphi) \tag{3.53}$$

$$E_2(\mathbf{r}, t) = |E_{01}| \cos(\mathbf{k}\cdot\mathbf{r} - \omega t + \varphi \pm \pi/2) = \mp |E_{01}| \sin(\mathbf{k}\cdot\mathbf{r} - \omega t + \varphi) \tag{3.54}$$

Equations 3.53 and 3.54 reveal that the time-varying electric field $\mathbf{E}$ is constant in magnitude since $|\mathbf{E}| = \sqrt{(E_1)^2 + (E_2)^2} = |E_{01}|$. However, the direction of the electric field rotates as the wave propagates forward in time (see the time evolution in Figure 3.4). In this case, the wave is said to be *circularly polarized* since the tip of the vector $\mathbf{E}$ traces out a circle on the plane perpendicular to the direction of propagation.

In the case of circularly polarized electromagnetic waves, the electric field vector rotates with an angular frequency ω, and after one complete rotation the wave has moved forward by a distance equal to the wavelength λ. Whether the rotation is clockwise or counterclockwise is determined by the signs in Equation 3.54.

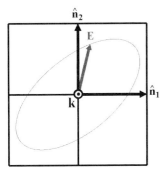

Fig. 3.5 An elliptically polarized electromagnetic plane wave.

3.5.3
Elliptically Polarized Electromagnetic Plane Waves

In the general case, the complex amplitudes E_{01} and E_{02} of the component waves can have arbitrary magnitudes and phases. In this case, the time-varying electric field **E** given by Equation 3.44 rotates and changes its magnitude as the wave moves forward in time. As a result, the tip of the vector **E** describes an ellipse instead of a circle, as viewed along **k**, and the wave is said to be *elliptically polarized* (Figure 3.5).

Note: circularly and linearly polarized plane waves can be viewed as particular cases of elliptical polarized plane waves. For example, a circularly polarized wave can be considered as an elliptically polarized wave for which the semi axes of the ellipse have the same magnitude. Similarly, a linearly polarized wave can be considered as an elliptically polarized wave for which the area of the ellipse tends to zero.[8]

3.6
Electromagnetic Energy

A fundamental aspect of wave propagation is the transport of energy. Waves have the significant feature that they can carry energy as they move forward in the material. In our particular case, it is electromagnetic energy what is transported from one point to another as the wave moves forward. In the following sections, we study how this electromagnetic energy is transported and conserved as the plane wave propagates in the material.

[8] In order to visualize in real time a continuous animation of the propagation of waves in different states of polarizations, we recommend the reader to download a freely available and user-friendly program called *EMANIM* from the following link: http://www.enzim.hu/~szia/emanim/emanim.htm

By considering Maxwell's curl equations (3.7 and 3.8) and taking the dot product with respect to **H** and **E**, respectively, we obtain

$$\mathbf{H} \cdot (\nabla \times \mathbf{E}) = -\mu \mathbf{H} \cdot \frac{\partial \mathbf{H}}{\partial t} \tag{3.55}$$

$$\mathbf{E} \cdot (\nabla \times \mathbf{H}) = \varepsilon \mathbf{E} \cdot \frac{\partial \mathbf{E}}{\partial t} \tag{3.56}$$

Subtracting Equation 3.56 from Equation 3.55, and using the vector identity $\mathbf{H} \cdot (\nabla \times \mathbf{E}) - \mathbf{E} \cdot (\nabla \times \mathbf{H}) = \nabla \cdot (\mathbf{E} \times \mathbf{H})$, we have

$$\nabla \cdot (\mathbf{E} \times \mathbf{H}) = -\varepsilon \mathbf{E} \cdot \frac{\partial \mathbf{E}}{\partial t} - \mu \mathbf{H} \cdot \frac{\partial \mathbf{H}}{\partial t} \tag{3.57}$$

or, equivalently,

$$\nabla \cdot (\mathbf{E} \times \mathbf{H}) = -\varepsilon \frac{1}{2} \frac{\partial}{\partial t}(\mathbf{E} \cdot \mathbf{E}) - \mu \frac{1}{2} \frac{\partial}{\partial t}(\mathbf{H} \cdot \mathbf{H}) \tag{3.58}$$

By defining

$$u = \varepsilon \frac{1}{2} \mathbf{E} \cdot \mathbf{E} + \mu \frac{1}{2} \mathbf{H} \cdot \mathbf{H} \tag{3.59}$$

and

$$\mathbf{S} = \mathbf{E} \times \mathbf{H} \tag{3.60}$$

Equation 3.58 takes the form of an energy conservation law

$$\nabla \cdot \mathbf{S} = -\frac{\partial u}{\partial t} \tag{3.61}$$

where $u = u(\mathbf{r}, t)$ is the instantaneous electromagnetic *energy density* (Joules per cubic meter), and $\mathbf{S} = \mathbf{S}(\mathbf{r}, t)$ is the instantaneous *Poynting vector*, which represents the amount of electromagnetic energy per unit area per unit time (Joules per square meter per second) crossing a surface perpendicular to the direction of **S**. Both quantities u and **S** are real functions of the position vector **r** and time t. Note that the energy density u given by Equation 3.59 is the sum of electric and magnetic energy densities.

To understand Equation 3.61 as an energy conservation law, we take the integral over an arbitrary volume V enclosed by a surface A, and use the divergence theorem to obtain

$$\iint_A \mathbf{S} \cdot d\mathbf{A} = -\frac{d}{dt} \int_V u \, dV \tag{3.62}$$

Equation 3.62 indicates that the integral of the Poynting vector over the closed three-dimensional surface A, which represents the energy per unit time flowing across the surface A (out of the volume V), is equal to the rate at which the energy decreases in the volume V. Therefore, Equation 3.62 shows that the electromagnetic energy is conserved.

3.6.1
Energy Density and Energy Flux for Electromagnetic Plane Waves

In the more general case, the electromagnetic energy density u is the sum of electric and magnetic energy densities (Equation 3.59). However, in the case of electromagnetic plane waves, the electromagnetic energy density u can be written in a more convenient form because of the fact that the electric and magnetic energy densities are equal.

For example, by using Equations 3.41 and 3.42 and the vector identity $(\hat{\mathbf{k}} \times \mathbf{H}) \cdot \mathbf{E} = (\mathbf{E} \times \hat{\mathbf{k}}) \cdot \mathbf{H}$, we have

$$\varepsilon \frac{1}{2} \mathbf{E} \cdot \mathbf{E} = -\varepsilon \frac{1}{2}\sqrt{\frac{\mu}{\varepsilon}}(\hat{\mathbf{k}} \times \mathbf{H}) \cdot \mathbf{E} = -\varepsilon \frac{1}{2}\sqrt{\frac{\mu}{\varepsilon}}(\mathbf{E} \times \hat{\mathbf{k}}) \cdot \mathbf{H} = \mu \frac{1}{2} \mathbf{H} \cdot \mathbf{H} \qquad (3.63)$$

Equation 3.63 shows that the instantaneous energy density u given by Equation 3.59 can therefore be written as

$$u = \varepsilon \mathbf{E} \cdot \mathbf{E} \qquad (3.64)$$

We showed in the previous section that the instantaneous Poynting vector $\mathbf{S}$ is given by the cross-product between the electric field $\mathbf{E}$ and the magnetic field $\mathbf{H}$. In the case of electromagnetic plane waves, however, the Poynting vector $\mathbf{S}$ can also be written in a more convenient form. For example, by using Equations 3.41 and 3.42, the vector identity $\mathbf{E} \times (\hat{\mathbf{k}} \times \mathbf{E}) = (\mathbf{E} \cdot \mathbf{E})\hat{\mathbf{k}} - (\mathbf{E} \cdot \hat{\mathbf{k}})\mathbf{E}$, and the fact that $\mathbf{E} \cdot \hat{\mathbf{k}} = 0$, we can write

$$\mathbf{S} = \mathbf{E} \times \mathbf{H} = \sqrt{\frac{\varepsilon}{\mu}} \mathbf{E} \times (\hat{\mathbf{k}} \times \mathbf{E}) = \sqrt{\frac{\varepsilon}{\mu}}(\mathbf{E} \cdot \mathbf{E})\hat{\mathbf{k}} - \sqrt{\frac{\varepsilon}{\mu}}(\mathbf{E} \cdot \hat{\mathbf{k}})\mathbf{E} = \sqrt{\frac{\varepsilon}{\mu}}(\mathbf{E} \cdot \mathbf{E})\hat{\mathbf{k}} \qquad (3.65)$$

Importantly, Equation 3.65 shows that the direction of the Poynting vector is parallel to the unit vector $\hat{\mathbf{k}}$. This means that the electromagnetic energy flows along the direction of the wave vector $\mathbf{k}$, which is the direction of propagation of the plane wave.

3.6.2
Time-averaged Values

The electric and magnetic fields $\mathbf{E}$ and $\mathbf{H}$ vary so rapidly in time (Table 3.1) that their instantaneous values, which are given by Equations 3.14 and 3.15, cannot be measured in practice. In fact, with currently available experimental techniques, only time-averaged values of the fields can be measured. The same occurs with the instantaneous values of the electromagnetic energy density u and the Poynting vector $\mathbf{S}$, which are given by Equations 3.64 and 3.65, respectively. Therefore, it is important to calculate time-averaged values for these relevant physical quantities, because these are the values that can be measured in practice.

By using Equations 3.64 and 3.65, the *time-averaged energy density* $\langle u \rangle$ and the *time-averaged energy crossing a unit area per unit time* $\langle \mathbf{S} \rangle$ of an electromagnetic plane wave are given by

$$\langle u \rangle = \varepsilon \langle \mathbf{E} \cdot \mathbf{E} \rangle \tag{3.66}$$

$$\langle \mathbf{S} \rangle = \sqrt{\frac{\varepsilon}{\mu}} \langle \mathbf{E} \cdot \mathbf{E} \rangle \hat{\mathbf{k}} \tag{3.67}$$

Since both physical quantities depend on the time-averaged value of the dot product of the electric field $\mathbf{E}$, we next calculate explicitly the value of $\langle \mathbf{E} \cdot \mathbf{E} \rangle$.

By using Equation 3.14, we have

$$\mathbf{E} \cdot \mathbf{E} = \mathrm{Re}(\mathbf{E}_0\, e^{i(\mathbf{k}\cdot\mathbf{r}-\omega t)}) \cdot \mathrm{Re}(\mathbf{E}_0\, e^{i(\mathbf{k}\cdot\mathbf{r}-\omega t)}) \tag{3.68}$$

By defining

$$\tilde{\mathbf{E}} = \mathbf{E}_0\, e^{i\mathbf{k}\cdot\mathbf{r}} \tag{3.69}$$

Equation 3.68 transforms into

$$\mathbf{E} \cdot \mathbf{E} = \mathrm{Re}(\tilde{\mathbf{E}}\, e^{-i\omega t}) \cdot \mathrm{Re}(\tilde{\mathbf{E}}\, e^{-i\omega t}) \tag{3.70}$$

By separating the complex vector $\tilde{\mathbf{E}}$ into its real and imaginary parts as $\tilde{\mathbf{E}} = \tilde{\mathbf{E}}_R + i\tilde{\mathbf{E}}_I$, and taking the real parts in Equation 3.70, we have

$$\begin{aligned}\mathbf{E} \cdot \mathbf{E} &= [\tilde{\mathbf{E}}_R \cos(\omega t) + \tilde{\mathbf{E}}_I \sin(\omega t)] \cdot [\tilde{\mathbf{E}}_R \cos(\omega t) + \tilde{\mathbf{E}}_I \sin(\omega t)] \\ &= \tilde{\mathbf{E}}_R \cdot \tilde{\mathbf{E}}_R \cos^2(\omega t) + \tilde{\mathbf{E}}_I \cdot \tilde{\mathbf{E}}_I \sin^2(\omega t) + 2\tilde{\mathbf{E}}_R \cdot \tilde{\mathbf{E}}_I \cos(\omega t)\sin(\omega t)\end{aligned} \tag{3.71}$$

Because $\tilde{\mathbf{E}}_R$ and $\tilde{\mathbf{E}}_I$ do not depend on the time t, the time-averaged value of Equation 3.71 is given by

$$\langle \mathbf{E} \cdot \mathbf{E} \rangle = \tilde{\mathbf{E}}_R \cdot \tilde{\mathbf{E}}_R \langle \cos^2(\omega t)\rangle + \tilde{\mathbf{E}}_I \cdot \tilde{\mathbf{E}}_I \langle \sin^2(\omega t)\rangle + 2\tilde{\mathbf{E}}_R \cdot \tilde{\mathbf{E}}_I \langle \cos(\omega t)\sin(\omega t)\rangle \tag{3.72}$$

By considering that

$$\langle \cos^2(\omega t)\rangle = \frac{1}{2}, \quad \langle \sin^2(\omega t)\rangle = \frac{1}{2}, \quad \langle \cos(\omega t)\sin(\omega t)\rangle = 0 \tag{3.73}$$

we finally have

$$\langle \mathbf{E} \cdot \mathbf{E} \rangle = \frac{1}{2}\tilde{\mathbf{E}}_R \cdot \tilde{\mathbf{E}}_R + \frac{1}{2}\tilde{\mathbf{E}}_I \cdot \tilde{\mathbf{E}}_I = \frac{1}{2}(\tilde{\mathbf{E}} \cdot \tilde{\mathbf{E}}^*) = \frac{1}{2}(\mathbf{E}_0 \cdot \mathbf{E}_0^*) = \frac{1}{2}|\mathbf{E}_0|^2 \tag{3.74}$$

where $\tilde{\mathbf{E}}^*$ indicates the vector conjugate of $\tilde{\mathbf{E}}$.

3.6 Electromagnetic Energy

By considering Equation 3.74, the time-averaged value of the energy density $\langle u \rangle$ given by Equation 3.66 and the time-averaged value of the energy per unit area per unit time $\langle \mathbf{S} \rangle$ given by Equation 3.67 can now be written as

$$\langle u \rangle = \frac{1}{2}\varepsilon |\mathbf{E}_0|^2 \tag{3.75}$$

$$\langle \mathbf{S} \rangle = \frac{1}{2}\sqrt{\frac{\varepsilon}{\mu}}|\mathbf{E}_0|^2 \hat{\mathbf{k}} \tag{3.76}$$

Importantly, since ε, μ, and $\mathbf{E}_0$ are constant quantities, Equations 3.75 and 3.76 determine that the time-averaged quantities $\langle u \rangle$ and $\langle \mathbf{S} \rangle$ do not depend on the position and have the same *constant* value throughout the material in which the plane wave propagates.

■ **Example**
Consider an electromagnetic plane wave propagating along the z direction, with a complex amplitude vector $\mathbf{E}_0 = (2 + 4i)\hat{\mathbf{x}} + (1 + 3i)\hat{\mathbf{y}}$, within a material characterized by the parameters ε and μ.
Because $|\mathbf{E}_0|^2 = \mathbf{E}_0 \cdot \mathbf{E}_0^* = [(2 + 4i)\hat{\mathbf{x}} + (1 + 3i)\hat{\mathbf{y}}] \cdot [(2 - 4i)\hat{\mathbf{x}} + (1 - 3i)\hat{\mathbf{y}}] = 20 + 10 = 30$, the time-averaged energy density and the time-averaged Poynting vector are given by

$$\langle u \rangle = \frac{1}{2}\varepsilon\, 30, \quad \langle \mathbf{S} \rangle = \frac{1}{2}\sqrt{\frac{\varepsilon}{\mu}}\, 30\,\hat{\mathbf{z}}$$

We now examine the relationship between the time-averaged energy density $\langle u \rangle$ and the time-averaged energy per unit area per unit time $\langle \mathbf{S} \rangle$.

Consider an electromagnetic plane wave with velocity v crossing a circularly shaped surface area A, which is aligned perpendicularly to the direction of propagation of the wave (Figure 3.6). Since the Poynting vector $\mathbf{S}$ is parallel to the direction of propagation of the wave, the amount of energy that passes over the surface area A, after an interval of time Δt, is given by

$$\text{Energy} = |\langle \mathbf{S} \rangle| A \Delta t \tag{3.77}$$

After an interval of time Δt, this amount of energy is contained within the cylinder shown in Figure 3.6, which has a volume V equal to $Av\Delta t$. The amount of energy within the volume is also given by

$$\text{Energy} = \langle u \rangle Av\Delta t \tag{3.78}$$

By comparing Equations 3.77 and 3.78, we have

$$|\langle \mathbf{S} \rangle| = \langle u \rangle v \tag{3.79}$$

This means that the Poynting vector $\mathbf{S}$ and the energy density u are not independent quantities. In fact, the time-averaged energy per unit area per unit

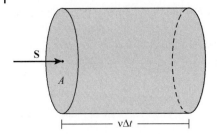

Fig. 3.6 The electromagnetic energy that crosses the surface area A during the time Δt is contained within the cylindrical volume of area A and length $v\Delta t$.

time (which is given by **S**) is equal to the time-averaged energy density u multiplied by the velocity v of the wave (which is the speed at which the energy is flowing).

3.6.3
Intensity

The magnitude of the time-averaged Poynting vector is also known as the *intensity* I (or irradiance) of the electromagnetic plane wave. This means that the intensity I of a plane wave, at any given point in the material, is the time-averaged energy per unit area per unit time (with units of Joule per square meter per second) crossing a surface perpendicular to the direction of the propagation of the plane wave.

For a *single* electromagnetic plane wave given by

$$\mathbf{E}(\mathbf{r}, t) = \text{Re}(\mathbf{E}_0 \, e^{i(\mathbf{k}\cdot\mathbf{r} - \omega t)}) \tag{3.80}$$

the intensity I of the wave is thus (Equation 3.67)

$$I = |\langle \mathbf{S} \rangle| = \sqrt{\frac{\varepsilon}{\mu}} \langle \mathbf{E} \cdot \mathbf{E} \rangle \tag{3.81}$$

which by using Equation 3.74 transforms into

$$I = \frac{1}{2}\sqrt{\frac{\varepsilon}{\mu}} |\mathbf{E}_0|^2 \tag{3.82}$$

This means that the intensity I of a single plane wave is proportional to the square of the modulus of the electric field amplitude vector $\mathbf{E}_0$. Note that since ε, μ, and $\mathbf{E}_0$ are constant quantities, the intensity of a plane wave has the same value at each point where the wave propagates.

For reasons that will be evident in the next section, it is useful to consider the intensity that corresponds to a more general type of electromagnetic wave given by the equation

$$\mathbf{E}(\mathbf{r}, t) = \text{Re}(\mathbf{A}(\mathbf{r}) \, e^{-i\omega t}) \tag{3.83}$$

where **A** is an arbitrary complex amplitude vector that depends on the position **r**.

For this type of wave, the intensity relation 3.81 holds at least as an approximation [3]. We can therefore measure the intensity of the more general wave (Equation 3.83) as

$$I = \sqrt{\frac{\varepsilon}{\mu}} \langle \mathbf{E} \cdot \mathbf{E} \rangle \tag{3.84}$$

By following the scheme presented in Section 3.6.2, we can calculate the time-averaged value $\langle \mathbf{E} \cdot \mathbf{E} \rangle$ for the more general electromagnetic wave given by Equation 3.83 and replace this value in Equation 3.84. The intensity I corresponding to the general type of wave given by Equation 3.83 is therefore

$$I(\mathbf{r}) = \frac{1}{2}\sqrt{\frac{\varepsilon}{\mu}} |\mathbf{A}(\mathbf{r})|^2 \tag{3.85}$$

Importantly, in contrast to a single plane wave, Equation 3.85 determines that the intensity $I(\mathbf{r})$ of the more general type of wave depends on the position vector **r**.

Finally, since it is common in optics to compare intensities of electromagnetic waves propagating in the same material, we can neglect the common factors in Equations 3.82 and 3.85 and simply write

$$I = |\mathbf{E}_0|^2 \quad \text{for a single plane wave} \tag{3.86}$$

and

$$I(\mathbf{r}) = |\mathbf{A}(\mathbf{r})|^2 \quad \text{for a more general type of wave} \tag{3.87}$$

3.7
Interference of Electromagnetic Plane Waves

In the previous sections, we considered the propagation of an electromagnetic plane wave given by Equations 3.14 and 3.15 and studied many of its physical properties (e.g. the transverse character, polarization states, energy density, energy flux, and intensity). It is important to remark that in the previous sections it was assumed that there was a *single* wave propagating within the material.

Conversely, in the following sections, we examine the propagation of many electromagnetic plane waves in the same material at the same time. We will see that various interesting physical effects arise from the superposition of many electromagnetic plane waves. In particular, we begin our study by examining the resultant effects on the distribution of the intensity in space when *two* electromagnetic plane waves, with the same angular frequency ω, propagate simultaneously in the same material.

The two plane waves are given by

$$\mathbf{E}_1(\mathbf{r}, t) = \mathrm{Re}(\mathbf{E}_{01}\, e^{i(\mathbf{k}_1 \cdot \mathbf{r} - \omega t)}) \tag{3.88}$$

$$\mathbf{E}_2(\mathbf{r}, t) = \mathrm{Re}(\mathbf{E}_{02}\, e^{i(\mathbf{k}_2 \cdot \mathbf{r} - \omega t)}) \tag{3.89}$$

where $\mathbf{k}_1$ and $\mathbf{k}_2$ are the corresponding wave vectors and $\mathbf{E}_{01}$ and $\mathbf{E}_{02}$ are the complex electric field amplitude vectors. According to the previous section, the intensities of the *individual* plane waves (Equations 3.88 and 3.89) at any given point in the material are given by

$$I_1 = |\mathbf{E}_{01}|^2 \tag{3.90}$$

$$I_2 = |\mathbf{E}_{02}|^2 \tag{3.91}$$

Because I_1 and I_2 are constant, Equations 3.90 and 3.91 determine that if the plane waves propagate separately, the electromagnetic energy of each wave is constant and therefore homogeneously distributed in space.

We now consider that the two plane waves propagate in the same material at the same time. Because the material is assumed to be *linear*, we can apply the principle of superposition. This principle states that at any given point in space the total electric field is the sum of the electric fields of the individual plane waves. That is,

$$\mathbf{E}_T(\mathbf{r}, t) = \mathbf{E}_1(\mathbf{r}, t) + \mathbf{E}_2(\mathbf{r}, t) \tag{3.92}$$

or, equivalently,

$$\begin{aligned}\mathbf{E}_T(\mathbf{r}, t) &= \mathrm{Re}(\mathbf{E}_{01}\, e^{i(\mathbf{k}_1 \cdot \mathbf{r} - \omega t)}) + \mathrm{Re}(\mathbf{E}_{02}\, e^{i(\mathbf{k}_2 \cdot \mathbf{r} - \omega t)}) \\ &= \mathrm{Re}[(\mathbf{E}_{01}\, e^{i\mathbf{k}_1 \cdot \mathbf{r}} + \mathbf{E}_{02}\, e^{i\mathbf{k}_2 \cdot \mathbf{r}})\, e^{-i\omega t}]\end{aligned} \tag{3.93}$$

The electric field in Equation 3.93, which describes the superposition of the two plane waves, is not a single plane wave but a more general type of wave. By considering Equations 3.83 and 3.87, the intensity of the *resultant* wave (Equation 3.93) is given by

$$I(\mathbf{r}) = |\mathbf{A}(\mathbf{r})|^2, \quad \text{where} \quad \mathbf{A}(\mathbf{r}) = \mathbf{E}_{01}\, e^{i\mathbf{k}_1 \cdot \mathbf{r}} + \mathbf{E}_{02}\, e^{i\mathbf{k}_2 \cdot \mathbf{r}} \tag{3.94}$$

By expanding Equation 3.94, we have

$$\begin{aligned}I(\mathbf{r}) = |\mathbf{A}(\mathbf{r})|^2 &= \mathbf{A}(\mathbf{r}) \cdot \mathbf{A}^*(\mathbf{r}) \\ &= (\mathbf{E}_{01}\, e^{i\mathbf{k}_1 \cdot \mathbf{r}} + \mathbf{E}_{02}\, e^{i\mathbf{k}_2 \cdot \mathbf{r}}) \cdot (\mathbf{E}_{01}^*\, e^{-i\mathbf{k}_1 \cdot \mathbf{r}} + \mathbf{E}_{02}^*\, e^{-i\mathbf{k}_2 \cdot \mathbf{r}})\end{aligned} \tag{3.95}$$

or, equivalently,

$$\begin{aligned}I(\mathbf{r}) = &|\mathbf{E}_{01}|^2 + |\mathbf{E}_{02}|^2 + 2\mathrm{Re}(\mathbf{E}_{01} \cdot \mathbf{E}_{02}^*) \cos[(\mathbf{k}_1 - \mathbf{k}_2) \cdot \mathbf{r}] \\ & - 2\mathrm{Im}(\mathbf{E}_{01} \cdot \mathbf{E}_{02}^*) \sin[(\mathbf{k}_1 - \mathbf{k}_2) \cdot \mathbf{r}]\end{aligned} \tag{3.96}$$

where Re and Im denote respectively the real and imaginary part of the corresponding complex numbers.

Equation 3.96 reveals one of the most important consequences of the simultaneous propagation of two electromagnetic plane waves with the same frequency in the same material. The intensity I of the total wave is *not* equal to the sum of the intensities $I_1 + I_2$ of the individual waves but consists of two additional trigonometric terms called the interference terms. These terms cause the intensity I to depend on the position $\mathbf{r}$. Therefore, *the superposition of electromagnetic plane waves creates a nonhomogeneous distribution of energy in space.*

This important physical phenomenon is known as *wave interference*. Note that there is no gain or loss of energy. As a consequence of the superposition of the plane waves, the resultant intensity I is distributed over regions where its value is larger than the sum of the intensities $I_1 + I_2$, but also distributed over regions where its value is less than $I_1 + I_2$. Since the spatial average of the intensity I is equal to $I_1 + I_2$ (Equation 3.96), the energy is conserved (although it is nonhomogeneously distributed in space).

Note that the magnitudes of the wave vectors $\mathbf{k}_1$ and $\mathbf{k}_2$ in Equation 3.96 are equal because the two waves that propagate in the same material have the same frequency (Equations 3.12 and 3.20). As a result, if the wave vectors $\mathbf{k}_1$ and $\mathbf{k}_2$ are collinear, $\mathbf{k}_1 - \mathbf{k}_2 = 0$ and the intensity I is homogeneously distributed in space. This means that interference effects are observed only if the directions of propagation of the waves (which are given by $\mathbf{k}_1$ and $\mathbf{k}_2$) are noncollinear. Also note that in order to obtain the interference terms, the dot product $\mathbf{E}_{01} \cdot \mathbf{E}_{02}^*$ must be different from zero. That is, the electric field amplitude vectors (i.e. the polarization states) of the individual plane waves must not be perpendicular.

In brief, interference effects between two electromagnetic plane waves are observed only when

- the wave vectors of the interfering waves are not collinear and
- the electric field vectors of the interfering waves are not perpendicular.

To illustrate the basic concepts on wave interference, we show an example where two linearly polarized electromagnetic plane waves propagate at right angles in the same material at the same time.

The linearly polarized plane waves are given by

$$\mathbf{E}_1(\mathbf{r}, t) = \mathrm{Re}(E_{01} \hat{\mathbf{n}}_1 \, e^{i \, (\mathbf{k}_1 \cdot \mathbf{r} - \omega t)}) \tag{3.97}$$

$$\mathbf{E}_2(\mathbf{r}, t) = \mathrm{Re}(E_{02} \hat{\mathbf{n}}_2 \, e^{i(\mathbf{k}_2 \cdot \mathbf{r} - \omega t)}) \tag{3.98}$$

whereas the electric field amplitude vectors and the wave vectors are given by

$$E_{01} = 1 \, e^{i\varphi_1}, \quad \hat{\mathbf{n}}_1 = \hat{\mathbf{z}} = (0, 0, 1), \quad \mathbf{k}_1 = \frac{2\pi}{\lambda}\hat{\mathbf{x}} = \left(\frac{2\pi}{\lambda}, 0, 0\right) \tag{3.99}$$

$$E_{02} = 1 \, e^{i\varphi_2}, \quad \hat{\mathbf{n}}_2 = \hat{\mathbf{z}} = (0, 0, 1), \quad \mathbf{k}_2 = \frac{2\pi}{\lambda}\hat{\mathbf{y}} = \left(0, \frac{2\pi}{\lambda}, 0\right) \tag{3.100}$$

where λ is the wavelength of the interfering electromagnetic plane waves.

After inserting Equations 3.99 and 3.100 into Equations 3.97 and 3.98, we have

$$\mathbf{E}_1(\mathbf{r}, t) = 1 \cos\left(\frac{2\pi}{\lambda}x - \omega t + \varphi_1\right) \hat{\mathbf{z}} \qquad (3.101)$$

$$\mathbf{E}_2(\mathbf{r}, t) = 1 \cos\left(\frac{2\pi}{\lambda}y - \omega t + \varphi_2\right) \hat{\mathbf{z}} \qquad (3.102)$$

and the total electric field corresponding to the superposition of the waves is given by

$$\mathbf{E}_T(\mathbf{r}, t) = 1 \cos\left(\frac{2\pi}{\lambda}x - \omega t + \varphi_1\right) \hat{\mathbf{z}} + 1 \cos\left(\frac{2\pi}{\lambda}y - \omega t + \varphi_2\right) \hat{\mathbf{z}} \qquad (3.103)$$

Figure 3.7 displays both the electric fields of the individual waves and the total electric field at a particular time t. High and low electric field values are illustrated in white and black, respectively. Owing to the principle of superposition, the total electric field is just the sum of the individual electric fields. It is important to remark that because the individual electric fields (Equations 3.101 and 3.102) and the total electric field (Equation 3.103) depend on the time t, the electric field patterns shown in Figure 3.7 vary with time.

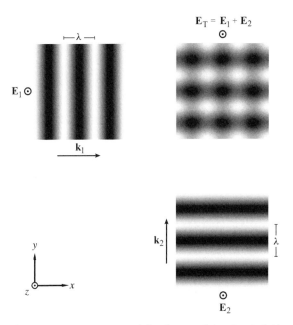

Fig. 3.7 Instantaneous spatial distribution of the electric field **E** corresponding to the superposition of two electromagnetic plane waves with perpendicular wave vectors $\mathbf{k}_1$ and $\mathbf{k}_2$, and parallel electric field vectors $\mathbf{E}_1$ and $\mathbf{E}_2$. The diagram on the right upper corner shows the superposition of the electric fields $\mathbf{E}_1$ and $\mathbf{E}_2$ at a particular time t.

On the other hand, the intensities of the *individual* plane waves are given by

$$I_1 = |\mathbf{E}_{01}|^2 = 1 \tag{3.104}$$
$$I_2 = |\mathbf{E}_{02}|^2 = 1 \tag{3.105}$$

Because I_1 and I_2 are constant, if the plane waves are allowed to propagate separately, the intensity of each wave is homogeneously distributed in space (gray regions in Figure 3.8). On the other hand, if the waves are allowed to propagate in the same region of space at the same time, the intensity $I(\mathbf{r})$ of the *total* wave is *not* homogeneously distributed in space (diagram in the right upper corner of Figure 3.8). The intensity $I(\mathbf{r})$ of the total wave can be obtained by inserting Equations 3.99 and 3.100 into Equation 3.96, which gives

$$I(\mathbf{r}) = 1^2 + 1^2 + 2\cos(\varphi_1 - \varphi_2)\cos\left[\frac{2\pi}{\lambda}(x - y)\right]$$
$$- 2\sin(\varphi_1 - \varphi_2)\sin\left[\frac{2\pi}{\lambda}(x - y)\right] \tag{3.106}$$

or, equivalently,

$$I(\mathbf{r}) = 2 + 2\cos\left[\frac{2\pi}{\lambda}(x - y) + \varphi_1 - \varphi_2\right] \tag{3.107}$$

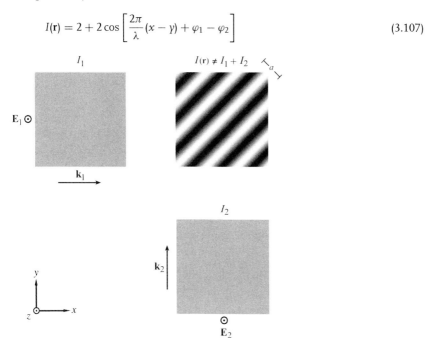

Fig. 3.8 Spatial distribution of the intensity (or energy) $I(\mathbf{r})$ corresponding to the superposition of two electromagnetic plane waves with perpendicular wave vectors $\mathbf{k}_1$ and $\mathbf{k}_2$, and with parallel electric field vectors $\mathbf{E}_1$ and $\mathbf{E}_2$. The diagram on the right upper corner shows the resultant distribution of the intensity $I(\mathbf{r})$ when the plane waves propagate in the same region of space at the same time. Note: the intensities I_1, I_2, and $I(\mathbf{r})$ do not vary with time.

Figure 3.8 shows the homogeneous intensities I_1 and I_2 of the individual waves and the nonhomogeneous intensity $I(\mathbf{r})$ of the total wave for the case $\varphi_1 = \varphi_2 = 0$. High and low intensity values are illustrated in white and black, respectively. We can see that the two linearly polarized waves interfere with each other, generating a one-dimensional periodic spatial distribution of the intensity $I(\mathbf{r})$ that varies along a single direction. This periodic variation of the intensity is called an *interference pattern*. Importantly, because Equations 3.104, 3.105, and 3.107 do not depend on the time t, the intensities I_1, I_2, and $I(\mathbf{r})$ shown in Figure 3.8 *do not change with time*.

The fringe-to-fringe spacing, a, of the periodic interference pattern $I(\mathbf{r})$ shown in Figure 3.8 is determined by the wavelength λ of the interfering plane waves (Equation 3.107). In the more general case, the spacing (or period) a of the one-dimensional interference pattern created by the superposition of two plane waves is given by the formula

$$a = \frac{\lambda}{2\sin(\theta/2)} \tag{3.108}$$

where θ is the angle between the wave vectors of the interfering plane waves. In the particular case shown in Figure 3.8 we have $\theta = \pi/2$, and therefore $a = \lambda/\sqrt{2}$.

Equation 3.107 also determines that the phase difference $\varphi_1 - \varphi_2$ between the complex amplitudes E_{01} and E_{02} does not alter the shape of the one-dimensional periodic interference pattern. In fact, the phase difference $\varphi_1 - \varphi_2$ only translates the interference pattern in space without changing its shape.

3.7.1
Three-dimensional Interference Patterns

In the previous section, we showed that a one-dimensional periodic interference pattern can be created by the superposition of two electromagnetic plane waves. However, if we want to create two- and three-dimensional periodic interference patterns, we need to consider more than two plane waves. This is because the dimensionality of the resultant periodic interference pattern is determined by the number of electromagnetic plane waves that interfere with each other. For example,

- the superposition of two noncollinear plane waves generates a one-dimensional periodic interference pattern;
- the superposition of three noncoplanar plane waves generates a two-dimensional periodic interference pattern;
- the superposition of four or more noncoplanar plane waves generates a three-dimensional periodic interference pattern.

In this section, we are interested in obtaining three-dimensional periodic interference patterns, and we therefore need to consider the simultaneous propagation of at least four electromagnetic plane waves. For simplicity, we next examine the distribution of the intensity $I(\mathbf{r})$ created by the simultaneous propagation of *four* plane waves with the same frequency ω.

3.7 Interference of Electromagnetic Plane Waves

The plane waves are given by

$$E_1(r, t) = \mathrm{Re}(E_{01}\, e^{i(k_1 \cdot r - \omega t)}) \tag{3.109}$$

$$E_2(r, t) = \mathrm{Re}(E_{02}\, e^{i(k_2 \cdot r - \omega t)}) \tag{3.110}$$

$$E_3(r, t) = \mathrm{Re}(E_{03}\, e^{i(k_3 \cdot r - \omega t)}) \tag{3.111}$$

$$E_4(r, t) = \mathrm{Re}(E_{04}\, e^{i(k_4 \cdot r - \omega t)}) \tag{3.112}$$

where $k_i (i = 1, \ldots, 4)$ are the corresponding wave vectors, and E_{0i} are the complex electric field amplitude vectors.

In complete analogy with the previous section, the total electric field E_T at any point is given by the sum of the electric fields of the individual waves. That is

$$E_T(r, t) = E_1(r, t) + E_2(r, t) + E_3(r, t) + E_4(r, t) \tag{3.113}$$

The intensities of the individual plane waves (Equations 3.109–3.112) are given by

$$I_1 = |E_{01}|^2 \tag{3.114}$$

$$I_2 = |E_{02}|^2 \tag{3.115}$$

$$I_3 = |E_{03}|^2 \tag{3.116}$$

$$I_4 = |E_{04}|^2 \tag{3.117}$$

whereas the intensity $I(r)$ of the total wave is given by (Equation 3.94)

$$I(r) = |A(r)|^2, \quad \text{where} \quad A(r) = E_{01}\, e^{ik_1 \cdot r} + E_{02}\, e^{ik_2 \cdot r} + E_{03}\, e^{ik_3 \cdot r} + E_{04}\, e^{ik_4 \cdot r} \tag{3.118}$$

After expanding Equation 3.118, we have

$$\begin{aligned}
I(r) = &\, |E_{01}|^2 + |E_{02}|^2 + |E_{03}|^2 + |E_{04}|^2 \\
&+ 2\mathrm{Re}(E_{01} \cdot E_{02}^*) \cos[(k_1 - k_2) \cdot r] - 2\,\mathrm{Im}(E_{01} \cdot E_{02}^*) \sin[(k_1 - k_2) \cdot r] \\
&+ 2\mathrm{Re}(E_{01} \cdot E_{03}^*) \cos[(k_1 - k_3) \cdot r] - 2\,\mathrm{Im}(E_{01} \cdot E_{03}^*) \sin[(k_1 - k_3) \cdot r] \\
&+ 2\mathrm{Re}(E_{01} \cdot E_{04}^*) \cos[(k_1 - k_4) \cdot r] - 2\,\mathrm{Im}(E_{01} \cdot E_{04}^*) \sin[(k_1 - k_4) \cdot r] \\
&+ 2\mathrm{Re}(E_{02} \cdot E_{03}^*) \cos[(k_2 - k_3) \cdot r] - 2\,\mathrm{Im}(E_{02} \cdot E_{03}^*) \sin[(k_2 - k_3) \cdot r] \\
&+ 2\mathrm{Re}(E_{02} \cdot E_{04}^*) \cos[(k_2 - k_4) \cdot r] - 2\,\mathrm{Im}(E_{02} \cdot E_{04}^*) \sin[(k_2 - k_4) \cdot r] \\
&+ 2\mathrm{Re}(E_{03} \cdot E_{04}^*) \cos[(k_3 - k_4) \cdot r] - 2\,\mathrm{Im}(E_{03} \cdot E_{04}^*) \sin[(k_3 - k_4) \cdot r]
\end{aligned} \tag{3.119}$$

Equation 3.119 shows that the superposition of four electromagnetic plane waves generates 16 terms in the expression of the intensity $I(r)$ (four constant terms $|E_{01}|^2$, $|E_{02}|^2$, $|E_{03}|^2$, and $|E_{04}|^2$, plus 12 interference terms in the form of cosine and sine terms). If the wave vectors k_1, k_2, k_3, and k_4 are noncoplanar, the intensity $I(r)$ forms a three-dimensional interference pattern.

To illustrate the basic concepts of wave interference in three dimensions, we examine a simple example where four linearly polarized electromagnetic plane waves propagate in the same material at the same time:

$$\mathbf{E}_1(\mathbf{r}, t) = \text{Re}(E_{01}\hat{\mathbf{n}}_1 \, e^{i(\mathbf{k}_1 \cdot \mathbf{r} - \omega t)}) \tag{3.120}$$

$$\mathbf{E}_2(\mathbf{r}, t) = \text{Re}(E_{02}\hat{\mathbf{n}}_2 \, e^{i(\mathbf{k}_2 \cdot \mathbf{r} - \omega t)}) \tag{3.121}$$

$$\mathbf{E}_3(\mathbf{r}, t) = \text{Re}(E_{03}\hat{\mathbf{n}}_3 \, e^{i(\mathbf{k}_3 \cdot \mathbf{r} - \omega t)}) \tag{3.122}$$

$$\mathbf{E}_4(\mathbf{r}, t) = \text{Re}(E_{04}\hat{\mathbf{n}}_4 \, e^{i(\mathbf{k}_4 \cdot \mathbf{r} - \omega t)}) \tag{3.123}$$

We carefully choose the electric field amplitude vectors and the wave vectors to be

$$E_{01} = 1.000, \quad \hat{\mathbf{n}}_1 = (+0.000, +0.707, -0.707),$$

$$\mathbf{k}_1 = \frac{2\pi}{\lambda}\left(-\frac{1}{\sqrt{3}}, -\frac{1}{\sqrt{3}}, -\frac{1}{\sqrt{3}}\right) \tag{3.124}$$

$$E_{02} = 0.632, \quad \hat{\mathbf{n}}_2 = (-0.500, +0.309, -0.809),$$

$$\mathbf{k}_2 = \frac{2\pi}{\lambda}\left(+\frac{1}{\sqrt{3}}, -\frac{1}{\sqrt{3}}, -\frac{1}{\sqrt{3}}\right) \tag{3.125}$$

$$E_{03} = 0.874, \quad \hat{\mathbf{n}}_3 = (+0.809, +0.500, -0.309),$$

$$\mathbf{k}_3 = \frac{2\pi}{\lambda}\left(-\frac{1}{\sqrt{3}}, +\frac{1}{\sqrt{3}}, -\frac{1}{\sqrt{3}}\right) \tag{3.126}$$

$$E_{04} = 2.288, \quad \hat{\mathbf{n}}_4 = (-0.309, +0.809, +0.500),$$

$$\mathbf{k}_4 = \frac{2\pi}{\lambda}\left(-\frac{1}{\sqrt{3}}, -\frac{1}{\sqrt{3}}, +\frac{1}{\sqrt{3}}\right) \tag{3.127}$$

where λ is the wavelength of the interfering waves.

Note that the magnitudes of the wave vectors $\mathbf{k}_1$, $\mathbf{k}_2$, $\mathbf{k}_3$, and $\mathbf{k}_4$ in Equations 3.124–3.127 are equal because the four waves propagate in the same material with the same frequency (Equations 3.12 and 3.20).

By inserting Equations 3.124–3.127 into 3.119, the spatial distribution of the intensity $I(\mathbf{r})$ is given by

$$I(\mathbf{r}) = 7.4 + \cos\left(\frac{4\pi x}{\sqrt{3}\lambda}\right) + \cos\left(\frac{4\pi y}{\sqrt{3}\lambda}\right) + \cos\left(\frac{4\pi z}{\sqrt{3}\lambda}\right) \tag{3.128}$$

Note that Equation 3.128 indicates that the distribution of the intensity $I(\mathbf{r})$ corresponding to the superposition of the four electromagnetic plane waves varies periodically in three dimensions. This means that the four plane waves given by Equations 3.120–3.127 generate a three-dimensional interference pattern. Equation 3.128 also determines that the spacing a (or period) of the three-dimensional interference pattern along the x, y, and z directions is the same and it is given by

$$a = \frac{\sqrt{3}\lambda}{2} \tag{3.129}$$

After substituting Equation 3.129 into Equation 3.128, we finally have

$$I(\mathbf{r}) = 7.4 + \cos\left(\frac{2\pi x}{a}\right) + \cos\left(\frac{2\pi y}{a}\right) + \cos\left(\frac{2\pi z}{a}\right) \tag{3.130}$$

It is important to note the similarity between the resultant intensity distribution (Equation 3.130) and the formula for the simple periodic function (2.33) that we developed in Chapter 2. As we show in the next section, this similarity is the basis for creating desired periodic structures by using the interference of electromagnetic plane waves.

3.8
Interference Lithography

Interference lithography is basically a method in which the nonhomogeneous distribution of energy generated by the superposition of electromagnetic plane waves is registered (or recorded) in a photoresist material. The most important aspect of this technique is that it allows the experimental realization of periodic structures such as those presented in Chapter 2. We show in this section the basic theoretical concepts on how periodic structures are created by using this technique. The experimental aspects associated with the realization of the structures are treated in the next chapters.

3.8.1
Photoresist Materials

The first step to fabricate periodic structures by interference lithography is to create a desired spatial distribution of electromagnetic energy within a photoresist material. In these materials, certain molecules absorb a fraction of the incident electromagnetic energy to initiate chemical reactions. When the amount of energy supplied by the electromagnetic waves in the surroundings of the point **r** is larger than a certain threshold value, the photoresist material is chemically modified at that point to distinguish it from nearby regions receiving lower amounts of energy which are not chemically modified.[9] In this manner, photoresist materials are able to 'record' the distribution of electromagnetic energy in chemically modified and nonmodified regions. In a subsequent process called *developing*, either the regions that are chemically modified or the nonmodified regions can be selectively dissolved and the complementary regions remain to form a desired periodic structure.

---

9) Strictly speaking, the photoresist material is chemically modified everywhere. However, we consider that when the supplied energy is larger than a certain threshold value, the photoresist material is chemically modified (i.e. sufficiently chemically modified), whereas when the supplied energy is below the threshold value the photoresist material is not chemically modified (although in this latter case, the photoresist material is in fact insufficiently chemically modified).

The amount of work $W(\mathbf{r})$ done by the superposition of electromagnetic plane waves at the position $\mathbf{r}$ on the photoresist material, during the time Δt, is proportional to the product of the resultant intensity $I(\mathbf{r})$ (Joule per square meter per second) and the time Δt (seconds). That is,

$$W(\mathbf{r}) \propto I(\mathbf{r})\Delta t \tag{3.131}$$

We take the quantity $I(\mathbf{r})\Delta t$ as a measure of the amount of work done on the material and note that the photoresist material is chemically modified when $I(\mathbf{r})\Delta t$ is larger than a fixed threshold value T_0, which is given by the specific type of photoresist material. That is, if the plane waves interfere in the material for a period of time Δt, the region for which $I(\mathbf{r})\Delta t > T_0$ is chemically modified, whereas the region for which $I(\mathbf{r})\Delta t < T_0$ is not changed.

To understand this better, consider the one-dimensional distribution of the intensity $I(\mathbf{r})$ generated by the interference of two plane waves in a photoresist material. Figure 3.9 shows the amount of work $I(\mathbf{r})\Delta t$ as a function of the position $\mathbf{r}$ at different intervals of times Δt_1, Δt_2, and Δt_3, where $\Delta t_1 < \Delta t_2 < \Delta t_3$.

After the interval of time Δt_1 (Figure 3.9a), we have

$$\begin{aligned} I(\mathbf{r}_A)\Delta t_1 &< T_0 \\ I(\mathbf{r}_B)\Delta t_1 &< T_0 \\ I(\mathbf{r}_C)\Delta t_1 &< T_0 \end{aligned} \tag{3.132}$$

This means that the amount of work done on the material at the points $\mathbf{r}_A$, $\mathbf{r}_B$, and $\mathbf{r}_C$ (and at any other point in the material) is less than that required to chemically modify the material. As a result, the photoresist material is not modified at any given point.

However, after the interval of time Δt_2, the work done on the material has increased (Figure 3.9b) and we have

$$\begin{aligned} I(\mathbf{r}_A)\Delta t_2 &> T_0 \\ I(\mathbf{r}_B)\Delta t_2 &< T_0 \\ I(\mathbf{r}_C)\Delta t_2 &< T_0 \end{aligned} \tag{3.133}$$

Now, there are some regions (one of which includes the point $\mathbf{r}_A$) where the work done on the material is larger than the value required to modify the material. These regions (shown in gray) are therefore able to be chemically distinguished from the other regions.

After the interval of time Δt_3, the work done on the material is even larger (Figure 3.9c), and we have

$$\begin{aligned} I(\mathbf{r}_A)\Delta t_3 &> T_0 \\ I(\mathbf{r}_B)\Delta t_3 &> T_0 \\ I(\mathbf{r}_C)\Delta t_3 &< T_0 \end{aligned} \tag{3.134}$$

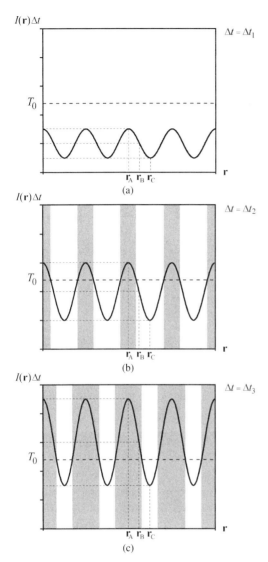

Fig. 3.9 The one-dimensional spatial distribution of $I(\mathbf{r})\Delta t$ that corresponds to the interference of two electromagnetic plane waves in a photoresist material as a function of $\mathbf{r}$. (a) $\Delta t = \Delta t_1$, (b) $\Delta t = \Delta t_2$, and (c) $\Delta t = \Delta t_3$, where $\Delta t_1 < \Delta t_2 < \Delta t_3$. The regions in the photoresist material where $I(\mathbf{r})\Delta t > T_0$ (shown in gray) are chemically modified so that they can be distinguished from the other regions.

As a result, the gray regions where the photoresist material is chemically modified are increased.

Note that if the amount of time Δt is large enough, the entire photoresist material is chemically modified. This is something undesired, since the objective of the interference lithography technique is to create periodic structures by recording the distribution of the intensity $I(\mathbf{r})$ in the photoresist material. This means that the time Δt is an important experimental variable that must be precisely controlled.

3.8.2
The Interference Lithography Technique

There exist two different types of photoresist materials: positive and negative resists. When a *positive* photoresist material is used, the regions that are chemically modified can be dissolved in the developing process because they become soluble in a photoresist developer. In this case, the regions that are not chemically modified remain intact after developing and form the resultant periodic structure.

On the other hand, when a *negative* photoresist material is used, the regions that are not chemically modified can be dissolved by the photoresist developer, whereas the regions that are chemically modified are insoluble and remain intact after developing and forming the resultant periodic structure.

That is, after developing a positive photoresist material, we have

$$\text{If } I(\mathbf{r})\Delta t > T_0 \Rightarrow \text{air} \tag{3.135}$$

$$\text{If } I(\mathbf{r})\Delta t < T_0 \Rightarrow \text{photoresist material} \tag{3.136}$$

whereas after developing a negative photoresist, we have

$$\text{If } I(\mathbf{r})\Delta t > T_0 \Rightarrow \text{photoresist material} \tag{3.137}$$

$$\text{If } I(\mathbf{r})\Delta t < T_0 \Rightarrow \text{air} \tag{3.138}$$

Thus, the intensity distribution generated by the interference of the beams is divided into two distinct regions (photoresist material and air regions) by the use of the photoresist material/developer system.

For example, if we create the distribution of intensity $I(\mathbf{r})$ given by Equation 3.130 within a *positive* photoresist material where $T_0 = 6.6$, during a time $\Delta t = 1$, after developing we have

$$\text{for } I(\mathbf{r}) = 7.4 + \cos\left(\frac{2\pi x}{a}\right) + \cos\left(\frac{2\pi y}{a}\right) + \cos\left(\frac{2\pi z}{a}\right) > 6.6$$

$$\Rightarrow \text{air} \tag{3.139}$$

$$\text{for } I(\mathbf{r}) = 7.4 + \cos\left(\frac{2\pi x}{a}\right) + \cos\left(\frac{2\pi y}{a}\right) + \cos\left(\frac{2\pi z}{a}\right) < 6.6$$

$$\Rightarrow \text{photoresist material} \tag{3.140}$$

Figure 3.10a shows the chemically modified regions (in black) and the nonmodified regions (in white) generated by this intensity distribution. Since the photoresist material is positive, the chemically modified regions (i.e. the regions of high intensity) are dissolved after developing and become air, whereas the nonmodified

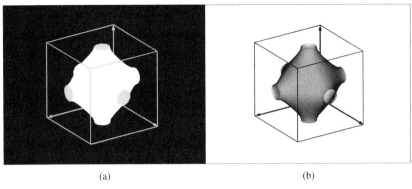

Fig. 3.10 Interference lithography technique. (a) One unit cell of the three-dimensional periodic spatial distribution of $I(\mathbf{r})\Delta t$ corresponding to the interference of four electromagnetic plane waves within a positive photoresist material. The chemically modified region where $I(\mathbf{r})\Delta t > T_0$ is shown in black, whereas the nonmodified region where $I(\mathbf{r})\Delta t < T_0$ is shown in white. (b) The unit cell of the corresponding bicontinuous three-dimensional periodic structure obtained after developing the photoresist material.

material forms the resultant periodic structure. Note that after developing, *we obtain the simple periodic structure* $f_1(x, y, z)$ (Figure 3.10b). This example clearly shows how the interference lithography technique allows us to create three-dimensional periodic structures.

It is important to note, however, that some particular structures might be impractical from the point of view of fabrication. For example, if the *insoluble* regions that form the structure are disconnected, the structure collapses after developing. In addition, if the *soluble* regions are disconnected, the solvent cannot penetrate to dissolve the chemically modified material. This means that it is highly desirable to create both soluble and insoluble regions that are both self-supporting and completely connected within themselves. These types of structures are called bicontinuous structures. Note that the periodic structure shown in Figure 3.10 is bicontinuous and it is therefore suitable for fabrication by interference lithography.

3.8.3
Designing Periodic Structures

By comparing the expression of the spatial distribution of the intensity given by Equation 3.130 and the analytical formula of the simple periodic structure $f_1(x, y, z)$ given by Equation 2.33, it is clear that in order to fabricate a desired periodic structure by using the interference lithography technique, *the expression of the intensity $I(\mathbf{r})$ must be similar to the formula $f(\mathbf{r})$ of the periodic structure we want to fabricate.*

In the previous sections, we showed how to obtain the spatial distribution of the intensity $I(\mathbf{r})$, given the wave vectors $\mathbf{k}$ and electric field amplitude vectors $\mathbf{E}_0$ of the individual electromagnetic plane waves that propagate in the same material at the same time (Equation 3.119). That is, we know how to solve the following problem:

> Given the parameters of the interfering plane waves
> $\longrightarrow$ determine the intensity distribution $I(\mathbf{r})$

However, in order to fabricate a desired periodic structure by interference lithography, we need to consider the inverse case. That is, given a specific desired spatial distribution of the intensity $I(\mathbf{r})$ (which resembles the formula of the periodic structure that we want to fabricate), we want to be able to determine the wave vectors $\mathbf{k}$ and electric field amplitude vectors $\mathbf{E}_0$ of the plane waves that we need to use to create the required distribution of intensity. That is

> Given a desired intensity distribution $I(\mathbf{r})$
> $\longrightarrow$ what are the parameters of the plane waves?

We develop this subject in detail in the next chapter, where we introduce a systematic method to obtain the plane-wave parameters (i.e. wave vectors $\mathbf{k}$ and electric field amplitude vectors $\mathbf{E}_0$) required to fabricate desired periodic structures by the interference of electromagnetic plane waves.

Further Reading

1 Chen, H.C. (**1983**) *Theory of Electromagnetic Waves*, McGraw-Hill, New York.
2 Staelin, D.H., Morgenthaler, A.W. and Kong, J.A. (**1994**) *Electromagnetic Waves*, Prentice-Hall, New Jersey.
3 Born, M. and Wolf, E. (**1959**) *Principles of Optics*, Pergamon Press, London.
4 Jackson, J.D. (**1975**) *Classical Electrodynamics*, John Wiley & Sons, Ltd, New York.
5 Hecht, E. and Zajac, A. (**1974**) *Optics*, Addison-Wesley, Massachusetts.
6 Campbell, M., Sharp, D.N., Harrison, M.T., Denning, R.G. and Turberfield, A.J. (**2000**) "Fabrication of photonic crystals for the visible spectrum by holographic lithography". *Nature*, **404**, 53–56.

Problems

3.1 An electromagnetic plane wave corresponding to visible green light (frequency $f = 5.9 \times 10^{14}$ s^{-1}) propagates in vacuum. What is the wavelength and velocity of the wave?

3.2 The same electromagnetic plane wave is now propagating within a photoresist material (e.g. SU-8) where $\varepsilon = 3\,\varepsilon_0$ and $\mu = 1\,\mu_0$. What is the corresponding wavelength and velocity of the wave? Note: the frequency of the wave does not change when the plane wave moves into different materials.

3.3 Derive the wave equation (3.13) for the magnetic field $\mathbf{H}$.

3.4 An electromagnetic plane wave is propagating along the z direction in a material characterized by the parameters ε and μ. Its instantaneous electric field vector is given by

$$E(\mathbf{r}, t) = \hat{x} E_0 \cos(kz - \omega t)$$

(a) Show that $\mathbf{E}(\mathbf{r}, t)$ satisfies the wave equation (3.11).

(b) By using Equation 3.41 find the corresponding magnetic field vector $\mathbf{H}(\mathbf{r}, t)$.

3.5 An electromagnetic plane wave, with an angular frequency ω, is propagating along the z direction in a material characterized by ε and μ. The complex electric field amplitude vector is given by $\mathbf{E}_0 = a\hat{x} + ia\hat{y}$, where i is the imaginary unit. Find the instantaneous electric and magnetic field vectors $\mathbf{E}(\mathbf{r}, t)$ and $\mathbf{H}(\mathbf{r}, t)$.

3.6 Consider an electromagnetic plane wave propagating in a material characterized by ε and μ. The instantaneous electric field vector of the plane wave is given by

$$\mathbf{E}(\mathbf{r}, t) = \mathrm{Re}\!\left(4\hat{x}\, e^{i\left(\frac{\pi}{a} x + \frac{\pi}{a} y - \omega t\right)}\right)$$

(a) Find the wave vector $\mathbf{k}$ and its absolute value k.

(b) What is the value of the wavelength λ and the frequency ω?

(c) Find the instantaneous magnetic field vector $\mathbf{H}(\mathbf{r}, t)$.

3.7 Prove the vector identities

$$\nabla \times (\nabla \times \mathbf{E}) = -\nabla^2 \mathbf{E} + \nabla(\nabla \cdot \mathbf{E})$$
$$\mathbf{H} \cdot (\nabla \times \mathbf{E}) - \mathbf{E} \cdot (\nabla \times \mathbf{H}) = \nabla \cdot (\mathbf{E} \times \mathbf{H})$$
$$(\hat{\mathbf{k}} \times \mathbf{H}) \cdot \mathbf{E} = (\mathbf{E} \times \hat{\mathbf{k}}) \cdot \mathbf{H}$$
$$\mathbf{E} \times (\hat{\mathbf{k}} \times \mathbf{E}) = (\mathbf{E} \cdot \mathbf{E})\hat{\mathbf{k}} - (\mathbf{E} \cdot \hat{\mathbf{k}})\mathbf{E}$$

3.8 Prove the following identities

$$\nabla^2 \left(\mathbf{E}_0\, e^{i(\mathbf{k}\cdot\mathbf{r} - \omega t)}\right) = -(\mathbf{k} \cdot \mathbf{k})\left(\mathbf{E}_0\, e^{i(\mathbf{k}\cdot\mathbf{r} - \omega t)}\right)$$
$$\nabla \times \left(\mathbf{E}_0\, e^{i(\mathbf{k}\cdot\mathbf{r} - \omega t)}\right) = i\mathbf{k} \times \left(\mathbf{E}_0\, e^{i(\mathbf{k}\cdot\mathbf{r} - \omega t)}\right)$$
$$\nabla \cdot \left(\mathbf{E}_0\, e^{i(\mathbf{k}\cdot\mathbf{r} - \omega t)}\right) = i\mathbf{k} \cdot \mathbf{E}_0\, e^{i(\mathbf{k}\cdot\mathbf{r} - \omega t)}$$

3.9 An electromagnetic plane wave, with an angular frequency ω, is propagating along the z direction in a material characterized by ε and μ. The complex electric field amplitude vector is given by $\mathbf{E}_0 = \hat{x} + (a + ib)\hat{y}$, where i is the imaginary unit. Determine the values of a and b for which the plane wave is

(a) linearly polarized

(b) circularly polarized

(c) elliptically polarized

3.10 An electromagnetic plane wave with a frequency $f = 500$ GHz propagates within a homogeneous medium characterized by $\varepsilon = 2\varepsilon_0$ and $\mu = \mu_0$. The electric field amplitude vector is given by $\mathbf{E}_0 = 5\hat{x}$.

(a) What is the instantaneous energy density of the wave?

(b) What is the time-averaged energy density of the wave?

(c) Find the Poynting vector S(r,t).

(d) What is the intensity of the wave?

3.11 Is the sum of two plane waves a plane wave?

3.12 Demonstrate Equation 3.108, which is valid for the superposition of two plane waves propagating with the wavevector magnitude k. Hint: $\cos\theta = 1 - 2\sin^2(\frac{\theta}{2})$.

3.13 Show that the wave vectors and electric field amplitude vectors given by Equations 3.124–3.127 generate the three-dimensional intensity pattern given by Equation 3.128.

3.14 Show that the linearly polarized electromagnetic waves given by the wave vectors and electric field amplitude vectors

$E_{01} = 0.877$, $\hat{n}_1 = (+0.612, -0.774, +0.161)$, $k_1 = \frac{\pi}{a}(+3, +3, +3)$

$E_{02} = 0.716$, $\hat{n}_2 = (+0.250, -0.905, -0.346)$, $k_2 = \frac{\pi}{a}(+5, +1, +1)$

$E_{03} = 1.036$, $\hat{n}_3 = (+0.346, -0.250, +0.905)$, $k_3 = \frac{\pi}{a}(+1, +5, +1)$

$E_{04} = 2.317$, $\hat{n}_4 = (+0.905, +0.346, -0.250)$, $k_4 = \frac{\pi}{a}(+1, +1, +5)$

generate the three-dimensional intensity pattern

$$I(r) = 7.73 + \cos\left[\frac{2\pi}{a}(-x+y+z)\right] + \cos\left[\frac{2\pi}{a}(x-y+z)\right] + \cos\left[\frac{2\pi}{a}(x+y-z)\right]$$

Suppose $T_0 = 8.5$ and $\Delta t = 1$. Show over a unit cell and over $2 \times 2 \times 2$ unit cells, the resultant three-dimensional periodic structure after development by using some mathematical software such as Maple, Mathematica, or Matlab.

4

Periodic Structures and Interference Lithography

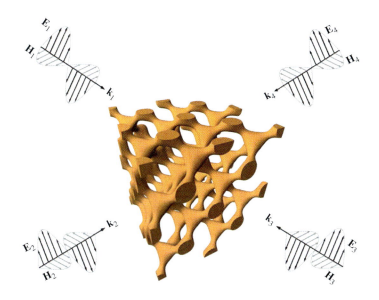

Interference lithography allows us to fabricate large-area, defect-free periodic structures. In this chapter we present a method to obtain the parameters of the electromagnetic plane waves (or laser beams) required to fabricate desired periodic structures by interference lithography. This method is based on the similarity between the spatial distribution of the intensity generated by the interference of electromagnetic waves and the Fourier series expansion of a periodic function representing a periodic structure. We also provide in this chapter all laser beam parameters required to construct the simple periodic structures presented in Chapter 2. Practical engineering aspects associated with the control of the laser beams and the experimental realization of the structures will be treated in the next chapter.

Periodic Materials and Interference Lithography. M. Maldovan and E. Thomas
Copyright © 2009 WILEY-VCH Verlag GmbH & Co. KGaA, Weinheim
ISBN: 978-3-527-31999-2

4.1
The Connection between the Interference of Plane Waves and Fourier Series

In the previous chapter, we examined the superposition of electromagnetic plane waves in the same material at the same time and showed that the intensity $I(\mathbf{r})$ of the total wave is nonhomogeneously distributed in space. In particular, we showed that the intensity $I(\mathbf{r})$ depends on the position vector $\mathbf{r}$ through a number of cosine and sine functions known as *the interference terms*. On the other hand, in Chapter 1 we established that periodic functions $f(\mathbf{r})$ representing periodic structures can be written as the sum of cosine and sine functions known as *Fourier series expansions*. In fact, the similarity between the expressions for $I(\mathbf{r})$ and $f(\mathbf{r})$ translates into the possibility of fabricating periodic structures by the interference of electromagnetic plane waves (Section 3.8).

In this section, we introduce a method to calculate the wave vectors $\mathbf{k}_j$ and electric field amplitude vectors $\mathbf{E}_{0j}$ of the electromagnetic plane waves required to fabricate a periodic structure represented by the formula $f(\mathbf{r})$. As we show next, the method is based on the comparison between the expression of the intensity $I(\mathbf{r})$ of the total wave and the formula $f(\mathbf{r})$ of the periodic structure.

We start by considering the intensity $I(\mathbf{r})$ generated by the superposition of N electromagnetic plane waves having the same frequency ω. Each plane wave is given by the formula

$$\mathbf{E}_j(\mathbf{r}, t) = \text{Re}(\mathbf{E}_{0j} e^{i(\mathbf{k}_j \cdot \mathbf{r} - \omega t)}), \quad j = 1, \ldots, N \tag{4.1}$$

As in the previous chapter, the principle of superposition states that the total electric field (at each point where the plane waves superpose) is equal to the sum of the N electric fields of the individual plane waves. That is

$$\mathbf{E}_T(\mathbf{r}, t) = \sum_{j=1}^{N} \mathbf{E}_j(\mathbf{r}, t) \tag{4.2}$$

According to Section 3.7, the intensity $I(\mathbf{r})$ of the total wave is thus given by

$$I(\mathbf{r}) = |\mathbf{A}(\mathbf{r})|^2 = \left(\sum_{j=1}^{N} \mathbf{E}_{0j} e^{i\mathbf{k}_j \cdot \mathbf{r}}\right) \cdot \left(\sum_{j=1}^{N} \mathbf{E}_{0j}^* e^{-i\mathbf{k}_j \cdot \mathbf{r}}\right) \tag{4.3}$$

which can also be written as

$$I(\mathbf{r}) = \sum_{j=1}^{N} |\mathbf{E}_{0j}|^2 + \sum_{1 \leq i < j \leq N} 2\,\text{Re}(\mathbf{E}_{0i} \cdot \mathbf{E}_{0j}^*) \cos[(\mathbf{k}_i - \mathbf{k}_j) \cdot \mathbf{r}]$$
$$- 2\,\text{Im}(\mathbf{E}_{0i} \cdot \mathbf{E}_{0j}^*) \sin[(\mathbf{k}_i - \mathbf{k}_j) \cdot \mathbf{r}] \tag{4.4}$$

4.1 The Connection between the Interference of Plane Waves and Fourier Series

Because the spatial distribution of the intensity $I(\mathbf{r})$ of the total wave must be similar to the formula $f(\mathbf{r})$ of the periodic structure that we want to fabricate, we compare Equation 4.4 with the Fourier series expansion of a periodic function $f(\mathbf{r})$ defined within an orthorhombic unit cell, which is given by

$$f(\mathbf{r}) = a_{000} + \sum_n \sum_m \sum_p \left\{ a_{nmp} \cos\left[2\pi \left(\frac{nx}{a} + \frac{my}{b} + \frac{pz}{c}\right)\right] \right.$$
$$\left. + b_{nmp} \sin\left[2\pi \left(\frac{nx}{a} + \frac{my}{b} + \frac{pz}{c}\right)\right] \right\} \tag{4.5}$$

The comparison of the trigonometric functions in Equations 4.4 and 4.5 determines that, in order to experimentally realize a specific periodic structure represented by the formula $f(\mathbf{r})$, the wave vectors in Equation 4.4 cannot have arbitrary values. They must satisfy

$$\mathbf{k}_i - \mathbf{k}_j = 2\pi \left(\frac{n}{a}\hat{\mathbf{x}} + \frac{m}{b}\hat{\mathbf{y}} + \frac{p}{c}\hat{\mathbf{z}}\right) \tag{4.6}$$

This means that in order to have a particular n, m, and p term in the Fourier series expansion (Equation 4.5), there must be a pair of plane waves in the interference set of waves (Equation 4.1) whose wave vectors $\mathbf{k}_i$ and $\mathbf{k}_j$ satisfy Equation 4.6.

At the same time, by comparing the coefficients corresponding to the trigonometric functions, we can see that the complex electric field amplitude vectors $\mathbf{E}_{0i}$ and $\mathbf{E}_{0j}$ and the Fourier coefficients a_{nmp} and b_{nmp} must satisfy the relations

$$2 \operatorname{Re}(\mathbf{E}_{0i} \cdot \mathbf{E}_{0j}^*) = a_{nmp} \tag{4.7}$$

$$-2 \operatorname{Im}(\mathbf{E}_{0i} \cdot \mathbf{E}_{0j}^*) = b_{nmp} \tag{4.8}$$

Therefore, Equations 4.6–4.8 establish the connection between the spatial distribution of the intensity $I(\mathbf{r})$ generated by the superposition of N electromagnetic plane waves and the Fourier series expansion of the periodic function $f(\mathbf{r})$ representing the periodic structure that we want to fabricate by the interference of these waves.

For example, if we are interested in making a periodic structure whose analytical formula $f(\mathbf{r})$ has a Fourier series term given by

$$a_{nmp} \cos\left[2\pi \left(\frac{nx}{a} + \frac{my}{b} + \frac{pz}{c}\right)\right] \tag{4.9}$$

we need a pair of plane waves in Equation 4.1 whose wave vectors satisfy Equation 4.6 and whose electric field amplitude vectors satisfy Equation 4.7.

In the next sections, we apply this method and obtain the wave vectors and electric field amplitude vectors required to fabricate the periodic structures presented in Chapter 2.

4.2
Simple Periodic Structures in Two Dimensions Via Interference Lithography

In this section, we are interested in creating *two-dimensional* periodic structures by the interference of electromagnetic plane waves. In order to do this, we consider the superposition of *three* noncoplanar plane waves (Section 3.7.1). We can obtain the spatial distribution of the intensity $I(\mathbf{r})$ of the total wave by replacing $N = 3$ in Equation 4.4. We thus have

$$I(\mathbf{r}) = |\mathbf{E}_{01}|^2 + |\mathbf{E}_{02}|^2 + |\mathbf{E}_{03}|^2$$
$$+ 2\text{Re}(\mathbf{E}_{01} \cdot \mathbf{E}_{02}^*) \cos[(\mathbf{k}_1 - \mathbf{k}_2) \cdot \mathbf{r}] - 2\text{Im}(\mathbf{E}_{01} \cdot \mathbf{E}_{02}^*) \sin[(\mathbf{k}_1 - \mathbf{k}_2) \cdot \mathbf{r}]$$
$$+ 2\text{Re}(\mathbf{E}_{01} \cdot \mathbf{E}_{03}^*) \cos[(\mathbf{k}_1 - \mathbf{k}_3) \cdot \mathbf{r}] - 2\text{Im}(\mathbf{E}_{01} \cdot \mathbf{E}_{03}^*) \sin[(\mathbf{k}_1 - \mathbf{k}_3) \cdot \mathbf{r}] \quad (4.10)$$
$$+ 2\text{Re}(\mathbf{E}_{02} \cdot \mathbf{E}_{03}^*) \cos[(\mathbf{k}_2 - \mathbf{k}_3) \cdot \mathbf{r}] - 2\text{Im}(\mathbf{E}_{02} \cdot \mathbf{E}_{03}^*) \sin[(\mathbf{k}_2 - \mathbf{k}_3) \cdot \mathbf{r}]$$

Note that in the case of three noncoplanar plane waves, the intensity $I(\mathbf{r})$ consists of the sum of nine terms: three constant terms $|\mathbf{E}_{01}|^2$, $|\mathbf{E}_{02}|^2$, and $|\mathbf{E}_{03}|^2$ plus six interference terms. This means that the superposition of three noncoplanar plane waves can create periodic structures whose Fourier series expansions $f(\mathbf{r})$ consist of the sum of *at most* six trigonometric terms.

As an illustrative example, consider that we are interested in fabricating the two-dimensional periodic structure shown in Figure 2.1b, which is defined in square unit cells and represented by the simple periodic function

$$f_1(\mathbf{r}) = +\cos\left(\frac{2\pi x}{a}\right) + \cos\left(\frac{2\pi y}{a}\right) \quad (4.11)$$

To fabricate this periodic structure by the interference of three plane waves, the spatial distribution of the intensity (Equation 4.10) must be similar to the formula of the structure (Equation 4.11). By comparing Equations 4.10 and 4.11, we see that the following must be satisfied:

$$\mathbf{k}_1 - \mathbf{k}_2 = \frac{2\pi}{a}(1, 0, 0) \quad (4.12)$$
$$\mathbf{k}_1 - \mathbf{k}_3 = \frac{2\pi}{a}(0, 1, 0)$$

and

$$+2\text{Re}(\mathbf{E}_{01} \cdot \mathbf{E}_{02}^*) = 1, \quad -2\text{Im}(\mathbf{E}_{01} \cdot \mathbf{E}_{02}^*) = 0$$
$$+2\text{Re}(\mathbf{E}_{01} \cdot \mathbf{E}_{03}^*) = 1, \quad -2\text{Im}(\mathbf{E}_{01} \cdot \mathbf{E}_{03}^*) = 0 \quad (4.13)$$
$$+2\text{Re}(\mathbf{E}_{02} \cdot \mathbf{E}_{03}^*) = 0, \quad -2\text{Im}(\mathbf{E}_{02} \cdot \mathbf{E}_{03}^*) = 0$$

In addition, because electromagnetic plane waves are transverse, the three interfering plane waves must satisfy

$$\mathbf{k}_1 \cdot \mathbf{E}_{01} = 0$$

$$\mathbf{k}_2 \cdot \mathbf{E}_{02} = 0 \tag{4.14}$$

$$\mathbf{k}_3 \cdot \mathbf{E}_{03} = 0$$

The solution of the Equations 4.12–4.14 determines the entire set of plane-wave parameters required to obtain a spatial distribution of the intensity $I(\mathbf{r})$ that resembles the simple periodic function $f_1(\mathbf{r})$. For example, by solving the set of Equations 4.12, we can obtain the values of the wave vectors $\mathbf{k}_1$, $\mathbf{k}_2$, and $\mathbf{k}_3$ of the three interfering plane waves. Once the wave vectors are found, we can then solve the set of Equations 4.13–4.14 and obtain the corresponding values for the electric field amplitude vectors $\mathbf{E}_{01}$, $\mathbf{E}_{02}$, and $\mathbf{E}_{03}$.

We first look at the set of wave vectors $\mathbf{k}_1$, $\mathbf{k}_2$, and $\mathbf{k}_3$ required to obtain the desired intensity $I(\mathbf{r})$. Note that Equation 4.12 generates six equations with nine unknowns (i.e. the nine Cartesian components of the three the wave vectors $\mathbf{k}_1$, $\mathbf{k}_2$, and $\mathbf{k}_3$) and we thus have infinite solutions. Also note that we need $|\mathbf{k}_1| = |\mathbf{k}_2| = |\mathbf{k}_3|$ because the interfering plane waves have the same frequency and propagate in the same material. One solution is the set of wave vectors

$$\begin{aligned}
\mathbf{k}_1 &= \frac{\pi}{a}(1, 1, 1) \\
\mathbf{k}_2 &= \frac{\pi}{a}(-1, 1, 1) \\
\mathbf{k}_3 &= \frac{\pi}{a}(1, -1, 1)
\end{aligned} \tag{4.15}$$

where $k = \frac{\pi}{a}\sqrt{3}$ is the magnitude of the wave vectors.

By considering Equation 3.20, we have

$$k = \frac{\pi}{a}\sqrt{3} = \frac{2\pi}{\lambda} \quad \text{or} \quad a = \frac{\sqrt{3}}{2}\lambda \tag{4.16}$$

This means that the periodicity a of the resultant two-dimensional structure is determined by the wavelength λ of the interfering plane waves (which is in turn determined by the magnitude of the wave vectors k). It is important to mention that the difference between the wave vectors $\mathbf{k}_i - \mathbf{k}_j$ determines the translational periodicity (and hence the point lattice) of the resultant two-dimensional periodic structure. This means that our choice of wave vectors (Equation 4.15) can be broadly used to obtain two-dimensional periodic structures defined in square unit cells.

We next concentrate on the solutions for the electric field amplitude vectors $\mathbf{E}_{01}$, $\mathbf{E}_{02}$, and $\mathbf{E}_{03}$, which are given by the set of Equations 4.13–4.14. Before finding the electric field amplitudes, we want to mention that the value of $|\mathbf{E}_{01}|^2 + |\mathbf{E}_{02}|^2 + |\mathbf{E}_{03}|^2$ in Equation 4.10, which corresponds to the background intensity, should be minimized in order to maximize the contrast of the interference pattern. The intensity contrast C is defined as

$$C = \frac{I_{max} - I_{min}}{I_{max} + I_{min}} \tag{4.17}$$

and can be seen to be maximal when $|\mathbf{E}_{01}|^2 + |\mathbf{E}_{02}|^2 + |\mathbf{E}_{03}|^2$ is minimal. As discussed in the next chapter, high intensity contrast is a beneficial experimental aspect, as it reduces constraints on the structure reproduction.

We then find the solutions of the Equations 4.13–4.14 while minimizing the value of $|\mathbf{E}_{01}|^2 + |\mathbf{E}_{02}|^2 + |\mathbf{E}_{03}|^2$ by using the MATLAB code described in Appendix A. It is important to mention that different solutions can be obtained if we use different initial conditions, but all of them should minimize the background intensity. In the case that the interfering plane waves are considered to be linearly polarized, we found the following solution for the electric field amplitude vectors

$$\mathbf{E}_{01} = E_{01}\,\hat{\mathbf{n}}_1 = 1.0000(0.5721, 0.2185, -0.7906)$$
$$\mathbf{E}_{02} = E_{02}\,\hat{\mathbf{n}}_2 = 0.7071(0.5000, 0.8091, -0.3090) \quad (4.18)$$
$$\mathbf{E}_{03} = E_{03}\,\hat{\mathbf{n}}_3 = 0.7071(0.3090, -0.5000, -0.8091)$$

We can test the correctness of our solution by inserting the wave vectors (Equation 4.15) and the electric field amplitude vectors (Equation 4.18) into Equation 4.10. By doing this, we obtain the following spatial distribution of intensity (which has a contrast $C = 1$)

$$I(\mathbf{r}) = 2.0 + \cos\left(\frac{2\pi x}{a}\right) + \cos\left(\frac{2\pi y}{a}\right) \quad (4.19)$$

We can see that the resultant intensity distribution $I(\mathbf{r})$ given by Equation 4.19 is identical to the formula of the periodic structure $f_1(\mathbf{r})$ given by Equation 4.11 except for the constant. This means that the periodic structure represented by the function $f_1(\mathbf{r})$ can be fabricated by the interference of the three noncoplanar beams. The wave parameters of these beams are determined by Equations 4.15 and 4.18. As we point out later, chemical additives to the photoresist material can be used to reduce the constant, which decreases the contrast C.

We now systematically apply the above scheme and find the beam parameters required to fabricate the two-dimensional periodic structures presented in Chapter 2. The results are shown in Table 4.1.

Note that in the first two cases the electric field amplitude vectors are real and the interfering beams are in phase and *linearly polarized*. On the other hand, in the case of the $f_{4/3}^{II}$ structure, the electric fields amplitudes are complex vectors and the interfering beams are not in phase and are *elliptically polarized*.

Before moving into the study of three-dimensional periodic structures, we want to show the comparison between the theoretical periodic structures in two dimensions and their experimental fabrication by interference lithography. For example, Figures 4.1 and 4.2 show the two-dimensional periodic structures represented by the formulas $f_1(x, y) - 0.6$ and $f_{4/3}^{I}(x, y) - 0.8$, respectively, together with the scanning electron microscopy (SEM) images of their experimental realization. Note the small length scale at which the periodic structures are fabricated since the structures are defined respectively in square and triangular unit cells of side $a \sim 1\,\mu\text{m}$.

4.2 Simple Periodic Structures in Two Dimensions Via Interference Lithography

Table 4.1 Beam parameters to fabricate the two-dimensional periodic structures f_1, $f_{4/3}^{I}$, and $f_{4/3}^{II}$. (Chapter 2 shows the 2D images of these structures)

Structure	Wave vectors	Electric field amplitude vectors
$f_1(x,y)$	$\mathbf{k}_1 = \dfrac{\pi}{a}(1,1,1)$ $\mathbf{k}_2 = \dfrac{\pi}{a}(-1,1,1)$ $\mathbf{k}_3 = \dfrac{\pi}{a}(1,-1,1)$	$\mathbf{E}_{01} = 1.0000(0.5721, 0.2185, -0.7906)$ $\mathbf{E}_{02} = 0.7071(0.5000, 0.8091, -0.3090)$ $\mathbf{E}_{03} = 0.7071(0.3090, -0.5000, -0.8091)$
$f_{4/3}^{I}(x,y)$	$\mathbf{k}_1 = \dfrac{2\pi}{a}\left(-\dfrac{2}{3}, 0, \dfrac{2}{3}\right)$ $\mathbf{k}_2 = \dfrac{2\pi}{a}\left(\dfrac{1}{3}, \dfrac{1}{\sqrt{3}}, \dfrac{2}{3}\right)$ $\mathbf{k}_3 = \dfrac{2\pi}{a}\left(\dfrac{1}{3}, -\dfrac{1}{\sqrt{3}}, \dfrac{2}{3}\right)$	$\mathbf{E}_{01} = 0.9788(0.7071, 0.0000, 0.7071)$ $\mathbf{E}_{02} = 0.8833(0.9019, -0.4238, -0.0840)$ $\mathbf{E}_{03} = 0.8833(0.9019, 0.4238, -0.0840)$
$f_{4/3}^{II}(x,y)$	$\mathbf{k}_1 = \dfrac{2\pi}{a}\left(-\dfrac{2}{3}, 0, \dfrac{2}{3}\right)$ $\mathbf{k}_2 = \dfrac{2\pi}{a}\left(\dfrac{1}{3}, \dfrac{1}{\sqrt{3}}, \dfrac{2}{3}\right)$ $\mathbf{k}_3 = \dfrac{2\pi}{a}\left(\dfrac{1}{3}, -\dfrac{1}{\sqrt{3}}, \dfrac{2}{3}\right)$	$\mathbf{E}_{01} = (0.0722 - i0.0050, -0.0596 + i0.9537, 0.0722 - i0.0050)$ $\mathbf{E}_{02} = (0.7274 + i0.3694, -0.4947 - i0.0896, 0.0647 - i0.1071)$ $\mathbf{E}_{03} = (-0.6157 + i0.4949, -0.5094 + i0.0263, -0.1333 - i0.2247)$

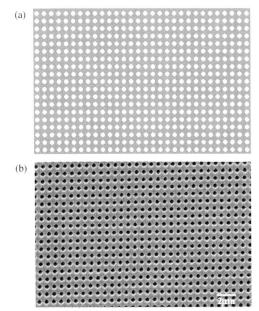

Fig. 4.1 (a) The two-dimensional periodic structure represented by the formula $f_1(x,y) - 0.6$, which is defined in square unit cells. (b) A scanning electron microscopy image of the structure fabricated by interference lithography using the laser beam parameters shown in Table 4.1

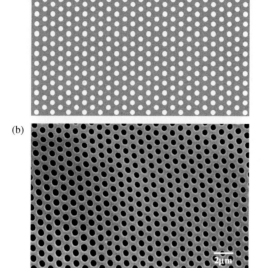

Fig. 4.2 (a) The two-dimensional periodic structure represented by the formula $f^1_{4/3}(x, y) - 0.8$, which is defined in triangular unit cells. (b) A scanning electron microscopy image of the structure fabricated by interference lithography using the laser beam parameters shown in Table 4.1

4.3
Simple Periodic Structures in Three Dimensions Via Interference Lithography

We now turn to the task of fabricating *three-dimensional* periodic structures by the interference of electromagnetic plane waves. As mentioned in Section 3.7.1, to fabricate a three-dimensional periodic structure via interference lithography, we need the superposition of at least *four* plane waves. However, there is a trade-off involving the number of interfering plane waves due to the following facts:

- *The smaller the number of plane waves, the easier the assembling and positioning of the required optical components to properly control the corresponding laser beams.*
- *The larger the number of plane waves, the larger the number of interference terms (and Fourier terms) that can be obtained and thus the larger the number of different periodic structures that can be fabricated.*

With concern for the fabrication process, we consider in this section the superposition of *four* electromagnetic plane waves as a means to obtain three-dimensional periodic structures by interference lithography. For $N = 4$, the spatial distribution of the intensity (Equation 4.4) transforms into

$$I(\mathbf{r}) = |\mathbf{E}_{01}|^2 + |\mathbf{E}_{02}|^2 + |\mathbf{E}_{03}|^2 + |\mathbf{E}_{04}|^2$$
$$+ 2\text{Re}(\mathbf{E}_{01} \cdot \mathbf{E}_{02}^*)\cos[(\mathbf{k}_1 - \mathbf{k}_2) \cdot \mathbf{r}] - 2\text{Im}(\mathbf{E}_{01} \cdot \mathbf{E}_{02}^*)\sin[(\mathbf{k}_1 - \mathbf{k}_2) \cdot \mathbf{r}]$$
$$+ 2\text{Re}(\mathbf{E}_{01} \cdot \mathbf{E}_{03}^*)\cos[(\mathbf{k}_1 - \mathbf{k}_3) \cdot \mathbf{r}] - 2\text{Im}(\mathbf{E}_{01} \cdot \mathbf{E}_{03}^*)\sin[(\mathbf{k}_1 - \mathbf{k}_3) \cdot \mathbf{r}]$$
$$+ 2\text{Re}(\mathbf{E}_{01} \cdot \mathbf{E}_{04}^*)\cos[(\mathbf{k}_1 - \mathbf{k}_4) \cdot \mathbf{r}] - 2\text{Im}(\mathbf{E}_{01} \cdot \mathbf{E}_{04}^*)\sin[(\mathbf{k}_1 - \mathbf{k}_4) \cdot \mathbf{r}] \quad (4.20)$$
$$+ 2\text{Re}(\mathbf{E}_{02} \cdot \mathbf{E}_{03}^*)\cos[(\mathbf{k}_2 - \mathbf{k}_3) \cdot \mathbf{r}] - 2\text{Im}(\mathbf{E}_{02} \cdot \mathbf{E}_{03}^*)\sin[(\mathbf{k}_2 - \mathbf{k}_3) \cdot \mathbf{r}]$$
$$+ 2\text{Re}(\mathbf{E}_{02} \cdot \mathbf{E}_{04}^*)\cos[(\mathbf{k}_2 - \mathbf{k}_4) \cdot \mathbf{r}] - 2\text{Im}(\mathbf{E}_{02} \cdot \mathbf{E}_{04}^*)\sin[(\mathbf{k}_2 - \mathbf{k}_4) \cdot \mathbf{r}]$$
$$+ 2\text{Re}(\mathbf{E}_{03} \cdot \mathbf{E}_{04}^*)\cos[(\mathbf{k}_3 - \mathbf{k}_4) \cdot \mathbf{r}] - 2\text{Im}(\mathbf{E}_{03} \cdot \mathbf{E}_{04}^*)\sin[(\mathbf{k}_3 - \mathbf{k}_4) \cdot \mathbf{r}]$$

As an illustrative example of our method in three dimensions, consider that we are interested in fabricating the three-dimensional simple periodic structure $f_1(\mathbf{r})$ (Figure 2.4) represented by the formula

$$f_1(\mathbf{r}) = +\cos\left(\frac{2\pi x}{a}\right) + \cos\left(\frac{2\pi y}{a}\right) + \cos\left(\frac{2\pi z}{a}\right) \quad (4.21)$$

In complete analogy with the two-dimensional case, the expression of the spatial distribution of the intensity (Equation 4.20) must be similar to the formula (4.21) representing the periodic structure that we want to fabricate. As a result, by comparing Equations 4.20 and 4.21, the following constraints must be satisfied:

$$\mathbf{k}_1 - \mathbf{k}_2 = \frac{2\pi}{a}(1, 0, 0)$$
$$\mathbf{k}_1 - \mathbf{k}_3 = \frac{2\pi}{a}(0, 1, 0) \quad (4.22)$$
$$\mathbf{k}_1 - \mathbf{k}_4 = \frac{2\pi}{a}(0, 0, 1)$$

and

$$+2\text{Re}(\mathbf{E}_{01} \cdot \mathbf{E}_{02}^*) = 1, \quad -2\text{Im}(\mathbf{E}_{01} \cdot \mathbf{E}_{02}^*) = 0$$
$$+2\text{Re}(\mathbf{E}_{01} \cdot \mathbf{E}_{03}^*) = 1, \quad -2\text{Im}(\mathbf{E}_{01} \cdot \mathbf{E}_{03}^*) = 0$$
$$+2\text{Re}(\mathbf{E}_{01} \cdot \mathbf{E}_{04}^*) = 1, \quad -2\text{Im}(\mathbf{E}_{01} \cdot \mathbf{E}_{04}^*) = 0 \quad (4.23)$$
$$+2\text{Re}(\mathbf{E}_{02} \cdot \mathbf{E}_{03}^*) = 0, \quad -2\text{Im}(\mathbf{E}_{02} \cdot \mathbf{E}_{03}^*) = 0$$
$$+2\text{Re}(\mathbf{E}_{02} \cdot \mathbf{E}_{04}^*) = 0, \quad -2\text{Im}(\mathbf{E}_{02} \cdot \mathbf{E}_{04}^*) = 0$$
$$+2\text{Re}(\mathbf{E}_{03} \cdot \mathbf{E}_{04}^*) = 0, \quad -2\text{Im}(\mathbf{E}_{03} \cdot \mathbf{E}_{04}^*) = 0$$

And, as before, since electromagnetic plane waves are transverse, we also have

$$\mathbf{k}_1 \cdot \mathbf{E}_{01} = 0$$
$$\mathbf{k}_2 \cdot \mathbf{E}_{02} = 0 \quad (4.24)$$
$$\mathbf{k}_3 \cdot \mathbf{E}_{03} = 0$$
$$\mathbf{k}_4 \cdot \mathbf{E}_{04} = 0$$

By solving the set of Equations 4.22, we obtain the following solution for the wave vectors of the four interfering waves:

$$\begin{aligned}\mathbf{k}_1 &= \frac{\pi}{a}(1,1,1)\\ \mathbf{k}_2 &= \frac{\pi}{a}(-1,1,1)\\ \mathbf{k}_3 &= \frac{\pi}{a}(1,-1,1)\\ \mathbf{k}_4 &= \frac{\pi}{a}(1,1,-1)\end{aligned} \qquad (4.25)$$

Note that all the wave vectors we have chosen have the same magnitude because the interfering plane waves have the same frequency and propagate in the same material. By considering Equations 4.25 and 3.20 we have

$$k = \frac{\pi}{a}\sqrt{3} = \frac{2\pi}{\lambda} \quad \text{or} \quad a = \frac{\sqrt{3}}{2}\lambda \qquad (4.26)$$

Therefore, as in the two-dimensional case, the periodicity a of the resultant structure is determined by the wavelength λ of the interfering electromagnetic plane waves. Since the difference between the wave vectors $\mathbf{k}_i - \mathbf{k}_j$ determines the point lattice of the resultant periodic structure, the set of wave vectors (Equation 4.25) can be broadly used to obtain three-dimensional periodic structures defined in simple cubic unit cells.

In order to find the electric field amplitude vectors given by Equations 4.23–4.24, we again employ the MATLAB code described in Appendix A. This code finds the solutions of Equations 4.23–4.24 while minimizing the value of $|\mathbf{E}_{01}|^2 + |\mathbf{E}_{02}|^2 + |\mathbf{E}_{03}|^2 + |\mathbf{E}_{04}|^2$ in Equation 4.20. As in the two-dimensional case, different solutions can be obtained by using different initial conditions for the amplitudes. In the case of employing a set of linearly polarized plane waves, we found the following electric field amplitude vectors:

$$\begin{aligned}\mathbf{E}_{01} &= 1.3554(-0.5825, -0.2041, 0.7867)\\ \mathbf{E}_{02} &= 0.9036(0.5000, -0.3090, 0.8090)\\ \mathbf{E}_{03} &= 0.4518(-0.8090, -0.5000, 0.3090)\\ \mathbf{E}_{04} &= 0.9036(-0.3090, 0.8090, 0.5000)\end{aligned} \qquad (4.27)$$

By replacing the wave vectors (Equation 4.25) and the electric field amplitude vectors (Equation 4.27) in Equation 4.20, we can test the correctness of our solution. The spatial distribution of the intensity becomes

$$I(\mathbf{r}) = 3.6742 + \cos\left(\frac{2\pi x}{a}\right) + \cos\left(\frac{2\pi y}{a}\right) + \cos\left(\frac{2\pi z}{a}\right) \qquad (4.28)$$

Table 4.2 Beam parameters to fabricate the simple periodic structure $f_1(x, y, z)$. (Chapter 2 shows the formula and 3D image of this structure)

Structure	Wave vectors	Electric field amplitude vectors
$f_1(x, y, z)$	$\mathbf{k}_1 = \pi/a\,(1, 1, 1)$ $\mathbf{k}_2 = \pi/a\,(-1, 1, 1)$ $\mathbf{k}_3 = \pi/a\,(1, -1, 1)$ $\mathbf{k}_4 = \pi/a\,(1, 1, -1)$	$\mathbf{E}_{01} = 1.3554(-0.5825, -0.2041, 0.7867)$ $\mathbf{E}_{02} = 0.9036(0.5000, -0.3090, 0.8090)$ $\mathbf{E}_{03} = 0.4518(-0.8090, -0.5000, 0.3090)$ $\mathbf{E}_{04} = 0.9036(-0.3090, 0.8090, 0.5000)$

Table 4.3 Beam parameters to fabricate the simple periodic structures $f_3(x, y, z)$. (Chapter 2 shows the formulas and 3D images of the structures)

Structure	Wave vectors	Electric field amplitude vectors
$f_3^{4-I}(x, y, z)$	$\mathbf{k}_1 = \pi/a\,(2, 0, 1)$ $\mathbf{k}_2 = \pi/a\,(-2, 0, 1)$ $\mathbf{k}_3 = \pi/a\,(0, 2, -1)$ $\mathbf{k}_4 = \pi/a\,(0, -2, -1)$	$\mathbf{E}_{01} = 1.0000(-0.3535, 0.6123, 0.7071)$ $\mathbf{E}_{02} = 1.0000(0.3535, -0.6123, 0.7071)$ $\mathbf{E}_{03} = 1.0000(0.6123, 0.3535, 0.7071)$ $\mathbf{E}_{04} = 1.0000(-0.6123, -0.3535, 0.7071)$
$f_3^{4-II}(x, y, z)$	$\mathbf{k}_1 = \pi/a\,(2, 0, 1)$ $\mathbf{k}_2 = \pi/a\,(-2, 0, 1)$ $\mathbf{k}_3 = \pi/a\,(0, 2, -1)$ $\mathbf{k}_4 = \pi/a\,(0, -2, -1)$	$\mathbf{E}_{01} = 1.6008(-0.0831, 0.9825, 0.1662)$ $\mathbf{E}_{02} = 0.6558(0.4444, -0.1127, 0.8887)$ $\mathbf{E}_{03} = 0.9149(0.7298, 0.3057, 0.6114)$ $\mathbf{E}_{04} = 1.4682(0.4899, 0.3899, -0.7797)$
$f_3^3(x, y, z)$	$\mathbf{k}_1 = \pi/a\,(2, 0, 1)$ $\mathbf{k}_2 = \pi/a\,(-2, 0, 1)$ $\mathbf{k}_3 = \pi/a\,(0, 2, -1)$ $\mathbf{k}_4 = \pi/a\,(0, -2, -1)$	$\mathbf{E}_{01} = 0.2827(-0.3332, 0.6667, 0.6667)$ $\mathbf{E}_{02} = 1.9039(0.3714, -0.5571, 0.7428)$ $\mathbf{E}_{03} = 1.9039(-0.5571, 0.3714, 0.7428)$ $\mathbf{E}_{04} = 0.2827(0.6667, -0.3332, 0.6667)$

We can see that the intensity distribution (Equation 4.28) is similar to the formula of the simple periodic function (Equation 4.21) and therefore the corresponding periodic structure can be fabricated by interference lithography. Note that as a result of optimizing the electric field amplitude vectors, the contrast generated by the intensity distribution (Equation 4.28) is greatly improved. Using the definition of contrast in Equation 4.17, the reader can show that the optimized solution (Equation 4.28) increases the contrast by a factor of 2 over the previous solution (Equation 3.130).

By systematically applying the above scheme to the formulas corresponding to the three-dimensional periodic structures presented in Chapter 2, we found all the beam parameters required to fabricate these periodic structures. Tables 4.2–4.4

Table 4.4 Beam parameters to fabricate the simple periodic structures $f_2(x, y, z)$. (Chapter 2 shows the formulas and 3D images of the structures)

Structure	Wave vectors	Electric field amplitude vectors
$f_2^{6-I}(x, y, z)$	$\mathbf{k}_1 = \pi/a\,(1, 1, 1)$ $\mathbf{k}_2 = \pi/a\,(1, -1, -1)$ $\mathbf{k}_3 = \pi/a\,(-1, 1, -1)$ $\mathbf{k}_4 = \pi/a\,(-1, -1, 1)$	$\mathbf{E}_{01} = 1.5811(-0.6325, -0.1310, 0.7634)$ $\mathbf{E}_{02} = 1.5811(0.6325, -0.1310, 0.7634)$ $\mathbf{E}_{03} = 0.7071(0.0000, 0.7071, 0.7071)$ $\mathbf{E}_{04} = 0.7071(0.0000, 0.7071, 0.7071)$
$f_2^{6-II}(x, y, z)$	$\mathbf{k}_1 = \pi/a\,(1, 1, 1)$ $\mathbf{k}_2 = \pi/a\,(1, -1, -1)$ $\mathbf{k}_3 = \pi/a\,(-1, 1, -1)$ $\mathbf{k}_4 = \pi/a\,(-1, -1, 1)$	$\mathbf{E}_{01} = 2.5956(0.4521, 0.3628, -0.8149)$ $\mathbf{E}_{02} = 0.7530(0.8136, 0.4677, 0.3454)$ $\mathbf{E}_{03} = 0.7466(0.4928, 0.8102, 0.3174)$ $\mathbf{E}_{04} = 2.8465(0.7014, -0.7126, -0.0111)$
$f_2^{6-III}(x, y, z)$	$\mathbf{k}_1 = \pi/a\,(1, 1, 1)$ $\mathbf{k}_2 = \pi/a\,(1, -1, -1)$ $\mathbf{k}_3 = \pi/a\,(-1, 1, -1)$ $\mathbf{k}_4 = \pi/a\,(-1, -1, 1)$	$\mathbf{E}_{01} = (-0.0408 + i0.6281, 0.6650 - i0.7091, -0.6242 + i0.0809)$ $\mathbf{E}_{02} = (0.8008 + i0.0498, 0.6608 + i0.4490, 0.1400 - i0.3992)$ $\mathbf{E}_{03} = (-0.3718 + i0.4487, 0.0522 + i0.0489, 0.4239 - i0.3997)$ $\mathbf{E}_{04} = (-0.3894 + i0.1652, 0.4904 + i0.6307, 0.1009 + i0.7959)$
$f_2^{6-IV}(x, y, z)$	$\mathbf{k}_1 = \pi/a\,(1, 1, 1)$ $\mathbf{k}_2 = \pi/a\,(1, -1, -1)$ $\mathbf{k}_3 = \pi/a\,(-1, 1, -1)$ $\mathbf{k}_4 = \pi/a\,(-1, -1, 1)$	$\mathbf{E}_{01} = (0.6303 + i0.2425, -0.1822 - i0.4667, -0.4480 + i0.2243)$ $\mathbf{E}_{02} = (-0.1834 + i0.4767, 0.3212 + i0.3972, -0.5046 + i0.0795)$ $\mathbf{E}_{03} = (-0.3131 + i0.3912, -0.4182 - i0.2759, -0.1051 - i0.6671)$ $\mathbf{E}_{04} = (-0.0298 + i0.5002, 0.5251 - i0.4246, 0.4953 + i0.0755)$
$f_2^{5-I}(x, y, z)$	$\mathbf{k}_1 = \pi/a\,(1, 1, 1)$ $\mathbf{k}_2 = \pi/a\,(1, -1, -1)$ $\mathbf{k}_3 = \pi/a\,(-1, 1, -1)$ $\mathbf{k}_4 = \pi/a\,(-1, -1, 1)$	$\mathbf{E}_{01} = 1.4573(0.7874, -0.5809, -0.2065)$ $\mathbf{E}_{02} = 0.8282(0.6800, -0.0515, 0.7315)$ $\mathbf{E}_{03} = 1.4870(0.7503, 0.6540, -0.0963)$ $\mathbf{E}_{04} = 0.6742(0.4681, 0.3453, 0.8134)$
$f_2^{5-II}(x, y, z)$	$\mathbf{k}_1 = \pi/a\,(1, 1, 1)$ $\mathbf{k}_2 = \pi/a\,(1, -1, -1)$ $\mathbf{k}_3 = \pi/a\,(-1, 1, -1)$ $\mathbf{k}_4 = \pi/a\,(-1, -1, 1)$	$\mathbf{E}_{01} = 0.5724(0.4082, 0.4082, -0.8165)$ $\mathbf{E}_{02} = 1.0133(-0.0187, 0.6975, -0.7163)$ $\mathbf{E}_{03} = 1.0133(0.6975, -0.0187, -0.7163)$ $\mathbf{E}_{04} = 0.9741(-0.7071, 0.7071, 0.0000)$
$f_2^{5-III}(x, y, z)$	$\mathbf{k}_1 = \pi/a\,(1, 1, 1)$ $\mathbf{k}_2 = \pi/a\,(1, -1, -1)$ $\mathbf{k}_3 = \pi/a\,(-1, 1, -1)$ $\mathbf{k}_4 = \pi/a\,(-1, -1, 1)$	$\mathbf{E}_{01} = (0.4178 + i0.2967, 0.0383 + i0.3068, -0.4562 - i0.6035)$ $\mathbf{E}_{02} = (-0.6154 - i0.3667, -0.6350 + i0.1954, 0.0196 - i0.5622)$ $\mathbf{E}_{03} = (-0.0629 + i0.0698, 0.4444 - i0.5995, 0.5073 - i0.6693)$ $\mathbf{E}_{04} = (0.4920 - i0.4944, 0.0193 + i0.4204, 0.5114 - i0.0739)$
$f_2^{5-IV}(x, y, z)$	$\mathbf{k}_1 = \pi/a\,(1, 1, 1)$ $\mathbf{k}_2 = \pi/a\,(1, -1, -1)$ $\mathbf{k}_3 = \pi/a\,(-1, 1, -1)$ $\mathbf{k}_4 = \pi/a\,(-1, -1, 1)$	$\mathbf{E}_{01} = (0.1785 - i0.3838, -0.3885 - i0.1681, 0.2099 + i0.5519)$ $\mathbf{E}_{02} = (-0.5023 - i0.0457, 0.2175 - i0.4624, -0.7197 + i0.4166)$ $\mathbf{E}_{03} = (0.7042 - i0.3212, 0.6396 - i0.4281, -0.0646 - i0.1068)$ $\mathbf{E}_{04} = (0.0462 - i0.4151, -0.4137 - i0.0573, -0.3675 - i0.4724)$
$f_2^{4-I}(x, y, z)$	$\mathbf{k}_1 = \pi/a\,(1, 1, 1)$ $\mathbf{k}_2 = \pi/a\,(1, -1, -1)$ $\mathbf{k}_3 = \pi/a\,(-1, 1, -1)$ $\mathbf{k}_4 = \pi/a\,(-1, -1, 1)$	$\mathbf{E}_{01} = 1.0000(0.7071, -0.7071, 0.0000)$ $\mathbf{E}_{02} = 1.0000(0.7071, 0.7071, 0.0000)$ $\mathbf{E}_{03} = 1.0000(0.7071, 0.0000, -0.7071)$ $\mathbf{E}_{04} = 1.0000(0.7071, 0.0000, 0.7071)$

4.3 Simple Periodic Structures in Three Dimensions Via Interference Lithography

Table 4.4 (continued)

Structure	Wave vectors	Electric field amplitude vectors
$f_2^{4-II}(x,y,z)$	$\mathbf{k}_1 = \pi/a\,(1, 1, 1)$ $\mathbf{k}_2 = \pi/a\,(1, -1, -1)$ $\mathbf{k}_3 = \pi/a\,(-1, 1, -1)$ $\mathbf{k}_4 = \pi/a\,(-1, -1, 1)$	$\mathbf{E}_{01} = 1.3900(-0.2477, 0.7976, -0.5500)$ $\mathbf{E}_{02} = 0.7819(0.8119, 0.4806, 0.3313)$ $\mathbf{E}_{03} = 1.3962(0.7577, 0.1155, -0.6422)$ $\mathbf{E}_{04} = 0.7705(-0.6280, -0.1378, -0.7659)$
$f_2^{4-III}(x,y,z)$	$\mathbf{k}_1 = \pi/a\,(1, 1, 1)$ $\mathbf{k}_2 = \pi/a\,(1, -1, -1)$ $\mathbf{k}_3 = \pi/a\,(-1, 1, -1)$ $\mathbf{k}_4 = \pi/a\,(-1, -1, 1)$	$\mathbf{E}_{01} = 0.8990(0.7388, -0.0683, -0.6704)$ $\mathbf{E}_{02} = 1.2492(0.0552, 0.7332, -0.6778)$ $\mathbf{E}_{03} = 0.6455(0.8162, 0.3908, -0.4254)$ $\mathbf{E}_{04} = 1.1241(-0.0599, -0.6752, -0.7352)$
$f_2^{4-IV}(x,y,z)$	$\mathbf{k}_1 = \pi/a\,(1, 1, 1)$ $\mathbf{k}_2 = \pi/a\,(1, -1, -1)$ $\mathbf{k}_3 = \pi/a\,(-1, 1, -1)$ $\mathbf{k}_4 = \pi/a\,(-1, -1, 1)$	$\mathbf{E}_{01} = (0.5468 + i0.6477, -0.4594 + i0.0071, -0.0874 - i0.6548)$ $\mathbf{E}_{02} = (-0.5139 - i0.0157, -0.7388 - i0.0587, 0.2249 + i0.0430)$ $\mathbf{E}_{03} = (-0.1031 - i0.2073, 0.1928 - i0.6808, 0.2960 - i0.4734)$ $\mathbf{E}_{04} = (-0.2091 + i0.5346, 0.1546 - i0.2848, -0.0545 + i0.2498)$
$f_2^{3-I}(x,y,z)$	$\mathbf{k}_1 = \pi/a\,(1, 1, 1)$ $\mathbf{k}_2 = \pi/a\,(1, -1, -1)$ $\mathbf{k}_3 = \pi/a\,(-1, 1, -1)$ $\mathbf{k}_4 = \pi/a\,(-1, -1, 1)$	$\mathbf{E}_{01} = 1.0172(-0.7801, 0.5988, 0.1813)$ $\mathbf{E}_{02} = 0.7645(0.5551, 0.7961, -0.2409)$ $\mathbf{E}_{03} = 1.1323(0.1815, 0.7802, 0.5987)$ $\mathbf{E}_{04} = 0.5804(-0.8141, 0.4607, -0.3535)$
$f_2^{3-II}(x,y,z)$	$\mathbf{k}_1 = \pi/a\,(1, 1, 1)$ $\mathbf{k}_2 = \pi/a\,(1, -1, -1)$ $\mathbf{k}_3 = \pi/a\,(-1, 1, -1)$ $\mathbf{k}_4 = \pi/a\,(-1, -1, 1)$	$\mathbf{E}_{01} = 1.3554(-0.5826, -0.2041, 0.7867)$ $\mathbf{E}_{02} = 0.9036(0.5000, -0.3090, 0.8090)$ $\mathbf{E}_{03} = 0.4518(-0.8090, -0.5000, 0.3090)$ $\mathbf{E}_{04} = 0.9036(-0.3090, 0.8090, 0.5000)$

summarize our results. Note that in some cases the electric field amplitude vectors are *linearly polarized* while in others they are *elliptically polarized*.

Figure 4.3 shows the comparison between the theoretical three-dimensional periodic structure represented by the formula $f_1(x, y, z)$ and its experimental fabrication by interference lithography. As before, note the small length scale of the experimental structure with a lattice constant $a \sim 2\,\mu\text{m}$.

In summary, in this chapter, we presented a method that allows us to obtain the wave vectors $\mathbf{k}$ and electric field amplitude vectors $\mathbf{E}_0$ of the electromagnetic plane waves (or laser beams) required to fabricate periodic structures by interference lithography. The method is based on the similarity between the intensity pattern generated by the superposition of electromagnetic plane waves and the Fourier series expansion of a function representing a periodic structure. By using this method, the parameters of the interfering electromagnetic plane waves needed to fabricate periodic structures can be obtained systematically.

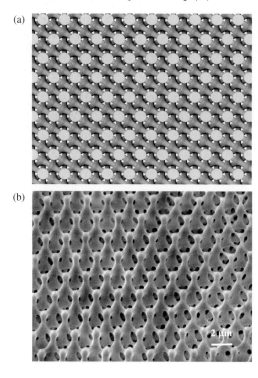

Fig. 4.3 (a) The three-dimensional periodic structure represented by the formula $f_1(x, y, z)$, which was slightly modified to obtain a better view of the structure. (b) Its fabrication by interference lithography by using the laser beam parameters shown in Table 4.2

Further Reading

1 Mei, D.B., Cheng, B.Y., Hu, W., Li, Z.L. and Zhang, D. (**1995**) "Three-dimensional ordered patterns by light interference". *Optics Letters*, **20**, 429–31.
2 Campbell, M., Sharp, D.N., Harrison, M.T., Denning, R.G. and Turberfield, A.J. (**2000**) "Fabrication of photonic crystals for the visible spectrum by holographic lithography". *Nature*, **404**, 53–56.
3 Turberfield, A.J. (**2001**) "Photonic crystals made by holographic lithography". *MRS Bulletin*, **26**, 632–36.
4 Cai, L.Z., Yang, X.L. and Wang, Y.R. (**2002**) "All fourteen Bravais lattices can be formed by interference of four noncoplanar beams". *Optics Letters*, **27**, 900–2.
5 Cai, L.Z., Yang, X.L. and Wang, Y.R. (**2002**) "Formation of three-dimensional periodic microstructures by interference of four noncoplanar beams". *Journal of the Optical Society of America A*, **19**, 2238–44.
6 Ullal, C.K., Maldovan, M., Wohlgemuth, M., White, C.A., Yang, S. and Thomas, E.L. (**2003**) "Triply periodic bicontinuous structures through interference lithography: a level-set approach". *Journal of the Optical Society of America A*, **20**, 948–54.
7 Sharp, D.N., Turberfield, A.J. and Denning, R.G. (**2003**) "Holographic photonic crystals with diamond symmetry". *Physical Review B*, **68**, 205102.
8 Toader, O., Chan, T.Y. and John, S. (**2004**) "Photonic band gap architectures for holographic lithography". *Physical Review Letters*, **92**, 043905.
9 Ao, X.Y. and He, S.L. (**2004**) "Two-stage design method for

realization of photonic band gap structures with desired symmetries by interference lithography". *Optics Express*, **12**, 978–83.

10 Ullal, C.K., Maldovan, M., Chen, G., Han, Y.-J., Yang, S. and Thomas, E.L. (**2004**) "Photonic crystals through holographic lithography: simple cubic, diamond-like, and gyroid-like structures". *Applied Physics Letters*, **84**, 5434–36.

11 Meisel, D.C., Wegener, M. and Busch, K. (**2004**) "Three-dimensional photonic crystals by holographic lithography using the umbrella configuration: symmetries and complete photonic band gaps". *Physical Review B*, **70**, 165104.

12 Chan, T.Y., Toader, O. and John, S. (**2005**) "Photonic band gap templating using optical interference lithography". *Physical Review E*, **71**, 046605.

13 Mao, W.D., Dong, J.W., Zhong, Y.C., Liang, G.Q. and Wang, H.Z. (**2005**) "Formation principles of two-dimensional compound photonic lattices by one-step holographic lithography". *Optics Express*, **13**, 2994–99.

14 Zhong, Y.C., Zhu, S.A., Su, H.M., Wang, H.Z., Chen, J.M., Zeng, Z.H. and Chen, H.L. (**2005**) "Photonic crystal with diamondlike structure fabricated by holographic lithography". *Applied Physics Letters*, **87**, 061103.

15 Moon, J.H., Yang, S., Pine, D.J. and Yang, S.M. (**2005**) "Translation of interference pattern by phase shift for diamond photonic crystals". *Optics Express*, **13**, 9841–46.

16 Lai, N.D., Liang, W.P., Lin, J.H., Hsu, C.C. and Lin, C.H. (**2005**) "Fabrication of two-and three-dimensional periodic structures by multi-exposure of two-beam interference technique". *Optics Express*, **13**, 9605–11.

17 Moon, J.H. and Yang, S. (**2005**) "Creating three-dimensional polymeric microstructures by multi-beam interference lithography". *Journal of Macromolecular Science-Polymer Reviews*, **C45**, 351–73.

18 Toader, O., Chan, T.Y.M. and John, S. (**2006**) "Diamond photonic band gap synthesis by umbrella holographic lithography". *Applied Physics Letters*, **89**, 101117.

19 Cai, L.Z., Dong, G.Y., Feng, C.S., Yang, X.L., Sheng, X.X. and Meng, X.F. (**2006**) "Holographic design of a two-dimensional photonic crystal of square lattice with a large two-dimensional complete gap". *Journal of the Optical Society of America B – Optical Physics*, **23**, 1708–11.

20 Meisel, D.C., Diem, M., Deubel, M., Perez-Williard, F., Linden, S., Gerthsen, D., Busch, K. and Wegener, M. (**2006**) "Shrinkage precompensation of holographic three-dimensional photonic-crystal templates". *Advanced Materials*, **18**, 2964–68.

21 Zhu, X.L., Xu, Y.G. and Yang, S. (**2007**) "Distortion of 3D SU8 photonic structures fabricated by four-beam holographic lithography with umbrella configuration". *Optics Express*, **15**, 16546–60.

Problems

4.1 Verify that the wave vectors and electric field amplitude vectors shown in Table 4.3 generate a spatial distribution of the intensity that is similar to the formulas of the simple periodic functions f_3^{4-I}, f_3^{4-II}, and f_3^3.

4.2 For the structures in Table 4.1, what is the relationship between the wavelength λ of the interfering electromagnetic plane waves and the spatial period a of the resultant periodic structures?

4.3 (a) By using the MATLAB program provided in Appendix A, find the wave vectors and electric field amplitude vectors required to fabricate the periodic structure represented by the formula

$$f(x, y, z) = 1\cos\left(\frac{2\pi x}{a}\right) + 2\cos\left(\frac{2\pi y}{a}\right) + 3\cos\left(\frac{2\pi z}{a}\right) \tag{4.29}$$

(b) Make a plot of $I(\mathbf{r})$.

Experimental

5
Fabrication of Periodic Structures

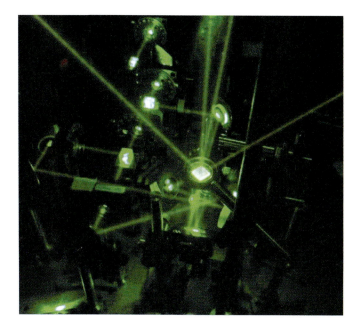

In the previous chapters, we have seen that an elegant and convenient way to create periodic structures represented by functions consisting of a few trigonometric terms is by the interference of electromagnetic plane waves (or laser beams). In this chapter, we explore some of the important considerations associated with the experimental realization of such structures.

Periodic Materials and Interference Lithography. M. Maldovan and E. Thomas
Copyright © 2009 WILEY-VCH Verlag GmbH & Co. KGaA, Weinheim
ISBN: 978-3-527-31999-2

5.1
Introduction

Since we are dealing with light, our task is essentially twofold: first, we want to obtain the desired beam configuration (i.e. wave vectors and electric field amplitude vectors for the electromagnetic plane waves) whose interference generates a spatial distribution of intensity $I(\mathbf{r})$ (or intensity pattern) that matches the function $f(\mathbf{r})$ representing the simple periodic structure that we want to fabricate. Second, we wish to record this intensity pattern into a material system, typically via a photosensitive medium, which will result in our final periodic structure. It is important to note that there already exist several variations and systems whereby these objectives can be achieved. Moreover, in the future it is reasonable to expect considerable progress in the ways an intensity pattern can be recorded into a photoresist material. We now undertake an examination of what can be done and how to make periodic materials.

5.2
Light Beams

The choice of light (or electromagnetic radiation) in the fabrication of periodic structures is in part driven by the convenience and reliability of lasers as a source of coherent electromagnetic plane waves. Throughout our derivations of three-dimensional intensity patterns, there was an implicit assumption that we were dealing with the interference of four monochromatic, coherent electromagnetic plane waves, which can be described by the formula

$$\mathbf{E}_j(\mathbf{r}, t) = \text{Re}(\mathbf{E}_{0j} e^{i(\mathbf{k}_j \cdot \mathbf{r} - \omega t)}), \quad j = 1, \ldots, 4 \tag{5.1}$$

In a monochromatic coherent plane wave, the electric field amplitude vector $\mathbf{E}_0$ at a particular point in space is constant and the phase varies linearly with time as ωt. In reality, light from real sources, including lasers, does not purely consist of a single frequency and is not coherent over arbitrary amounts of time. This means that for a particular point in space, the amplitude $\mathbf{E}_0$ and the phase ωt can fluctuate arbitrarily with time. How often such fluctuations occur depends on the spectral width of the light (in other words, the extent of the range of frequencies) in use and the mean frequency. The light source must be such that the light is sufficiently coherent over both the spatial extent of the sample and the duration of time for which the intensity pattern is being recorded into the chosen photoresist material. If this condition is not met, the resultant intensity pattern will become smoothened due to the random fluctuations in amplitude and phase of the interfering plane waves.

Typical light sources are regular lamps (often with filters to narrow the frequency spectrum), continuous wave lasers (which have a continuous constant amplitude of the electric field), and pulsed laser systems (in which very high peak powers can be achieved for each pulse). Each system will have a particular coherence length and time that must satisfy the requirements imposed by the manner in which the beam configuration is generated and the manner in which the intensity pattern is recorded.

The first task while creating simple periodic structures via interference lithography is to achieve the appropriate beam configuration. Once a particular periodic structure represented by an analytical function $f(\mathbf{r})$ is chosen, an optimum set of experimental parameters associated with the required interfering beams can be calculated as described in Chapter 4. In the simplest case, to obtain a three-dimensional periodic structure, we choose to interfere four beams of light. The experimental parameters that must be fixed for each beam are its direction (which is given by the wave vector $\mathbf{k}_j$), and the amplitude $|\mathbf{E}_{0j}|$ and state of polarization of its electric field (which are determined by the complex electric field amplitude vector $\mathbf{E}_{0j}$). In addition, the intensity of each beam is given by the square of the amplitude of the electric field as $I_j = |\mathbf{E}_{0j}|^2$ (Section 3.6.3).

An important consideration (which can only be demonstrated through a long derivation) is the fact that in the case of the interference of *four* noncoplanar beams, the addition of an arbitrary overall phase to each beam does not change the intensity pattern that the beams generate. For example, the interference of four beams given by

$$\mathbf{E}_j(\mathbf{r}, t) = \mathrm{Re}(\mathbf{E}_{0j} e^{i(\mathbf{k}_j \cdot \mathbf{r} - \omega t + \varphi_j)}), \quad j = 1, \ldots, 4 \tag{5.2}$$

(where φ_j is the arbitrary overall phase added to the j beam) generates the same intensity pattern $I(\mathbf{r})$ as the one generated by the original beams given by Equation 5.1. The only effect of the overall phases φ_j is the translation of the intensity pattern $I(\mathbf{r})$ in space.

The absence of any specification on the overall values of the phases φ_j of the interfering beams in the case of four-beam interference is a considerable and important simplification from an experimental standpoint. *The overall phases of the four beams are not relevant in terms of the fabrication of the periodic structures because their only effect is to translate the intensity pattern (and therefore the resultant periodic structure) in space.* That is, in the case of the interference of four laser beams, we can neglect the overall phases of the beams because different values of the phases create the *same* periodic structure, but located at different positions in space.

As an example, consider the interference of four linearly polarized beams. The Cartesian components of each complex amplitude vector $\mathbf{E}_{0j}$ have the same phase φ_j (Section 3.5). As a result, linearly polarized beams can be described by Equation 5.2, where the electric field amplitude vectors $\mathbf{E}_{0j}$ are *real* vectors. Because the overall phases φ_j only translate the resultant intensity pattern $I(\mathbf{r})$ in space, the four phases φ_j corresponding to the four linearly polarized beams can therefore be neglected.

In addition, circularly and elliptically polarized beams can also be described by Equation 5.2 and the overall phases can likewise be neglected. In these cases, however, the electric field amplitude vectors $\mathbf{E}_{0j}$ are still complex vectors, whose complex components have specific phases that determine the type of polarization. These phases must be taken into account to obtain the appropriate polarizations of the beams (Section 3.5).

5.3
Multiple Gratings and the Registration Challenge

In the most general case (which includes the interference of more than four beams), the first issue that must be addressed while creating structures via interference lithography is that of achieving registration of the multiple interference terms corresponding to the intensity equation (remember: each term is associated with the interference of two beams). This challenge is closely related to the ability to control the phases φ_j of the beams, in the region of interference, reliably and repeatedly. We have seen in the previous section that using four beams inherently solves this problem. It is, nevertheless, useful to understand the need for registration in the general case.

To appreciate this more fully, consider an arbitrary periodic structure. This structure can be represented by a sum of sinusoidal terms as given by its Fourier series expansion. The interference of two beams of light results in a one-dimensional sinusoidal variation of the intensity in space, which we will, for the purpose of distinction with the Fourier terms, temporarily call a grating. The direction and periodicity of this grating are determined by the difference $\mathbf{k}_i - \mathbf{k}_j$ between the wave vectors of the two interfering beams. We could theoretically fabricate our structure by individually registering in the photoresist material each sinusoidal Fourier term as a grating created by two interfering beams. For a grating, a change in the phase difference $\varphi_i - \varphi_j$ between the two interfering beams just causes a spatial translation of the intensity pattern in the direction of the periodicity of the grating. Once the phase difference for two interfering beams is fixed, to describe this grating by a particular sinusoidal function, we must fix our coordinate system. As we add the additional Fourier terms of our particular periodic structure, we introduce further gratings each of which has a coordinate system that matches it to a desired sinusoidal function. To fabricate our structure, we must ensure that the additional gratings are placed at the right locations with respect to the previously introduced gratings. To do this repeatedly and precisely in three dimensions taking both rotation and translation of the gratings into account is currently experimentally challenging.

Herein lies both the beauty and constraint of using four beams to create periodic structures by interference lithography. As explained in the previous section, when using four beams, the correct registration between the gratings we have selected is automatically guaranteed. This constraint on the number of beams is not overly restrictive. By using four beams, we have 12 interference terms in the intensity equation, and therefore 12 Fourier terms that we can work with to obtain desired periodic structures. As discussed in Chapter 4, the interference of four beams enables the fabrication of a large number of simple periodic structures.

5.4 Beam Configuration

5.4.1 Using Four Beams

From the perspective of beam configuration, the experimental task is thus reduced to ensuring the specified beam directions, amplitudes (or intensities), and polarizations of the interfering beams. The first scheme that we will examine to achieve this is by a 'free space' approach since it is the easiest to grasp. A typical beam path is shown in Figure 5.1, where four beams interfere within a recording medium (such as a photoresist material) to fabricate a three-dimensional periodic structure.

Referring to Figure 5.1, we see that the original beam is split into the requisite number of beams by three beam splitters. Since lasers typically produce a light beam with a specific linear polarization, the original beam is in general linearly polarized, and therefore the electric field oscillates along a fixed direction (Section 3.5.1).

The *intensity* of each beam can be controlled by using a specific type of beam splitter known as *a polarizing beam splitter*. Such a beam splitter splits the incident linearly polarized beam into two constituent linearly polarized beams (i.e. two linearly polarized beams for which the directions of oscillations of the electric field vectors are separated by 90°). By controlling the direction of oscillation of the electric field of the incoming linearly polarized beam (which is equivalent to controlling the angle of polarization θ as discussed in Section 3.5.1), the splitter allows us to control the relative amplitudes of the two constituent linearly polarized beams, and thus the intensity of the two resultant linearly polarized beams that

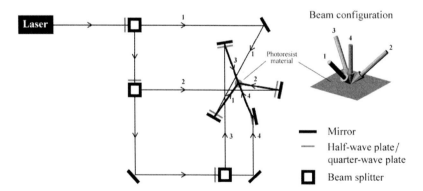

Fig. 5.1 Free space lithographic setup showing beam paths for four interfering beams where the final intersection of the four beams into the photoresist material is shown in thick lines. The inset in the upper-right corner shows the directions of the beams with respect to the photoresist material.

come out of the beam splitter. The direction of oscillation of the electric field of the incoming linearly polarized beam is controlled by using a half-wave plate[10] before the beam is incident on the beam splitter (Figure 5.1).

On the other hand, the beam *directions* are ensured by launching the beams toward the recording medium by bouncing them off mirrors at the appropriate angle. Note that a favorable experimental configuration is when all beams are incident from the same side of the photoresist material (also called the *umbrella* configuration). This is because in the case that the beams are incident from both sides of the photoresist material, it would be required to use a transparent and nonabsorbent substrate.

Finally, before the beams are incident on the photoresist material, the required type of polarization must be adjusted. Light beams with linear polarizations along specific directions are achieved by rotating the direction of oscillation of the electric field by means of a half-wave plate. On the other hand, light beams with specific elliptical polarizations are achieved by using a half-wave plate and a quarter-wave plate in sequence.

All considerations to obtain the right directions, intensities, and polarizations of the four interfering beams within the photoresist material are treated in subsequent sections. It is important to mention that vibrations of the photoresist material, mirrors, lenses, and so on, can result in the spatial motion of the intensity pattern, causing a large decrease in its contrast and the inability to make high fidelity structures.

5.4.2
Using a Single Beam (Phase Mask Lithography)

Another approach for achieving the desired beam configuration to fabricate periodic structures by the interference of waves is to create the interfering beams by using diffraction. For example, when a beam of light is incident on a periodic variation of the dielectric constant created by a particular dielectric structure, the light is diffracted along specific directions. The relative intensity and phase of the light that goes along each particular direction are determined by various factors such as the geometry and the refractive index of the dielectric structure. Now suppose we are looking to create a particular three-dimensional intensity pattern and suppose we know the beam parameters, which would give us our desired pattern. One can now create an appropriate two- or three-dimensional periodic dielectric structure,

---

10) A wave plate is an optical device made of a birefringent material with a specific thickness. As a consequence of the birefringence, the refractive indices along the two in-plane axes of the wave plate are different. This results in an optical path difference that light sees along the two axes. This optical path difference depends on the thickness of the wave plate, the respective refractive indices, and the wavelength λ of the light beam. In a half-wave plate, this path difference is set to $\lambda/2$ and in a quarter-wave plate it is set to $\lambda/4$. The polarization state of the emerging light on the other side of the wave plate depends on the angle that the incident polarization makes with the two axes. A half-wave plate is frequently used to rotate the orientation of linearly polarized light while a quarter-wave plate is used to convert light from being linearly polarized to elliptically polarized.

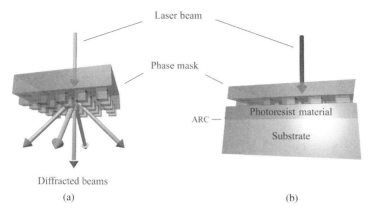

Fig. 5.2 Phase mask lithography. (a) A single monochromatic beam is incident onto a conformable phase mask. As a result of diffraction, the original beam divides into five diffracted beams. (b) The intensity pattern generated by the interference of the diffracted beams is recorded by placing a photoresist material next to the phase mask. Either an antireflection coating (ARC) between the substrate and the photoresist or an index-matched substrate is used to prevent interference with back reflected beams.

known as a *phase mask*, which, when exposed to a single beam of light, would give rise to diffracted beams that are traveling in the specific directions with the correct electric field amplitude vectors such that they match the beam parameters that we desire. The diffracted beams can be recombined either in the far field by using a lens to focus the beams onto the photoresist, or, as in Figure 5.2, the intensity pattern generated by the beams can be recorded by placing a photosensitive material in close proximity to the phase mask, where the diffracted beams overlap in space (near field). The advantage with this phase mask approach is that it simplifies the experimental setup tremendously because there is no need to arrange the four-beam configuration shown in Figure 5.1. More importantly, since the phases of the diffracted beams are fixed by the phase mask, in the case of placing the photoresist in the near field, it allows us to interfere more than four beams without worrying about the registration problem. Each time that a structure is transferred into a photoresist using the phase mask, all the beam parameters (including the relative phases) of the diffracted beams interfering in the photoresist will be the same. The final structure arising from a given phase mask would therefore be repeatedly reproduced each time. Unfortunately, there is presently no simple analytical scheme to obtain the required phase mask that would generate the correct beams that in turn will create a specific, desired, three-dimensional periodic structure. It is, however, possible to calculate the resultant intensity pattern (thereby the resultant structure) for a given phase mask. The process of targeting a desired three-dimensional structure, therefore, currently involves a trial and error process that is provided by knowledge of basic principles of diffraction. The development of such an analytical scheme that would directly generate the required phase mask for a desired three-dimensional structure is a reasonable possibility, and would constitute an important advance in the technology.

5.5
Pattern Transfer: Material Platforms and Photoresists

The second step in fabricating structures by interference lithography is in the transfer of the intensity pattern into a permanent structure via selective chemical changes of a photosensitive material. Once again, there are many material platforms that will facilitate such a transfer. Initial interference lithographic structures were fabricated in a class of materials known as *photoresists*, which were developed primarily for use in the semiconductor industry for printing two-dimensional patterns on circuit boards by use of transparency masks. Photoresists are organic materials, typically used in the form of thin films (often about 1–20-µm thick), which are designed to change their solubility in certain solvents upon exposure to light. An extensive research effort was involved in the development of this class of photosensitive materials. One of the most important goals in the semiconductor industry was to improve the speed of transistors, which depends on the time it takes for an electron to travel between the transistors components. There was thus a strong push for the miniaturization of individual circuit elements, thereby reducing this travel time, and increasing the speed of the device. An increased density in the pattern of the integrated circuits implied a corresponding demand of increased resolution that could be supported by the photoresists. This improvement in resolution was supported by the use of increasingly shorter wavelengths of light. Each downward shift in the exposure wavelength was accompanied by the development of a photoresist material platform that met the appropriate requirements such as optical transparency. This development of photoresist platforms across a wide spectrum of wavelengths of light is particularly relevant to our efforts given that the size of the resultant structures fabricated by interference lithography scales with the wavelength of the interfering light. Figure 5.3 presents the lithographic road map showing the development of photoresists as a function of the resolution (or wavelength of radiation).

The commonality of a large number of the properties required of photoresists used in the semiconductor industry and those of photosensitive materials used in interference lithography means that photoresists (with suitable modifications where necessary) constitute good candidates as materials for the intensity pattern transfer process.

The transfer of intensity patterns in photoresists takes advantage of the property that exposing the photoresist to an intensity of light above a certain threshold (i.e. overexposure) causes a chemical change in the heavily exposed regions.

There exist two types of photoresist materials: positive and negative. In a positive photoresist, the regions in the material that are overexposed to light are chemically changed and become soluble to the developing solvent. This means that the material in the regions that are overexposed is removed by the solvent. On the other hand, the underexposed regions, which are exposed to light intensities below the threshold intensity, remain insoluble to the solvent and form the resultant structure. One strategy for a positive photoresist material is to use a *polymer* photoresist and design the light induced chemical change such that in the overexposed regions either the

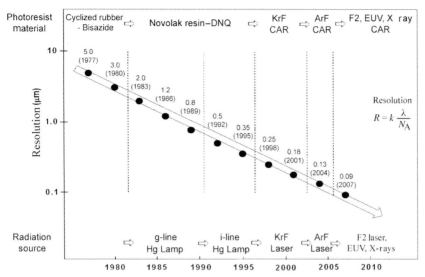

Fig. 5.3 Lithographic road map showing different types of photoresists for different radiation sources together with the resolution achieved and the corresponding date. Radiation sources and the associated wavelengths: Hg lamp (g-line) $\lambda = 436$ nm, Hg lamp (i-line) $\lambda = 365$ nm, krypton fluoride (KrF) excimer laser $\lambda = 248$ nm, argon fluoride (ArF) excimer laser $\lambda = 193$ nm, fluorine (F2) laser $\lambda = 157$ nm, extreme ultraviolet (EUV) $\lambda = 30-1$ nm, and X rays $\lambda = 10-0.01$ nm. The resolution R (or minimum feature size) is proportional to the ratio between the wavelength λ of the radiation used and the numerical aperture N_A of the lens. CAR is an abbreviation for chemically amplified resist. [Adapted from Thompson, L.F., Willson, C.G. and Bowden, M.J. (1994), *Introduction to Microlithography*, American Chemical Society, Washington, DC.]

bonds of the polymer are broken (thereby allowing these regions to be washed away by an appropriate solvent) or some other chemical change is induced that renders the polymer in these regions soluble in an appropriate solvent.

In a negative photoresist, the regions in the material that are overexposed to light are chemically changed and become difficult to dissolve. Therefore, the material in the overexposed regions remains intact forming the structure and the solvent removes the material in the underexposed regions. One strategy for a negative photoresist is to use a *monomeric* system in which the overexposed regions become polymerized and cross-link such that they form an insoluble network for the developing solvent.

Note that in the case of positive resists, the material in regions of high intensity of light is removed, whereas in the case of negative resists, the material in regions of low intensity light is removed. This means that the resultant structure in a positive (negative) resist is formed by the material located in the low (high) intensity regions. By changing from a negative to a positive resist, we can exchange the solid structure in one for the air structure in the other and vice versa. This allows access to inverse structures.

5.5.1
Negative Photoresists

The insolubility required in a negative resist can be achieved either by an increase in molecular weight or by photochemical rearrangements to form new insoluble products. To increase molecular weight, photoinitiators are generally used, which upon the absorption of light can generate free radicals or strong acids. These highly reactive chemical species facilitate cross-linking of the monomeric or oligomeric species that constitute the photoresist, and result in the formation of a high molecular weight polymer, which is difficult to dissolve. Without an increase in molecular weight, negative resists can also be achieved by the photochemical formation of hydrophobic or hydrophilic groups, which provide differential solubility between the overexposed and underexposed regions of the resist film.

One common negative photoresist for interference lithography is SU-8, an epoxy-based monomer that undergoes cationic polymerization. The reaction mechanism for SU-8 is shown in Figure 5.4 and serves to exemplify some of the considerations involved in choosing an appropriate photosensitive material platform. SU-8 belongs to a class of photoresists known as *chemically amplified resists* (CAR). In such photoresists, exposure to light results in the creation of a catalyst for a subsequent chemical reaction. In the first step, light is absorbed by the photoacid generator (PAG). This results in the release of a proton or H^+ ion. These photoacids initiate ring-opening reactions of the epoxy groups and catalyze the polymerization reaction in the SU-8 material, resulting in a highly cross-linked film in regions of high light intensity. During the polymerization reaction, the H^+ ion is regenerated, thereby allowing it to participate once more in further polymerization reactions. This recycling of H^+ ions implies that each absorbed photon gives rise to a large number of polymerization reactions. This chemical amplification results in an increased sensitivity of the photoresist material, which is an important consideration for the material platform.

Depending on the absorption spectrum of the PAG used, the photoresist will be responsive over a specific wavelength range. To extend the use of the photoresist to other wavelengths, an additive known as a *photosensitizer* is often incorporated. The photosensitizer serves to absorb the appropriate wavelength of light and transfer the energy to the PAG. For example, in the case of the SU-8 platform, diaryliodionium salts are typically chosen as the PAGs. The excited photosensitizer transfers an electron to the onium salt, which then generates the acid. Thus, the addition of a suitable photosensitizer allows for use of the photoresist at other wavelengths. A consideration in the extension of photoresists to other wavelengths is that the photoresist itself must not absorb the exposing light. The inherent broad wavelength transparency of SU-8 allows chemical routes to enable a wide processing window with respect to the wavelengths of the incident radiation.

The SU-8 photoresist serves to illustrate another consideration in the choice of a material platform for creating structures via interference lithography. It is important that the chemical change induced by the exposure of the photoresist material to the intensity pattern should not result in a change of refractive index

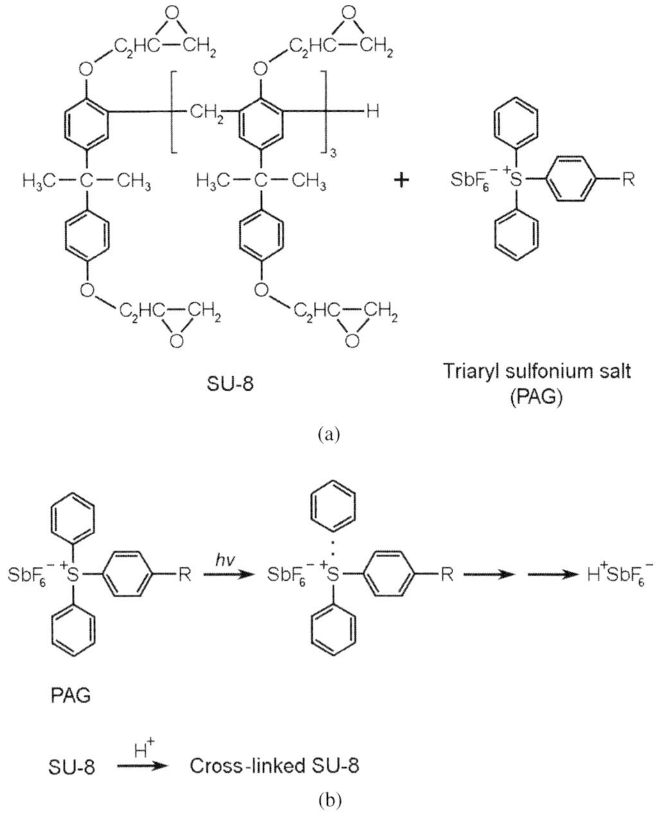

Fig. 5.4 SU-8 negative resist. (a) Chemical composition. Note the eight-epoxide groups per monomeric unit. (b) Photochemical reaction mechanism. The triaryl sulfonium salt acts as the photoacid generator upon irradiation with light.

of the photoresist material. If such a change occurs during the exposure process, then the interfering beams will begin to diffract off the pattern that is being written, which creates new beams that disrupt the desired intensity pattern. CAR avoid this problem since the creation of the H^+ ion does not significantly affect the refractive index of the photoresist material, and the intensity pattern is, therefore, effectively transferred to a spatial variation in concentration of the generated acid. Since the photoresist material is glassy at room temperature, the photo generated acid cannot diffuse. Once the exposure process is complete, the photoresist is then heated to a certain temperature at which the acid can diffuse to a limited extent and cause the appropriate chemical reaction, in this case the polymerization of the SU-8 photoresist. During polymerization, a change in the refractive index of the photoresist material occurs, but this change takes place after the exposure.

5.5.2
Positive Photoresists

As defined earlier, a positive resist is a photoresist in which regions overexposed to light become soluble to the developer, while underexposed regions remain insoluble. One common positive resist for ultraviolet (UV) light is composed of diazonaphthoquinone (DNQ) and Novolac resin (a phenol formaldehyde resin) (Figure 5.5). This combination is normally used with i-(365 nm), g-(436 nm), and

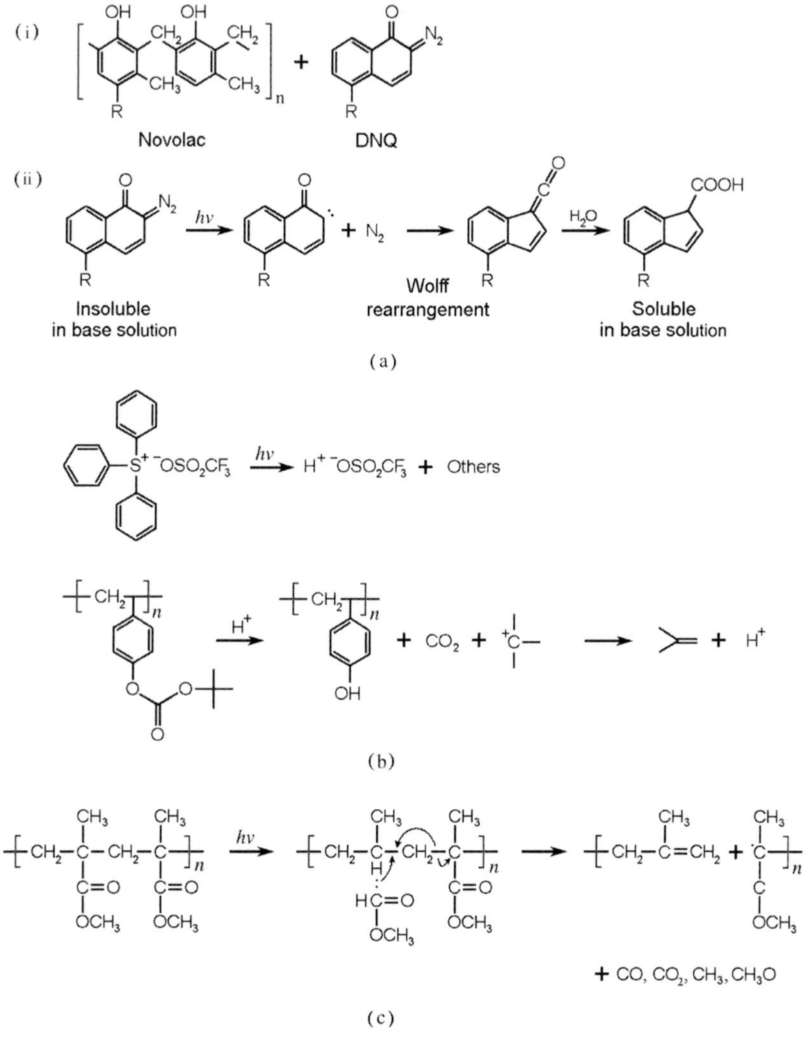

Fig. 5.5 Positive resist systems. (a) AZ type resist system (AZ5214E): (i) chemical composition, Novolac resin + DNQ (sensitizer and dissolution inhibitor) and (ii) photochemical reaction. (b) PHOST system. (c) PMMA resist system.

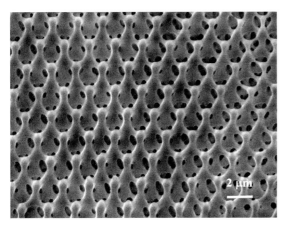

Fig. 5.6 A scanning electron microscope (SEM) image of the experimental realization of the simple periodic structure $f_1(x, y, z)$ in a photoresist material. The structure view is along the (1, 1, 1) diagonal direction of the cubic unit cell. [Source: Maldovan, M. et al. (2007) *Advanced Materials*, **19**, 3809.]

h-(404 nm) lines from a mercury lamp, and it is typically a nonchemically amplified resist (non-CAR). The phenolic resin is highly soluble in basic solution and has excellent film forming properties. DNQ acts both as a photosensitizer and dissolution inhibitor. Upon exposure, DNQ undergoes molecular rearrangement generating a carboxylic acid, and the overexposed area becomes soluble in basic developers, resulting in a positive image (empty regions) in areas of high light intensity. The amount of light absorbed by DNQ systems in the UV region decreases with increasing exposure due to the conversion of the compound into indene–carboxylic acid photoproducts. This 'photobleaching' effect allows for the propagation of the light through thicker resist films with minimum loss as the exposure proceeds (Figure 5.5).

Such a system cannot be used while employing wavelengths in the deep UV both due to the strong absorbance of the Novolac resins and the unbleachable characteristics of the DNQ chromophore in this spectral region. For exposure wavelengths shorter than 300 nm, systems based on polyhydroxystyrene (PHOST) polymers or methacrylate polymers (PMMA) are used. These positive photoresists consist of the corresponding polymeric backbone and contain *t*-butyloxycarbonyl (*t*-BOC) or tetrahydropyranyl groups, particular examples of which are shown in Figure 5.5. These groups behave as protecting groups. In the presence of an acid, which can be created by the exposure of a PAG to light of an appropriate wavelength and intensity, a deprotection reaction takes place. The consequence of this deprotection reaction is that a change in solubility of the polymer toward a developing agent is brought about. The deprotection reaction is also such that the acid is regenerated. These photoresists therefore fall into the category of CAR, and display a correspondingly higher exposure sensitivity.

5.5.3
Organic–Inorganic Hybrids Resists

Direct patterning of an organic–inorganic hybrid photoresist is interesting since hybrid materials are expected to have higher refractive index and improved mechanical properties as compared to a polymer-only structure. For example, two-dimensional organic–inorganic hybrid arrays were patterned by interference lithography via photo-cross-linking of the transition metal containing acrylate monomer obtained from the sol–gel reaction of methacryloxypropyl trimethoxysilane (MPTMS) and titanium alkoxide. In this case, the resultant structure is made of a material with an increased refractive index up to 2.0. This is an important consideration since one of the key applications of a periodic structure is a photonic crystal (Chapter 6), where the difference between the refractive indices of the material forming the periodic structure and air is crucial to achieve desired performances.

5.6
Practical Considerations for Interference Lithography

Having established the basic procedure by which one can fabricate our simple periodic structures, there are several additional practical considerations that are useful to keep in mind.

5.6.1
Preserving Polarizations and Directions

The periodic spatial distribution of the intensity generated by the interference of the multiple coherent beams of light allows us to create periodic structures. Since we want to transfer the intensity pattern into a photoresist material, it is important to ensure that the interference pattern *inside* our photoresist is the one that we desire. Three important experimental parameters that must be preserved as the light enters the photoresist material are the directions of propagation, the intensities, and the polarization states of the interfering beams.

Any time a beam of light is incident on an interface that presents a change in the dielectric constant ε or the magnetic permeability μ, its wave vector changes direction and magnitude. For example, Figure 5.7 shows incident and transmitted wave vectors $\mathbf{k}_i$ and $\mathbf{k}_t$, respectively, for a beam of light, which is incident from air on the photoresist material at an angle of incidence θ_1. The air and the photoresist material are characterized by the parameters ε_1, μ_1 and ε_2, μ_2, respectively, and their indices of refraction are defined as

$$n_1 = \sqrt{\frac{\varepsilon_1 \mu_1}{\varepsilon_0 \mu_0}}, \quad n_2 = \sqrt{\frac{\varepsilon_2 \mu_2}{\varepsilon_0 \mu_0}} \tag{5.3}$$

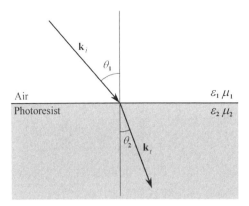

Fig. 5.7 Refraction of a light beam at the interface between air and the photoresist material.

where ε_0 and μ_0 are the parameters corresponding to vacuum, which satisfy $c = \sqrt{1/(\varepsilon_0 \mu_0)}$ where c is the speed of light in vacuum (Chapter 3). Note: since the air and the photoresist material are nonmagnetic, it can be assumed that $\mu_1 = \mu_2 = \mu_0$.

According to Equations 3.12 and 3.20, the change in the *magnitude* of the wave vector as the beam of light enters the photoresist material is given by

$$|\mathbf{k}_i| = k_i = \omega\sqrt{\varepsilon_1 \mu_1} = \frac{\omega}{c} n_1 \qquad (5.4)$$

$$|\mathbf{k}_t| = k_t = \omega\sqrt{\varepsilon_2 \mu_2} = \frac{\omega}{c} n_2 \qquad (5.5)$$

while the change in *direction* is determined by Snell's law, which establishes the relationship between the angle of incidence θ_1 and the angle of refraction θ_2.

$$\frac{\sin \theta_1}{\sin \theta_2} = \sqrt{\frac{\varepsilon_2 \mu_2}{\varepsilon_1 \mu_1}} = \frac{n_2}{n_1} \qquad (5.6)$$

In addition, since the electric field must always be orthogonal to the wave vector, the electric field also changes its direction as the beam of light is transmitted into the photoresist material. To account for this change, it is convenient to separate the electric field of the incident beam into its two independent perpendicular components (Section 3.5). In one component, the electric field is perpendicular to the plane of incidence (which is formed by the wave vectors $\mathbf{k}_i$ and $\mathbf{k}_t$), whereas in the other component, the electric field is parallel to the plane of incidence. These two perpendicular components are called *transverse electric* (TE) and *transverse magnetic* (TM) components, respectively, and their refraction is illustrated in Figure 5.8.

Moreover, the magnitude of the TE and TM electric fields also change as the beam of light enters the photoresist material. The intensity of the TE and TM

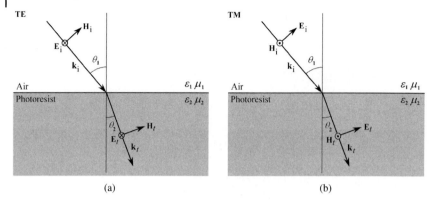

Fig. 5.8 (a) Refraction of the transverse electric (TE) component of the incident light beam at the interface between air and the photoresist material. (b) Refraction of the transverse magnetic (TM) component of the incident light beam.

components in the photoresist material can be obtained by using the formulas

$$I_{TE} = 1 - \left(\frac{n_1 \cos\theta_1 - \frac{\mu_1}{\mu_2} n_2 \cos\theta_2}{n_1 \cos\theta_1 + \frac{\mu_1}{\mu_2} n_2 \cos\theta_2} \right)^2$$

$$I_{TM} = 1 - \left(\frac{n_1 \cos\theta_2 - \frac{\mu_1}{\mu_2} n_2 \cos\theta_1}{n_1 \cos\theta_2 + \frac{\mu_1}{\mu_2} n_2 \cos\theta_1} \right)^2 \quad (5.7)$$

which can be derived by applying Maxwell's equations at an interface between two different materials having parameters ε_1, μ_1 and ε_2, μ_2, and requires knowledge of the angle of incidence θ_1. Note: the angle of refraction θ_2 is obtained by using Snell's law (Equation 5.6).

To preserve the correct beam directions, intensities, and polarizations within the final photoresist material, one has to solve this free space to photoresist transmission problem. That is, one has to obtain the starting parameters for the beams outside of the photoresist material such that when these beams are refracted into the material, the parameters equal the values required to fabricate the particular periodic structure. There is, however, an additional inconvenience presented by the fact that the photoresist material possesses a higher refractive index (typically $n_2 \sim 1.6$) than air ($n_1 = 1.0$). This means that the angle of incidence θ_1 is larger than the angle of refraction θ_2 and beam directions within the photoresist material that correspond to refraction angles θ_2 that are higher than the limit angle $\theta_L = \arcsin(n_1/n_2)$ for that particular photoresist material cannot be accessed.

An alternative solution to this free space to photoresist transmission problem is to construct a coupling prism whose refractive index matches that of the photoresist and whose shape is such that each incoming beam is presented with an interface that is normal to the direction of propagation for that beam (Figure 5.9).

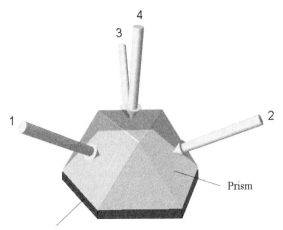

Fig. 5.9 Prism configuration to fabricate the periodic structure $f_3^3(x,y,z)$ by the use of four interfering beams. The prism is located on top of the photoresist material, where the base of the prism is defined by the light gray hexagon. The arrows represent the directions of the laser beams, which are numbered from 1 to 4. The fourth beam is normally incident on the base of the prism (and the photoresist material) from the triangular face located at the top of the prism. The other beams are symmetrically incident on the photoresist material from the three square faces of the prism.

A related experimental concern is the back reflection of light that arises from the interfaces generated by the substrate that supports the photoresist material. Since the reflected light is coherent with the incoming beams, if the intensity of the back reflected light is large enough, it would result in a distortion of the interference pattern. An obvious solution to minimize the reflection from the photoresist–substrate interface is to use a refractive index-matched substrate. In addition, the backside of the substrate can be contacted with a prism that has the same shape as the coupling prism so as to allow the beams to exit with a minimum amount of back reflection from the substrate–air interface. In the case of an opaque substrate, it would be necessary to utilize either an absorbing medium or an antireflection coating (ARC) underneath the photoresist material. Since, in general, all the beams are not symmetrically arranged in space with respect to the substrate, the antireflection coating would have to be chosen in a manner such that it minimizes the net back reflection intensity.

5.6.2
Contrast

Even when due precaution is taken to ensure that the optics results in the desired spatial distribution of the intensity within the photoresist material, low contrast in the intensity distribution could prevent the realization of the actual structure.

The intensity equation that describes the interference of multiple beams of light always contains a constant spatially invariant term. Mathematically, this arises from the dot product of each plane-wave term with the conjugate of itself (e.g. $|E_{01}|^2 + |E_{02}|^2 + |E_{03}|^2 + |E_{04}|^2$). Physically, this manifests itself as a dc offset to the spatially varying interference terms. When we transfer the intensity pattern into a photoresist material, the manner in which it is transferred depends on the response of the particular photoresist material in use. Therefore, in the case of our example acid catalyzed negative resist SU-8, exposure to a particular combination of intensity of light and time of exposure results in a given amount of polymerization. Since we seek to create a polymer–air structure, a material that has a binary response would be ideal: exposure above a certain threshold value would result in a constant degree of polymerization, while areas that receive exposure below this value would result in no polymerization. In reality, however, this is not the case and the response is monotonically increasing with dose (the SU-8 monomers react to form a multifunctional cross-linked network). If the magnitude of the spatially varying interference terms in the intensity equation is not large enough in comparison to the constant spatially invariant term, in other words for an insufficiently large contrast, then it becomes experimentally challenging to find an exposure condition for which the maxima of the intensity pattern results in network polymerization while the minima results in no polymerization.

We can define the contrast C present in a particular interference pattern as

$$C = \frac{I_{max} - I_{min}}{I_{max} + I_{min}} \tag{5.8}$$

where I_{max} and I_{min} are the maximum and minimum intensity values in the intensity pattern.

There are two independent approaches by which the contrast can be increased. The first is to recognize that even for a particular set of beam directions and a specific target intensity equation there are frequently multiple beam polarization states that will achieve the desired intensity pattern. As discussed in Chapter 4, one can therefore optimize the beam polarization states for the maximum contrast within this set of conditions. A complimentary approach to dealing with low contrast for a particular set of optical parameters is to effectively modify the response of the photoresist material by chemical means. For example, with SU-8, the spatial variation in polymerization is caused via the intermediate transfer of the light intensity pattern into a spatial variation of H^+ ions or acid. A certain, constant, spatially invariant, amount of the acid that is generated can be neutralized by adding a base to the photoresist material. This in effect lowers the dc offset while retaining the magnitude of the difference in the values of the maxima and minima of the intensity pattern, thereby raising the realizable contrast of the system.

5.6.3
Drying

Once the intensity pattern is transferred into the photoresist material, it is important to ensure that subsequent steps in the fabrication process do not distort the final structure. A key step is the development stage. After the solid, highly cross-linked structure is revealed by the development process, the developing solvent must be removed by drying. Forces that are a consequence of the surface tension of the solvent can be very large during drying. The distortion of the structure created by these forces can be circumvented by using a drying technique called *supercritical drying*. In this technique, the developing solvent is exchanged with liquid CO_2. If the developing solvent is miscible in liquid CO_2, no surface tension related effects are seen during this solvent exchange. Subsequently, the liquid CO_2 is heated under pressure until it is in a supercritical state. In the supercritical state, the densities of the liquid and vapor phases are the same. This state is characterized by low viscosity, high diffusion rates, and no surface tension. If the supercritical fluid is now held at a temperature above its supercritical temperature and its pressure reduced to one at which the equilibrium state is a gas, then the CO_2 is taken directly from a supercritical fluid to a gas. Since the supercritical fluid has no surface tension, the collapse of the structure that often accompanies the drying procedure is avoided.

5.6.4
Shrinkage

Another source of structural distortion is the stress created during the fabrication procedure. Such distortions are typically pronounced in negative photoresist materials. This is due to the fact that the transfer of the interference pattern into the photoresist material is associated with the polymerization and cross-linking of chains in the regions exposed to higher intensity of light. The formation of a cross-linked network is typically accompanied by a reduction in volume (increase in density). The stress associated with this change in volume is dissipated through distortions in the structure. Moreover, since the substrate presents, from a mechanical standpoint, a hard boundary condition that is not present at the upper free surface of the film, the shrinkage forces are not distributed in an isotropic manner. An approach to circumventing distortions arising out of volume shrinkage is to provide additional rigid boundary conditions that can help support the stress. Additionally and alternatively, one can calculate the distortions that a particular set of boundary conditions are likely to result in, modify the beam conditions to give a structure, which when subject to the accompanying distortions would yield the final desired target structure. In terms of the shrinkage of the resultant structure, the use of a positive photoresist material is more beneficial. In a positive photoresist material, one starts with a non-cross-linked polymer that is broken down into a soluble form in the regions exposed to higher intensity of light. This process is not accompanied by the volume change in the retained material that is seen in the negative resist platforms.

5.6.5
Backfilling – Creating Inverse Periodic Structures

The absence of a cross-linked network in positive photoresist materials is advantageous from the perspective of the use of the polymeric structures as templates for forming structures made of other materials. For example, simple periodic structures have a whole host of interesting properties ranging from their optical to their mechanical properties. The materials needed to optimally exploit each of these properties can be very different. It is therefore important to have a photoresist material that lends itself to being replaced by the more appropriate material for each targeted property. For example, in the case of photonic crystals and negative index materials working in the infrared, we might want to replace the polymeric photoresist material with a high refractive index, low absorption, dielectric material such as silicon. One can thus imagine back filling the air phase of the simple periodic structures with silicon using a process such as CVD, while using the polymeric structure as a template. The task would then be to remove the polymeric structure while retaining the structural integrity of the silicon structure. While a negative cross-linked photoresist has some advantages such as higher mechanical integrity, it is more difficult to replace with another material as compared to a positive photoresist. This is because the conditions under which the cross-linked material can be removed tend to be quite harsh, such as pyrolysis at high temperatures. This limits the materials that the cross-linked polymeric templates can be substituted with. Positive photoresists, on the other hand, can be removed more easily after backfilling by simple flood exposure and subsequent dissolution.

5.6.6
Volume Fraction Control

One of the primary advantages of using holographic interference lithography is the ability to control the geometry of the structure. In addition to affording control over the symmetries present in the structures, this technique also allows for a facile means by which the relative volume fractions of the two constituent materials of the structure (e.g. photoresist material and air) can be controlled. This is quite important since the properties of simple periodic structures can be very strongly dependent on the relative volume fractions of the constituent materials. In interference lithography, volume fraction control is achieved by controlling the value of the dc offset in the intensity equation. There are a number of experimental parameters through which this value can be controlled, including the intensity of the exposure, the duration of the exposure, the chemistry of the material platform, and the nature and temperature of the developing solvent. The first two parameters clearly affect the extent of exposure. In the case of a negative photoresist material, a larger exposure value (either due to a higher intensity or due to longer exposure duration) results in a larger volume fraction within the unit cell that is polymerized, and therefore a larger volume fraction of photoresist material in the developed structure. As we have seen earlier, modifying the chemistry of the photoresist also allows us to control the magnitude of the dc offset, and thereby the volume fraction of photoresist material in the resultant structure. The extent to which the photoresist dissolves depends on the strength of the developing solvent used. In the case of a negative resist, a stronger solvent will dissolve more of the photoresist material leading to a smaller volume fraction of this material in the resultant structure.

The ability to control volume fraction by any particular experimental parameter (including the ones previously mentioned) implies that an unintentional variation of such a parameter will result in an undesirable variation in volume fraction. For example, if the periodic spatial distribution of intensity decays along a particular direction in space (i.e. the periodic intensity maxima become weaker along a particular direction), the resultant periodic structure will have a spatial variation of its volume fraction. A case in which such a situation may arise is owing to the presence of absorption in the photoresist material. In fact, to transfer the intensity pattern into the photoresist material, the photoresist must of course necessarily be absorptive. For a thick film, this means that points at increasing distance from the surface of the photoresist material receive less light since an increasing amount of light is absorbed as it traverses the thickness of the film. Therefore, the periodic spatial distribution of intensity decays from the surface of the film and this translates into a resultant periodic structure with a spatial variation of its volume fraction. Minimizing this variation requires modification of the absorption characteristics of the photosensitive component of the photoresist material.

A variation in volume fraction may also occur within the plane of the film as a consequence of a variation in the intensity profile of the individual interfering beams of light. Lasers are typically the light sources that are used for interference lithography and it is not uncommon to have a Gaussian beam profile. To avoid this variation, the beam profile must be converted to a flat top beam, which can be achieved either by an absorptive element with the appropriate absorption profile or an appropriately shaped refractive element that redistributes the laser power within the beam.

5.7
Closing Remarks

The perception of a periodic structure as a sum of its constituent Fourier elements, and therefore as a sum of gratings that can be accessed by the interference of light, is an elegant one. Coupled with the fact that the first few Fourier elements that describe a periodic structure capture the essential nature of the properties that the structure would possess makes this approach potentially very powerful. Along with the promise of this technique there remain several key challenges that if addressed would truly unleash its potential. The primary challenge on the optics side is to develop a technique that will easily allow for the spatial registration of independent gratings. This would then allow us to write each grating separately, thereby allowing for easy access to virtually any geometry that would be readily scalable in size. Several approaches hold the promise toward this end including, aligning gratings using Moire interference patterns and using phase masks to guarantee specific phase relations. The challenge on the materials side is to deliver a materials platform that allows for the facile transfer into any material system that is optimum for any targeted property. These are not unreasonable challenges, and we can expect to see the promise of this approach toward creating functional materials that exploit structure property relations delivered in the near future.

Further Reading

1. Moore, G.E. (1965) "Cramming more components onto integrated circuits". *Electronics*, **38**, 114–17.
2. Deforest, W. (1975) *Photoresist: Materials and Processes*, McGraw-Hill, New York.
3. Taylor, G.N., Stillwagon, L.E., Houlihan, F.M., Wolf, T.M., Sogah, D.Y. and Hertler, W.R. (1991) "A positive, chemically amplified, aromatic methacrylate resist employing the tetrahydropyranyl protecting group". *Chemistry of Materials*, **3**, 1031–40.
4. Hanson, J.E., Reichmanis, E., Houlihan, F.M. and Neenan, T.X. (1992) "Synthesis and evaluation of copolymers of (tert-butoxycarbonyloxy)styrene and (2-nitrobenzyl)styrene sulfonates – single component chemically amplified deep-UV imaging materials". *Chemistry of Materials*, **4**, 837–42.
5. Thompson, L.F., Willson, C.G. and Bowden, M.J. (1994) *Introduction to Microlithography*, American Chemical Society, Washington, DC.
6. Rogers, J.A., Paul, K.E., Jackman, R.J. and Whitesides, G.M. (1998) "Generating similar to 90 nanometer features using near-field contact-mode photolithography with an elastomeric phase mask". *Journal of Vacuum Science and Technology B*, **16**, 59–68.
7. Goldfarb, D.L., de Pablo, J.J., Nealey, P.F., Simons, J.P., Moreau, W.M. and Angelopoulos, M. (2000) "Aqueous-based photoresist drying using supercritical carbon dioxide to prevent pattern collapse". *Journal of Vacuum Science and Technology B*, **18**, 3313–17.
8. Medeiros, D.R., Aviram, A., Guarnieri, C.R., Huang, W.S., Kwong, R., Magg, C.K., Mahorowala, A.P., Moreau, W.M., Petrillo, K.E. and Angelopoulos, M. (2001) " Recent progress in electron-beam a resists for advanced mask-making". *IBM Journal of Research and Development*, **45**, 639–50.
9. Shishido, A., Diviliansky, I.B., Khoo, I.C., Mayer, T.S., Nishimura, S., Egan, G.L. and Mallouk, T.E. (2001) "Direct fabrication of two-dimensional titania arrays using interference photolithography". *Applied Physics Letters*, **79**, 3332–34.
10. Wong, W.H. and Pun, E.Y.B. (2001) "Exposure characteristics and three-dimensional profiling of SU8C resist using electron beam lithography". *Journal of Vacuum Science and Technology B*, **19**, 732–35.
11. Sharp, D.N., Campbell, M., Dedman, E.R., Harrison, M.T., Denning, R.G. and Turberfield, A.J. (2002) "Photonic crystals for the visible spectrum by holographic lithography". *Optical and Quantum Electronics*, **34**, 3–12.
12. Yang, S., Megens, M., Aizenberg, J., Wiltzius, P., Chaikin, P.M. and Russel, W.B. (2002) "Creating periodic three-dimensional structures by multibeam interference of visible laser". *Chemistry of Materials*, **14**, 2831–33.
13. Malek, C.G.K. (2002) "SU8 resist for low-cost X-ray patterning of high-resolution, high-aspect-ratio MEMS". *Microelectronics Journal*, **33**, 101–5.
14. Stewart, M.D., Tran, H.V., Schmid, G.M., Stachowiak, T.B., Becker, D.J. and Willson, C.G. (2002) "Acid catalyst mobility in resist resins". *Journal of Vacuum Science and Technology B*, **20**, 2946–52.
15. Feng, R. and Farris, R.J. (2003) "Influence of processing conditions on the thermal and mechanical properties of SU8 negative photoresist coatings". *Journal of Micromechanics and Microengineering*, **13**, 80–88.
16. Saravanamuttu, K., Blanford, C.F., Sharp, D.N., Dedman, E.R., Turberfield, A.J. and Denning, R.G. (2003) "Sol-gel organic-inorganic composites for 3-D holographic lithography of photonic crystals with submicron periodicity". *Chemistry of Materials*, **15**, 2301–4.

17 Deubel, M., Von Freymann, G., Wegener, M., Pereira, S., Busch, K. and Soukoulis, C.M. (**2004**) "Direct laser writing of three-dimensional photonic-crystal templates for telecommunications". *Nature Materials*, **3**, 444–47.

18 Rumpf, R.C. and Johnson, E.G. (**2004**) "Fully three-dimensional modeling of the fabrication and behavior of photonic crystals formed by holographic lithography". *Journal of the Optical Society of America A-Optics Image Science and Vision*, **21**, 1703–13.

19 Jeon, S., Park, J.U., Cirelli, R., Yang, S., Heitzman, C.E., Braun, P.V., Kenis, P.J.A. and Rogers, J.A. (**2004**) "Fabricating complex three-dimensional nanostructures with high-resolution conformable phase mask". *Proceedings of the National Academy of Sciences*, **101**, 12428–33.

20 Felix, N.M., Tsuchiya, K. and Ober, C.K. (**2006**) "High-resolution patterning of molecular glasses using supercritical carbon dioxide". *Advanced Materials*, **18**, 442–46.

21 Jang, J.H., Ullal, C.K., Gorishnyy, T., Tsukruk, V.V. and Thomas, E.L. (**2006**) "Mechanically tunable three-dimensional elastomeric network/air structures via interference lithography". *Nano Letters*, **6**, 740–43.

22 Tetreault, N., von Freymann, G., Deubel, M., Hermatschweiler, M., Perez-Willard, F., John, S., Wegener, M. and Ozin, G.A. (**2006**) "New route to three-dimensional photonic bandgap materials: silicon double inversion of polymer templates". *Advanced Materials*, **18**, 457–60.

23 Meisel, D.C., Diem, M., Deubel, M., Perez-Williard, F., Linden, S., Gerthsen, D., Busch, K. and Wegener, M. (**2006**) "Shrinkage precompensation of holographic three-dimensional photonic-crystal templates". *Advanced Materials*, **18**, 2964–68.

Applications

6

Photonic Crystals

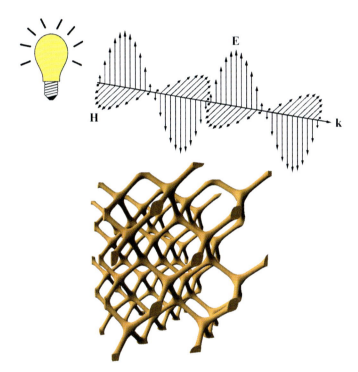

Photonic crystals are periodic structures made of different dielectric materials designed to control the propagation of electromagnetic waves. In this chapter, we study the most fundamental property of photonic crystals, that is, the existence of a photonic band gap. We also show that simple periodic structures such as those presented in the previous chapters can be usefully employed as photonic crystals. Some additional optical properties exhibited by photonic crystals are introduced in subsequent chapters, where we discuss important applications of these periodic dielectric structures.

6.1
Introduction

Photonic crystals are periodic structures designed to control the propagation of electromagnetic waves. These periodic structures are made of two materials with different dielectric constants (Figure 6.1); one of the materials has a high dielectric constant ε_1 and the other has a low dielectric constant ε_2. One of the basic properties exhibited by photonic crystals is that electromagnetic waves having frequencies within a specific range are not allowed to propagate within the periodic structure. This range of forbidden frequencies is called a *photonic band gap* and it is the main subject of this chapter.

The origin of photonic band gaps lies in the multiple scattering of a propagating wave at the interfaces between the different dielectric materials. To get an intuitive understanding of this effect, consider, for example, an electromagnetic wave propagating normal to the layers of a one-dimensional photonic crystal (Figure 6.1). As a consequence of scattering, secondary waves are created at each interface between different materials, which are reflected backwards and interfere with each other. If these secondary waves interfere constructively, the original electromagnetic wave is not allowed to propagate within the structure, as all the electromagnetic energy is reflected backwards. On the other hand, if the secondary waves do *not* interfere constructively, the electromagnetic wave is allowed to propagate in the structure.

The existence of photonic band gaps creates interesting physical phenomena that can help us to control the propagation of electromagnetic waves. For example, consider a periodic structure having a photonic band gap. If an electromagnetic wave with frequency within the photonic band gap is incident on the surface of the photonic crystal, the wave is totally reflected by the crystal. That is, the photonic crystal acts as a perfect mirror for the external wave because the wave is not allowed to propagate within the crystal. On the other hand, if the electromagnetic wave is generated *inside* the photonic crystal (by emission from an internal source), its propagation (or existence) within the photonic crystal is absolutely prohibited.

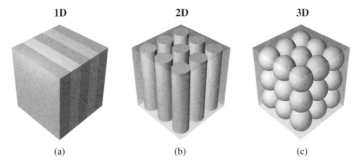

Fig. 6.1 Examples of photonic crystals with periodicities in one, two, and three dimensions. (a) A one-dimensional photonic crystal consisting of dielectric layers made of different dielectric materials. (b) A two-dimensional photonic crystal consisting of dielectric cylinders in a background dielectric material. (c) A three-dimensional photonic crystal consisting of dielectric spheres in a background dielectric material.

Because the formation of photonic band gaps is based on the phenomenon of diffraction, the wavelengths of the electromagnetic waves that are not allowed to propagate within photonic crystals are, in general, on the order of the spatial periodicity of the structure, which is determined by the thickness of the layers in the one-dimensional photonic crystal shown in Figure 6.1, or by the distance between the cylinders or the spheres in the case of the two- and three-dimensional photonic crystals shown in Figure 6.1. Thus, according to the electromagnetic spectrum shown in Chapter 3, photonic crystals with spatial periods on the order of centimeters to millimeters will forbid the propagation of microwaves ($3 \times 10^9 - 3 \times 10^{11}$ Hz); those with spatial periods on the order of millimeters to hundreds of nanometers will forbid infrared waves ($3 \times 10^{11} - 4 \times 10^{14}$ Hz); whereas those with 400–700 nm will forbid the propagation of visible light ($4 \times 10^{14} - 7.5 \times 10^{14}$ Hz).

As a result, fabrication and characterization of photonic crystals that forbid the propagation of microwaves is not difficult because the required length scales of the structures are relatively large. The great challenge, however, is the fabrication of photonic crystals at submicron length scales in order to forbid the propagation of visible light.

6.2
One-dimensional Photonic Crystals

6.2.1
Finite Periodic Structures

The simplest photonic crystal is a one-dimensional periodic structure made of two different dielectric materials such as the multilayer structure shown in Figure 6.2. The periodic structure consists of alternating homogeneous dielectric layers with dielectric constants ε_1 and ε_2 and magnetic permeabilities μ_1 and μ_2. Because the structure is one dimensional, the layers extend infinitely in the yz plane and the dielectric constant of the photonic crystal $\varepsilon = \varepsilon(\mathbf{r})$ varies periodically along a single direction in space (x direction in our case). The thicknesses of the layers along the x direction are given by t_1 and t_2, respectively, and the lattice constant of the structure is $a = t_1 + t_2$. For simplicity, we assume that the surrounding medium is made of air ($\varepsilon_0 = 8.85 \times 10^{-12}$ s^{2}C^{2}m^{-3}kg^{-1} and $\mu_0 = 4\pi \times 10^{-7}$ m kg C^{-2}) and that the dielectric layers are made of nonmagnetic materials ($\mu_1 = \mu_2 = \mu_0$).

We consider that the electromagnetic waves are normally incident on the surface of the one-dimensional photonic crystal. In general, when an electromagnetic wave with frequency ω (and the corresponding wavelength λ) is normally incident on the surface of the photonic crystal, part of the energy of the incident wave is reflected by the structure, while the remaining part is transmitted through the array of layers and emerges on the other side of the structure (Figure 6.2a). However, if the frequency ω of the incident wave is within the photonic band gap of the structure, the wave is not allowed to propagate within the structure and it is therefore totally

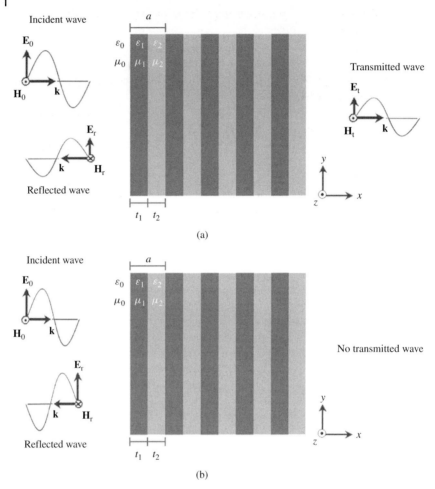

Fig. 6.2 (a) An electromagnetic wave with frequency *outside* the photonic band gap is normally incident on a one-dimensional photonic crystal. Part of the incident wave is reflected by the structure and part of the wave is transmitted through the structure. (b) An electromagnetic wave with frequency *within* the photonic band gap is normally incident on the same one-dimensional photonic crystal. In this case, the incident wave is not allowed to propagate within the structure and it is therefore *totally* reflected. As a result, no wave is transmitted on the other side of the structure.

reflected. In this case, there is no transmitted wave on the other side of the structure (Figure 6.2b).

The ratio between the intensity of the reflected wave and that of the incident wave is called the *reflectance R*, and represents the percentage of energy reflected by the photonic crystal. As we show next, the reflectance R will help us to determine whether or not an incident electromagnetic wave is allowed to propagate within the structure. In other words, it will help us to determine whether the structure possesses a photonic band gap.

The reflectance R varies between 0 and 1 and is given by the formula

$$R = \frac{|\mathbf{E}_r|^2}{|\mathbf{E}_0|^2} \tag{6.1}$$

where $\mathbf{E}_r$ and $\mathbf{E}_0$ are the electric field amplitude vectors of the reflected and incident waves, respectively (Chapter 3).

It is important to note that the reflectance R depends on the set of structural parameters of the photonic crystal such as the values of the dielectric constants ε_1 and ε_2, the thicknesses t_1 and t_2, and the number of layers. For a given photonic crystal, the reflectance R also depends on the frequency ω of the normally incident plane wave (Figure 6.3).

An interesting characteristic of the propagation of waves in periodic structures is that there is no fundamental length scale. For example, if the wavelength λ of the incident wave and the thicknesses t_1 and t_2 of the layers are increased by a factor r, the phases of the multiple reflected waves generated at the interfaces between the different materials are the same, and thus the same reflectance is obtained. Therefore, since the wavelength λ and the frequency ω of the incident wave are inversely proportional, the reflectance of a wave with frequency ω from a photonic crystal with thicknesses t_1 and t_2 is the same as the reflectance of a wave with frequency ω/r from a photonic crystal with thicknesses rt_1 and rt_2 (provided that the dielectric constant values ε_1 and ε_2 and the number of layers remain the same).

Because of this scaling property, it is convenient and fairly well established to plot the reflectance R as a function of the nondimensional variable $\omega a/2\pi c$, where c is the speed of light and $a = t_1 + t_2$ is the *lattice constant* of the one-dimensional

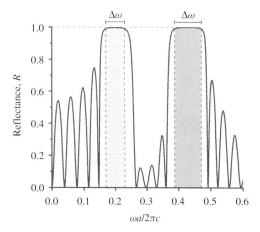

Fig. 6.3 The Reflectance R as a function of the nondimensional frequency $\omega a/2\pi c$ for normally incident electromagnetic waves on a one-dimensional photonic crystal consisting of 10 dielectric layers with $\varepsilon_1 = 11\varepsilon_0$, $\varepsilon_2 = 2\varepsilon_0$, and $t_1 = t_2 = 0.5a$ (Figure 6.2). For frequencies within the ranges $\omega a/2\pi c = 0.170 - 0.230$ (light gray) and $\omega a/2\pi c = 0.387 - 0.471$ (dark gray), the reflectance R is larger than 0.99 and the incident waves are considered to be totally reflected by the photonic crystal. These frequency ranges $\Delta \omega$ are called the *photonic band gaps*.

photonic crystal. This allows us to obtain the reflectance R for different lattice constants a (i.e. different structure periodicities) from a single graph. Figure 6.3 shows the reflectance R as a function of $\omega a/2\pi c$ for normally incident plane waves on a photonic crystal consisting of 10 layers with dielectric constants $\varepsilon_1 = 11\varepsilon_0$ and $\varepsilon_2 = 2\varepsilon_0$ and thicknesses $t_1 = t_2 = 0.5a$ (Figure 6.2). These results were obtained by using the MATLAB code provided in Appendix B.

As we can see from Figure 6.3, the reflectance R is strongly dependent on the frequency ω of the normally incident plane wave. Nevertheless, there exist some frequency ranges $\Delta\omega$ for which the reflectance of the incident wave is nearly equal to 1. This means that for these frequency ranges the incident wave is nearly totally reflected and therefore not allowed to propagate in the structure. *The frequency ranges $\Delta\omega$ for which electromagnetic waves are not allowed to propagate within the structure are called the photonic band gaps.*

In particular, for our example, the reflectance R is larger than 0.99 for frequencies within the ranges $\omega a/2\pi c = 0.170 - 0.230$ (light gray) and $\omega a/2\pi c = 0.387 - 0.471$ (dark gray). In addition, Figure 6.3 shows that $R \approx 1$ at $\omega a/2\pi c = 0.2$. This means that if the lattice constant of the photonic crystal is $a = 1\,\mu\text{m}$, an incident wave with frequency $\omega = 0.2 \times 2\pi c/a = 3.77 \times 10^{14}$ Hz is totally reflected by the photonic crystal.

On the other hand, incident waves with frequencies outside the photonic band gaps $\Delta\omega$ are partially reflected and partially transmitted through the photonic crystal. The percentage of energy reflected by the crystal is given by the value of the reflectance R, whereas the percentage of energy transmitted through the crystal is given by the transmittance $T = 1 - R$. Note that $T + R$ must be equal to 1 because the materials forming the photonic crystal are assumed to be nonabsorptive and the energy is therefore conserved.

The frequency ranges $\Delta\omega$ of the photonic band gaps depend on the particular structural parameters of the one-dimensional photonic crystal. However, in general, the following statements are valid:
- The larger the number of layers, the larger the width $\Delta\omega$ of the photonic band gaps;
- The larger the ratio $\varepsilon_1/\varepsilon_2$ between the dielectric constants of the layers, the larger the width $\Delta\omega$ of the photonic band gaps;
- For a one-dimensional photonic crystal with lattice constant a, the low-frequency photonic band gap reaches its maximum width when t_1 is approximately equal to

$$\frac{t_1}{a} = \frac{n_2}{n_1 + n_2} \tag{6.2}$$

where $n_1 = \sqrt{\dfrac{\varepsilon_1}{\varepsilon_0}}$ and $n_2 = \sqrt{\dfrac{\varepsilon_2}{\varepsilon_0}}$ are the refractive indices of the layers.

We encourage the reader to verify the above statements by using the MATLAB program provided in Appendix B (see Problems section).

6.2.2
Infinite Periodic Structures

An alternative and powerful approach for the theoretical evaluation of photonic band gaps is to assume that the photonic crystal is infinite. That is, *all* space is considered to be filled with the one-dimensional periodic dielectric structure. This assumption allows us to use relatively simple numerical techniques to establish the existence of photonic band gaps (especially in the case of three-dimensional photonic crystals) and it is also more appropriate because it deals directly with the propagation of waves within the photonic crystal. Note that, by definition, photonic band gaps are frequency ranges $\Delta\omega$ for which electromagnetic waves are not allowed to propagate *within* the structure.

In this section, we are thus interested in determining whether or not an electromagnetic wave with frequency ω is allowed to propagate within one-dimensional photonic crystals that extend infinitely in space. Note that since it is assumed that the photonic crystal extends infinitely in space, the electromagnetic wave is considered to be *inside* the photonic crystal (and not externally incident as in the previous section).

We showed in Chapter 3 that the relation between the frequency ω and the wave vector $\mathbf{k}$ of an electromagnetic wave propagating within a *homogeneous* material is given by $\omega = kv$, where k is the magnitude of the wave vector $\mathbf{k}$, and v is the velocity of the wave (Equation 3.20). On the other hand, in the case of an electromagnetic wave propagating within a *nonhomogeneous* material (such as a photonic crystal), the relation between the frequency ω and the wave vector $\mathbf{k}$ is a complicated function $\omega = \omega(\mathbf{k})$ that needs to be calculated numerically. This relation is called the *dispersion relation*. We thus have

$$\omega = kv \quad \text{for homogeneous materials} \tag{6.3}$$

$$\omega = \omega(\mathbf{k}) \text{ for photonic crystals}$$

Importantly, by plotting the dispersion relation $\omega = \omega(\mathbf{k})$ it is easy to determine if a particular infinite periodic structure possesses a photonic band gap. Or equivalently, it is easy to determine whether or not an electromagnetic wave with frequency ω is allowed to propagate within the photonic crystal. The graph of the dispersion relation $\omega = \omega(\mathbf{k})$ as a function of the wave vector $\mathbf{k}$ is called the *band diagram* or *band structure* of the photonic crystal. Next we plot the dispersion relation $\omega = \omega(\mathbf{k})$ for one-dimensional photonic crystals and show how to identify photonic band gaps.

In infinite photonic crystals (including one-, two-, and three-dimensional periodic structures), electromagnetic waves do not propagate as plane waves but as a more general type of wave called a *Bloch wave*. An electromagnetic Bloch wave, propagating with wave vector $\mathbf{k}$ within an infinite photonic crystal, can be described by the formula

$$\mathbf{H}_\mathbf{k}(\mathbf{r}, t) = \text{Re}\left[\mathbf{u}_\mathbf{k}(\mathbf{r}) e^{i(\mathbf{k}\cdot\mathbf{r} - \omega(\mathbf{k})t)}\right] \tag{6.4}$$

where $\mathbf{u}_\mathbf{k}(\mathbf{r})$ is a periodic vector function, with the same spatial period as the photonic crystal, that depends on the particular value of the wave vector $\mathbf{k}$. Note that Equation 6.4 indicates that a Bloch wave can be understood as a plane wave modulated by the periodic function $\mathbf{u_k(r)}$.

The Bloch wave (Equation 6.4) can also be written with its spatial and temporal parts separated. That is,

$$\mathbf{H_k(r}, t) = \mathrm{Re}\left[\mathbf{H_k(r)}e^{-i\omega(k)t}\right], \quad \text{where} \quad \mathbf{H_k(r)} = \mathbf{u_k(r)}e^{i\mathbf{k}\cdot\mathbf{r}} \tag{6.5}$$

The dispersion relation $\omega = \omega(\mathbf{k})$ is calculated by considering the electromagnetic wave equation for nonhomogeneous materials:

$$\nabla \times \left(\frac{1}{\varepsilon(\mathbf{r})} \nabla \times \mathbf{H_k(r)}\right) = \mu(\omega(\mathbf{k}))^2 \mathbf{H_k(r)} \tag{6.6}$$

where $\varepsilon = \varepsilon(\mathbf{r})$ is the nonhomogeneous periodic dielectric constant that describes the photonic crystal, μ is the magnetic permeability (which is, in general, assumed to be constant and equal to μ_0), and $\mathbf{H_k(r)}$ is the spatial part of the magnetic field vector of an electromagnetic Bloch wave propagating with wave vector $\mathbf{k}$ within the photonic crystal.

One of the most widely used numerical techniques to calculate the dispersion relation $\omega = \omega(\mathbf{k})$ using the wave equation (6.6) is the so-called plane-wave method (see References [14–17]). For example, for a given electromagnetic Bloch wave with wave vector $\mathbf{k}$ propagating within a photonic crystal characterized by $\varepsilon = \varepsilon(\mathbf{r})$, the numerical solution of Equation 6.6 determines the corresponding frequency $\omega = \omega(\mathbf{k})$. In addition, the solution of Equation 6.6 determines the spatial magnetic field $\mathbf{H_k(r)}$ and therefore the instantaneous magnetic field vector $\mathbf{H_k(r}, t)$ of the electromagnetic Bloch wave within the crystal.

We numerically solve the wave equation (6.6) by using the plane-wave method and show in Figure 6.4 the dispersion relation $\omega = \omega(\mathbf{k})$ for electromagnetic waves propagating within an infinite version of the one-dimensional photonic crystal presented in Figures 6.2 and 6.3. For this calculation, it is assumed that the electromagnetic waves propagate with wave vectors $\mathbf{k} = k_x \hat{\mathbf{x}}$ perpendicular to the dielectric layers. Because of the previously mentioned scaling property, it is convenient and fairly established to plot the nondimensional frequency $\omega a/2\pi c$ as a function of k_x, where c is the speed of light, and $a = t_1 + t_2$ is the lattice constant of the infinite one-dimensional photonic crystal. This allows us to obtain the dispersion relation for structures with different lattice constants from a single graph.

By examining the dispersion relation in Figure 6.4b, we can easily notice that there are some frequency ranges $\Delta\omega$ on the vertical axis (shown in light and dark gray) for which there is no associated wave vector $\mathbf{k}$. This means that an electromagnetic wave with frequency ω within these frequency ranges will not be allowed to propagate within the photonic crystal because there is no wave vector $\mathbf{k}$ that can sustain that frequency. These gray frequency ranges $\Delta\omega$ are thus the *photonic band gaps*. In particular, the photonic band gaps shown in Figure 6.4 correspond to frequency ranges $\omega a/2\pi c = 0.162 - 0.251$ (light gray) and $\omega a/2\pi c = 0.373 - 0.478$ (dark gray).

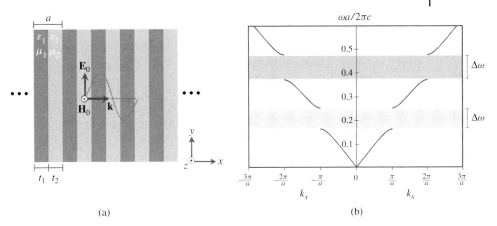

Fig. 6.4 (a) An infinite one-dimensional photonic crystal, where the black dots indicate that the sequence of dielectric layers extends infinitely in space. (b) The dispersion relation: a plot of nondimensional frequencies $\omega a/2\pi c$ versus the wave vector k_x for electromagnetic waves propagating within an infinite one-dimensional photonic crystal with dielectric constants of the layers $\varepsilon_1 = 11\varepsilon_0$, $\varepsilon_2 = 2\varepsilon_0$, and thicknesses $t_1 = t_2 = 0.5a$. In this example, it is assumed that the waves propagate perpendicularly to the layers with wave vectors $\mathbf{k} = k_x\hat{\mathbf{x}}$. The photonic band gaps $\Delta\omega$ are shown in light and dark gray.

It is interesting to note the similarity between the photonic band gaps $\Delta\omega$ obtained by considering an infinite periodic structure (Figure 6.4b) and those obtained by considering a similar but finite periodic structure (Figure 6.3). Note that the photonic band gaps for the infinite structure are wider than those corresponding to the finite structure, which is made of only 10 layers. For example, in the case of the infinite periodic structure the width of the low-frequency gap is $\Delta\omega = 0.089$, whereas in the case of the finite structure $\Delta\omega = 0.060$. Nevertheless, note that a structure made of only 10 layers presents a photonic band gap $\Delta\omega$ whose width is 67% of the width corresponding to the infinite structure. The required number of layers to obtain a photonic band gap comparable with an infinite structure is strongly dependent on the ratio $\varepsilon_1/\varepsilon_2$ between the dielectric constants of the layers. The smaller the dielectric ratio, the larger the number of required layers.

Figure 6.4 also shows that the dispersion relation for electromagnetic waves propagating within the infinite photonic crystal is discontinuous for values of $k_x = n\pi/a$, where n is an integer. In fact, these discontinuities signal the existence of the photonic band gaps. What is special about these k_x values? The photonic band gaps $\Delta\omega$ are created at $k_x = n\pi/a$, because at these k_x values the secondary waves reflected from an interface located at $x = x_0$ interfere constructively (with phase difference 2π) with those reflected from an interface located at $x = x_0 + a$. As a result, the reflected waves generated at different interfaces reinforce each other and the energy of the wave is reflected backwards.

6.2.3
Finite versus Infinite Periodic Structures

In the previous sections, we showed that the existence of photonic band gaps for electromagnetic waves propagating normally to the layers can be studied by considering either finite or infinite one-dimensional photonic crystals. In this section, we remark on some basic differences concerning the propagation of waves inside/outside of finite/infinite one-dimensional photonic crystals.

In the case that the one-dimensional photonic crystal is considered to be a finite periodic structure, the photonic band gaps are calculated by considering that an *external* electromagnetic wave is normally incident on the surface of the crystal. When the reflectance of the normally incident external wave is equal to 1, the wave is totally reflected and we can therefore conclude that it is not allowed to propagate within the crystal. In this case, we can infer that if the size of the one-dimensional photonic crystal is large enough, an electromagnetic wave generated *within* the photonic crystal (with frequency within the photonic band gap and wave vector normal to the layers) will also not be allowed to propagate within the crystal.

On the other hand, in the case that the one-dimensional photonic crystal is considered to be an infinite periodic structure, the photonic band gaps are obtained by calculating and plotting the dispersion relation for electromagnetic waves propagating *within* the photonic crystal normal to the layers (note that there is no outside since the photonic crystal is infinite). The frequency gaps in the dispersion relation determine that an electromagnetic wave generated within the infinite photonic crystal having a wave vector normal to the layers and with frequency within the photonic band gaps is not allowed to propagate within the structure. In this case, we can infer that if the photonic crystal is no longer infinite but its size is large enough, a normally incident *external* electromagnetic wave with frequency within the photonic band gaps will be totally reflected by the crystal (because it is not allowed to propagate within the crystal).

The above analysis is based on the fact that the frequency and the propagation direction of a normally incident electromagnetic wave do not vary when the wave moves into different dielectric layers.

Our discussion on photonic band gaps has so far been limited to the case in which the electromagnetic waves propagate normally to the layers of the photonic crystal. If this constraint is removed, it can be demonstrated that specific, finite, one-dimensional photonic crystals possess frequency ranges for which the reflectance of *external* waves is equal to 1 regardless of the angle of incidence of the waves. This property is called *omnidirectional reflection*. However, for *internal* waves, which are generated within the crystal, one-dimensional photonic crystals (either finite or infinite) can never prevent the propagation of waves parallel to the layers (e.g. along the y and z directions in Figure 6.2). This is due to the fact that there is no variation of the dielectric constant along any direction in the yz plane and there are thus no interfaces between different dielectric materials that can create secondary waves, which in turn can interfere with each other to create maximal reflectance and prevent an electromagnetic wave from propagating.

In order to obtain photonic crystals that forbid the propagation of electromagnetic waves along additional directions in space, we next consider that the nonhomogeneous dielectric constant $\varepsilon = \varepsilon(\mathbf{r})$ describing the photonic crystal varies periodically in two dimensions.

6.3
Two-dimensional Photonic Crystals

We show in Figure 6.5 an example of a two-dimensional photonic crystal consisting of cylinders with radius r arranged on a square lattice with lattice constant a, where the background material and the cylinders are made of different dielectric materials with dielectric constants ε_1 and ε_2, respectively. The cylinders extend infinitely along the z direction and therefore the dielectric constant $\varepsilon = \varepsilon(\mathbf{r})$ of the photonic crystal varies periodically in the xy plane.

In this section, we consider two-dimensional photonic crystals that extend infinitely in space, and we are therefore interested in calculating the dispersion relation $\omega = \omega(\mathbf{k})$ for electromagnetic waves propagating within the structures in order to examine the presence of photonic band gaps. In contrast to the one-dimensional case, we are here interested in investigating the existence of photonic band gaps for waves propagating along many directions. For example, we assume that the electromagnetic waves propagate within the infinite photonic crystal with wave vectors $\mathbf{k}$ lying in the xy plane (Figure 6.5). This means that the electromagnetic waves can have arbitrary directions in the xy plane. As we

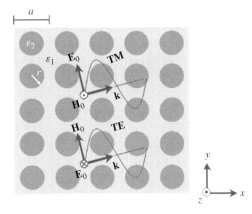

Fig. 6.5 A two-dimensional photonic crystal consisting of cylinders with radius r arranged in a square lattice with lattice constant a. The background material and the cylinders have dielectric constants ε_1 and ε_2, respectively. Therefore, the dielectric constant of the photonic crystal $\varepsilon = \varepsilon(x, y)$ varies periodically in two dimensions. Schematics of transverse electric (TE) and transverse magnetic (TM) waves are shown in the diagram. For specific values of the radius r and the dielectric constants ε_1 and ε_2, the periodic structure possesses photonic band gaps for which a general electromagnetic wave with specific frequencies is not allowed to propagate within the photonic crystal regardless of the direction of the wave vector $\mathbf{k}$ in the xy plane.

demonstrate in the following sections, one of the most important physical properties exhibited by two-dimensional photonic crystals is the fact that for certain frequencies these structures can prevent the propagation of electromagnetic waves regardless of their propagation direction in the plane of periodicity (i.e. the xy plane).

Since the electric field amplitude vector $\mathbf{E}_0$ of an electromagnetic wave is always in a plane perpendicular to the wave vector $\mathbf{k}$ (Chapter 3), we can divide a general electromagnetic wave propagating within the photonic crystal into two constituent waves: a transverse electric (TE) wave, in which the electric field amplitude vector is along the z direction, and a transverse magnetic (TM) wave, in which the electric field amplitude vector is along the direction perpendicular to both the z direction and the wave vector $\mathbf{k}$ (Figure 6.5). A general electromagnetic wave can thus be considered as formed by the sum of the TE and TM waves. In the case of two-dimensional photonic crystals, in order to examine the existence of photonic band gaps, the dispersion relation $\omega = \omega(\mathbf{k})$ is calculated separately for TE and TM waves. Importantly, the dispersion relations for TE and TM waves are in general different.

The calculation and plotting of the dispersion relation $\omega = \omega(\mathbf{k})$ for electromagnetic waves propagating within two-dimensional photonic crystals are more difficult than in the one-dimensional case. This is due to the fact that in this case we consider electromagnetic waves propagating with wave vectors $\mathbf{k}$ with different magnitudes and arbitrary directions in the xy plane (Figure 6.5). This means that the dispersion relation $\omega = \omega(\mathbf{k})$ is now a scalar function depending on the vector variable $\mathbf{k}$. In order to calculate the dispersion relation and investigate the existence of photonic band gaps, we first need to determine the domain of the function $\omega = \omega(\mathbf{k})$. That is, we need to define the set of wave vectors $\mathbf{k}$ required to completely describe the propagation of electromagnetic waves in two-dimensional photonic crystals. This set of wave vectors $\mathbf{k}$ is called the *Brillouin zone*.

The specific spatial periodicity of the photonic crystal plays a crucial role in the determination of the Brillouin zones. We therefore dedicate the following section to study the Brillouin zones associated with two-dimensional photonic crystals, which will allow us to calculate and plot the dispersion relation $\omega = \omega(\mathbf{k})$, which in turn will help us to identify the existence of photonic band gaps in two dimensions.

6.3.1
Reciprocal Lattices and Brillouin Zones in Two Dimensions

In this section, we introduce the reciprocal lattice and Brillouin zone concepts, noting that more detailed explanations can be found in References 60 and 61. Readers familiar with these concepts can skip this section and move to Section 6.3.2. Note that the reciprocal lattice and Brillouin zone concepts are essential in order to correctly plot the dispersion relation for electromagnetic waves propagating within photonic crystals.

In Chapter 1, we have seen that *any* periodic structure (such as a photonic crystal) has an associated particular point lattice. In addition, we show in the following sections that each point lattice has an associated *reciprocal lattice*. This means that we have the following relationship:

$$\text{periodic structure} \rightarrow \text{point lattice} \rightarrow \text{reciprocal lattice}$$

which is valid for two- and three-dimensional point lattices.

Consider, in particular, a point lattice made up of a periodic arrangement of points in two dimensions (Figure 1.5). An arbitrary plane wave propagating in the plane of the lattice with wave vector **k** does not necessarily have the periodicity of the lattice; that is, $e^{i\mathbf{k}\cdot\mathbf{r}} \neq e^{i\mathbf{k}\cdot(\mathbf{r}+\mathbf{R})}$, where **R** is the vector that generates the point lattice (Equation 1.1). However, for specific values of the wave vectors **k** (which are denoted by **G**), the plane wave actually satisfies the periodicity of the lattice and we have $e^{i\mathbf{G}\cdot\mathbf{r}} = e^{i\mathbf{G}\cdot(\mathbf{r}+\mathbf{R})}$. These particular wave vectors are called the *reciprocal lattice vectors* and they form the associated *reciprocal lattice*. In other words, the reciprocal lattice of a given point lattice can be defined as the set of wave vectors **G** that creates plane waves that satisfy the spatial periodicity of the point lattice.

We next show how to obtain the reciprocal lattice vectors **G** that correspond to a particular two-dimensional point lattice. Consider a point lattice defined by the primitive vectors $\mathbf{a}_1$ and $\mathbf{a}_2$ (Chapter 1), which are assumed to be in the xy plane. The set of reciprocal lattice wave vectors **G** associated with the point lattice can be obtained by using the formula

$$\mathbf{G} = m_1 \mathbf{b}_1 + m_2 \mathbf{b}_2 \tag{6.7}$$

where m_1 and m_2 are arbitrary integers and $\mathbf{b}_1$ and $\mathbf{b}_2$ are the *reciprocal lattice primitive vectors*, which are given by the formulas

$$\begin{aligned}\mathbf{b}_1 &= 2\pi \frac{\mathbf{a}_2 \times \hat{\mathbf{z}}}{\mathbf{a}_1 \cdot (\mathbf{a}_2 \times \hat{\mathbf{z}})} \\ \mathbf{b}_2 &= 2\pi \frac{\hat{\mathbf{z}} \times \mathbf{a}_1}{\mathbf{a}_1 \cdot (\mathbf{a}_2 \times \hat{\mathbf{z}})}\end{aligned} \tag{6.8}$$

Equation 6.7 determines that for each arbitrary set of m_1 and m_2 integers there is a corresponding reciprocal lattice vector **G**. Therefore, Equation 6.7 generates a two-dimensional periodic arrangement of reciprocal lattice vectors **G**, which defines the corresponding *reciprocal lattice*.

To illustrate these concepts, Table 6.1 shows the real-space primitive vectors $\mathbf{a}_1$ and $\mathbf{a}_2$ corresponding to the square and triangular point lattices and the associated reciprocal lattice primitive vectors $\mathbf{b}_1$ and $\mathbf{b}_2$ calculated by using Equation 6.8. In addition, Figure 6.6 shows the square and triangular point lattices and the associated reciprocal lattices. Note that the real-space primitive vectors $\mathbf{a}_1$ and $\mathbf{a}_2$ have units of meters because they represent the locations of the lattice points in real space, whereas the reciprocal lattice primitive vectors $\mathbf{b}_1$ and $\mathbf{b}_2$ have units of inverse meters because they represent wave vectors in the reciprocal space.

One of the most important features of the propagation of electromagnetic waves in infinite photonic crystals is the fact that the dispersion relation $\omega = \omega(\mathbf{k})$ is

Table 6.1 Real-space and reciprocal lattice primitive vectors for the square and triangular lattices.

	Primitive vectors	
	Real space	Reciprocal lattice
Square lattice	$\mathbf{a}_1 = a\hat{\mathbf{x}}$	$\mathbf{b}_1 = \dfrac{2\pi}{a}\hat{\mathbf{x}}$
	$\mathbf{a}_2 = a\hat{\mathbf{y}}$	$\mathbf{b}_2 = \dfrac{2\pi}{a}\hat{\mathbf{y}}$
Triangular lattice	$\mathbf{a}_1 = a\hat{\mathbf{x}}$	$\mathbf{b}_1 = \dfrac{4\pi}{\sqrt{3}a}\left(\dfrac{\sqrt{3}}{2}\hat{\mathbf{x}} - \dfrac{1}{2}\hat{\mathbf{y}}\right)$
	$\mathbf{a}_2 = \dfrac{a}{2}\hat{\mathbf{x}} + \dfrac{\sqrt{3}a}{2}\hat{\mathbf{y}}$	$\mathbf{b}_2 = \dfrac{4\pi}{\sqrt{3}a}\hat{\mathbf{y}}$

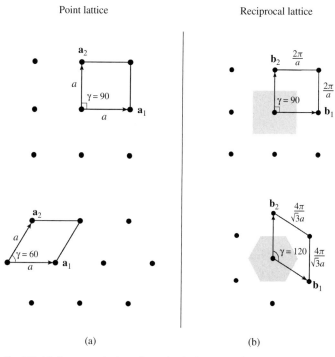

Fig. 6.6 (a) Square and triangular point lattices in real space and their corresponding unit cells (Chapter 1). (b) Associated reciprocal lattices (black dots) in the reciprocal space and Brillouin zones (gray regions). The Brillouin zones determine the set of wave vectors **k** required to plot the dispersion relation $\omega = \omega(\mathbf{k})$.

periodic and therefore it needs to be calculated only for a specific subset of wave vectors **k**. Indeed, the dispersion relation satisfies

$$\omega(\mathbf{k}) = \omega(\mathbf{k} + \mathbf{G}) \tag{6.9}$$

where **G** is an arbitrary reciprocal lattice vector of the reciprocal lattice associated with the point lattice that corresponds to the infinite photonic crystal.

Equation 6.9 basically determines that for each frequency ω the assignment of a wave vector **k** is not unique. Moreover, Equation 6.9 indicates that the value of the frequency ω corresponding to the wave vector **k** is the same as the frequency value ω that corresponds to the wave vector **k** + **G**. Because of this periodicity, only a specific subset of wave vectors **k** is needed to plot the dispersion relation $\omega = \omega(\mathbf{k})$. The subset of wave vectors **k** that contains all the information about the dispersion relation is called the *Brillouin zone*. The examination of wave vectors **k** outside of the Brillouin zone is unnecessary since the dispersion relation is periodic.

As a result, in order to plot the dispersion relation $\omega = \omega(\mathbf{k})$ for electromagnetic waves propagating within a specific photonic crystal (which will allow us to determine the existence of photonic band gaps), we have to identity the following:

periodic structure → point lattice → reciprocal lattice → Brillouin zone

Figure 6.6 shows the Brillouin zones that correspond to the square and triangular point lattices, where all the wave vectors **k** enclosed within the gray regions form the Brillouin zones. Note that the limits of the Brillouin zones are determined by the locus of points that are equidistant to nearby reciprocal lattice points.

In spite of the fact that only wave vectors within the Brillouin zone need to be considered to plot the dispersion relation $\omega = \omega(\mathbf{k})$, the large number of wave vectors still makes the plotting of the dispersion relation very difficult. However, the presence of symmetries in the photonic crystal can make the plotting of the dispersion relation easier. For example, photonic crystals are periodic structures and therefore are invariant under certain translations in space. In addition, some photonic crystals may be invariant after a rotation, or when a mirror reflection is performed on the periodic structure. If the photonic crystal is invariant after a rotation, the dispersion relation $\omega = \omega(\mathbf{k})$ is invariant as well. This means that the presence of rotational symmetries in the photonic crystal can reduce the number of wave vectors needed to plot the dispersion relation since the invariance of the dispersion relation after a rotation means that the values of $\omega = \omega(\mathbf{k})$ in some regions of the Brillouin zone are the same as those in other regions. Significantly, the invariance of the dispersion relation $\omega = \omega(\mathbf{k})$ applies for the presence of all the following symmetries in the photonic crystal: rotations and mirror reflections in two dimensions, and rotations, mirror reflections, and inversions in three dimensions.

In the case that the two-dimensional photonic crystal is *not* invariant under any rotation or mirror reflection, we need to consider the entire Brillouin zone to plot the dispersion relation $\omega = \omega(\mathbf{k})$. In this case, it is common practice to plot $\omega = \omega(\mathbf{k})$ for those wave vectors **k** in the Brillouin zone whose tips are located

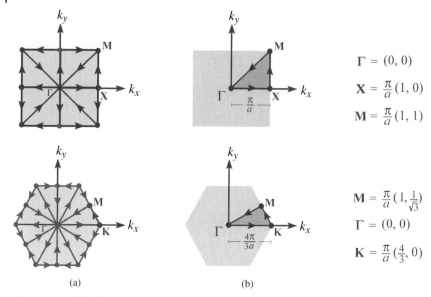

Fig. 6.7 (a) Brillouin zones (shaded gray polygons) for the square and triangular point lattices, where black solid lines with arrows indicate the wave vectors **k** commonly used to plot the dispersion relation $\omega = \omega(\mathbf{k})$. The black arrows show the standard convention for the direction in which the dispersion relation is plotted. (b) Irreducible Brillouin zones (shaded dark gray polygons) for the square and triangular point lattices, which take into account the presence of rotation and mirror symmetries in the corresponding photonic crystal. In this case, the dispersion relation $\omega = \omega(\mathbf{k})$ is plotted for wave vectors **k** along the paths determined by the $\Gamma - X - M$ and $\Gamma - K - M$ points, respectively (see black solid lines with arrows).

within the solid black lines shown in Figure 6.7a and whose origins are located at the Γ point. The black arrows in Figure 6.7a indicate the standard convention for the direction in which the dispersion relation is plotted.

On the other hand, in the case that the two-dimensional photonic crystal is invariant under rotations or mirror reflections, only part of the Brillouin zone needs to be considered. In general, the larger the number of symmetries present in the photonic crystal, the smaller the region of the Brillouin zone that needs to be considered. However, there exists a smallest region of the Brillouin zone, which cannot be reduced by the presence of additional symmetries, called the *irreducible Brillouin zone* (see dark gray regions in Figure 6.7b). For photonic crystals having symmetries such that only the irreducible Brillouin zone needs to be considered, it is common practice to plot the dispersion relation $\omega = \omega(\mathbf{k})$ for wave vectors **k** whose tips are located within the solid black lines defining the perimeter of the irreducible Brillouin zone (Figure 6.7b). Note that the center of the Brillouin zone is called the gamma point $\Gamma = (0, 0)$, while the vertices of the irreducible Brillouin zones have special names **X, M, K**, and so on.

In summary, we established the wave vectors **k** needed to calculate and plot the dispersion relation $\omega = \omega(\mathbf{k})$ for electromagnetic waves propagating within two-dimensional photonic crystals. The set of required wave vectors **k** depends on the translational symmetry of the photonic crystal and it is obtained by a procedure that includes the identification of the point lattice of the photonic crystal, the associated reciprocal lattice, and the corresponding Brillouin zone. In addition, the presence of additional symmetries may reduce the area of the Brillouin zone that needs to be considered. In the next section, we investigate the existence of photonic band gaps in two-dimensional photonic crystals by plotting the dispersion relation $\omega = \omega(\mathbf{k})$. The set of wave vectors **k** used to plot the dispersion relations is thus given by Figure 6.7.

6.3.2
Band Diagrams and Photonic Band Gaps in Two Dimensions

Now that we have established the set of wave vectors **k** required to plot the dispersion relation $\omega = \omega(\mathbf{k})$, we calculate and plot the dispersion relation for electromagnetic waves propagating in two-dimensional photonic crystals. We previously mentioned that certain two-dimensional photonic crystals can prevent the propagation of electromagnetic waves with specific frequencies independently of the direction of propagation of the waves in the plane of periodicity. In this section, we demonstrate this important property by plotting the dispersion relation $\omega = \omega(\mathbf{k})$ for wave vectors **k** along different directions in the plane of periodicity. These directions are determined by the edges of the Brillouin zones shown in Figure 6.7. From the plot of the dispersion relation, we will be able to determine the existence of photonic band gaps for these *in-plane* electromagnetic waves.

We consider two-dimensional photonic crystals composed of cylinders arranged in the square and triangular lattices (see insets in Figure 6.8). In particular, the cylinders are assumed to be made of air ($\varepsilon_2 = \varepsilon_0$), whereas the background material is made of a solid material with dielectric constant $\varepsilon_1 = 13\varepsilon_0$. The radius of the cylinders is $r = 0.45a$ in both cases, where a is the lattice constant of the structures. The dispersion relations $\omega = \omega(\mathbf{k})$ for these photonic crystals are shown in Figure 6.8, where the nondimensional frequencies $\omega a/2\pi c$ are plotted as a function of the wave vector **k** for electromagnetic waves propagating within the photonic crystals. Owing to the presence of rotational and mirror symmetries, in these particular photonic crystals the dispersion relation is plotted only for wave vectors **k** in the irreducible Brillouin zone. (Note: the dispersion relations $\omega = \omega(\mathbf{k})$ are obtained by numerically solving Equation 6.6).

On the horizontal axis, the segments $\Gamma - X, X - M, M - \Gamma$ and $M - \Gamma, \Gamma - K, K - M$ represent electromagnetic waves propagating with wave vectors **k** along the direction paths shown in the insets. The dispersion relations for TE and TM waves are calculated separately and are shown as solid and dashed lines, respectively.

In the case of the square photonic crystal, the dispersion relations show that all frequencies ω have associated with them at least one wave vector **k**. This means that this photonic crystal does not possess a photonic band gap for TE *and* TM waves propagating along arbitrary directions in the *xy* plane.

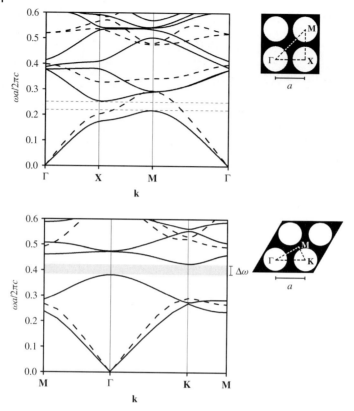

Fig. 6.8 The dispersion relation $\omega = \omega(\mathbf{k})$ for two-dimensional photonic crystals. Nondimensional frequencies $\omega a/2\pi c$ are plotted versus the wave vector $\mathbf{k}$ for TE (solid) and TM (dashed) electromagnetic waves propagating within photonic crystals made of cylinders with $r = 0.45a$ arranged in the square and triangular lattices, respectively. The cylinders are made of air ($\varepsilon_2 = \varepsilon_0$), and the background is a solid material with $\varepsilon_1 = 13\varepsilon_0$. The electromagnetic waves are assumed to propagate with wave vectors $\mathbf{k}$ in the xy plane. The triangular photonic crystal possesses a complete photonic band gap for both TE and TM waves (see gray frequency range). For these frequencies, neither TE nor TM waves can propagate in the photonic crystal irrespectively of the propagation direction of the waves in the plane of periodicity.

However, if we consider an electromagnetic wave with frequency ω within the range $\omega a/2\pi c = 0.22 - 0.25$ (see horizontal dashed lines), the dispersion relation for TE waves (solid lines) shows that there are no associated wave vectors $\mathbf{k}$ for this particular frequency. This means that a TE wave with frequency ω within this range is not allowed to propagate along any direction in the xy plane. In this case, it is said that the photonic crystal possesses a *complete* photonic band gap for TE waves.

On the other hand, the dispersion relation for TM waves (dashed lines) shows that a TM electromagnetic wave with frequency ω within the range $\omega a/2\pi c = 0.22 - 0.25$ has no associated wave vectors along the $\Gamma - X$ path but it has associated wave vectors along the $X - M$ and $M - \Gamma$ paths. This means that a TM wave with that frequency is not allowed to propagate along the $\Gamma - X$ direction but it can certainly propagate along other specific directions in the *xy* plane. Therefore, for the frequency range $\omega a/2\pi c = 0.22 - 0.25$, the photonic crystal does *not* possesses a complete photonic band gap for TM waves.

Since a general electromagnetic wave is made of the sum of TE and TM waves, this means that there is at least a direction in the *xy* plane along which a general electromagnetic wave with frequency ω within the range $\omega a/2\pi c = 0.22 - 0.25$ is allowed to propagate within the structure. Along this direction, the TE component of the wave will not be allowed to propagate but the TM component will propagate.

In contrast, in the case of the triangular photonic crystal, the dispersion relations show a range of frequencies $\Delta\omega$ (see thin rectangular gray region) for which there is no associated wave vector **k** for both TE and TM waves regardless of the direction of propagation of the waves. In this case, the photonic crystal possesses a *complete* photonic band gap for both TE and TM waves and neither TE nor TM waves can propagate along any direction in the *xy* plane. *In this significant case, a general electromagnetic wave with frequency within the photonic band gap $\Delta\omega$ is not allowed to propagate within the structure along any direction in the xy plane.*

Important note: In this book, in the case of two-dimensional photonic crystals, the word *complete* refers to the fact that the photonic band gap holds for arbitrary directions of propagation of the wave in the plane of periodicity. That is:
- If for a certain frequency range $\Delta\omega$, TE waves cannot propagate along any arbitrary direction in the plane of periodicity, the photonic crystal has a complete TE photonic band gap.
- If for a certain frequency range $\Delta\omega$, both TE and TM waves cannot propagate along any arbitrary direction in the plane of periodicity, the photonic crystal has a complete photonic band gap for both TE and TM waves.

It is important to clarify this point because in the case of two-dimensional photonic crystals the word *complete* is alternatively used to refer to the fact that the photonic band gap holds for both polarizations. For example, we could alternatively define (but we don't use these definitions in this book):
- If for a certain frequency range $\Delta\omega$, TE waves cannot propagate along any arbitrary direction in the plane of periodicity, the photonic crystal has a TE photonic band gap.
- If for a certain frequency range $\Delta\omega$, both TE and TM waves cannot propagate along any arbitrary direction in the plane of periodicity, the photonic crystal has a complete photonic band gap.

Note that regardless of the convention used, by using two-dimensional photonic crystals we are able to prevent for certain frequencies the propagation of electromagnetic waves of *any* polarization along *any* direction in the plane of periodicity of the photonic crystal (i.e. *xy* plane). This important result would have never been possible by considering one-dimensional photonic crystals for the reasons mentioned at the end of Section 6.2.3. Novel and intriguing physical effects created by the existence of these complete photonic band gaps are shown in Chapter 9.

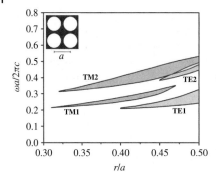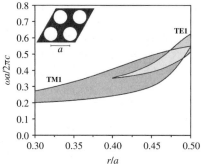

Fig. 6.9 Photonic band gap maps for air cylinders ($\varepsilon_2 = \varepsilon_0$) in a solid background ($\varepsilon_1 = 13\varepsilon_0$) arranged in the square and triangular lattices. Nondimensional frequencies $\omega a/2\pi c$ corresponding to complete TE and TM photonic band gaps are separately plotted as a function of r/a, where a is the lattice constant of the structure and c is the speed of light. When complete TE and TM photonic band gaps overlap (light gray), a general electromagnetic wave with frequency within the light gray region is not allowed to propagate along any direction in the xy plane.

To conclude this section, we want to show the dependence of the complete photonic band gaps for TE and TM waves on the radius of the cylinders. For example, the band diagrams shown in Figure 6.8 correspond to photonic crystals in which the radius of the cylinders is $r = 0.45a$. The complete photonic band gaps for TE and TM waves, however, vary as the radius of the cylinders is changed. Figure 6.9 shows separately the complete photonic band gaps for TE and TM waves as a function of the radius r of the cylinders. For these frequency ranges, TE or TM waves cannot propagate within the photonic crystal regardless of their propagation direction in the xy plane.

Importantly, for those frequencies for which the complete TE and TM photonic band gaps overlap (see light gray regions), the photonic crystal possesses a complete photonic band gap for *both* TE and TM waves, and an electromagnetic wave of *any* polarization is not allowed to propagate within the crystal along *any* direction in the xy plane. The graphs displayed in Figure 6.9 are called *photonic band gap maps* since the existence of complete photonic band gaps can be identified at a glance.[11]

6.3.3
Photonic Band Gaps in Two-dimensional Simple Periodic Structures

In the previous section, we considered in particular two-dimensional photonic crystals consisting of cylinders arranged in the square and triangular lattices and demonstrated the existence of a complete photonic band gap for both TE and TM waves. In this section, we show that two-dimensional simple periodic structures,

---

11) We could also show the complete photonic band gap maps for the inverse case in which the photonic crystals are made of solid cylinders arranged in air. However, these photonic crystals do not show a complete photonic band gap for both TE and TM waves.

such as those introduced in Chapter 2, also present a complete photonic band gap for both polarizations and can be usefully employed as photonic crystals. Importantly, as discussed in Chapter 4, these simple periodic structures can be fabricated at small length scales using interference lithography. In particular, if fabricated at submicron length scales, the structures present photonic band gaps for electromagnetic waves with frequencies in the visible range.

Consider, for example, the periodic structure represented by the simple periodic function

$$f_{4/3}^I(x,y) = \cos\left[\frac{2\pi}{a}\left(-x+\frac{1}{\sqrt{3}}y\right)\right] + \cos\left[\frac{2\pi}{a}\left(\frac{2}{\sqrt{3}}y\right)\right]$$
$$+ \cos\left[\frac{2\pi}{a}\left(-x-\frac{1}{\sqrt{3}}y\right)\right] + 0.95 \qquad (6.10)$$

which is shown in Figure 6.10a, where white and gray regions represent air and a solid material, respectively. The beam parameters required to fabricate this structure by interference lithography are given in Table 4.1.

Note that this simple periodic structure is very similar to a periodic structure made of air cylinders in the triangular lattice. In particular, Figure 6.10b shows the dispersion relation for electromagnetic waves propagating within the structure, where we can observe the existence of a complete photonic band gap $\Delta\omega$ for both TE and TM waves.

This example shows that the simple periodic structures presented in Chapter 2 together with the interference lithography technique offer an interesting method to fabricate two-dimensional photonic crystals with complete photonic band gaps

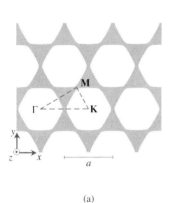

(a)

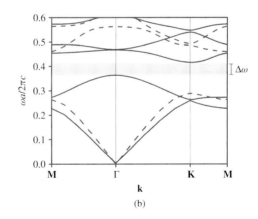

(b)

Fig. 6.10 (a) The simple periodic structure given by Equation 6.10, where white and gray regions are considered to be air ($\varepsilon_2 = \varepsilon_0$) and solid material ($\varepsilon_1 = 13\varepsilon_0$), respectively. This periodic structure can be fabricated by interference lithography.
(b) The dispersion relation $\omega = \omega(\mathbf{k})$, where nondimensional frequencies $\omega a/2\pi c$ are plotted versus the wave vector $\mathbf{k}$ for TE (solid) and TM (dashed) waves propagating within the structure. This simple periodic structure possesses a complete photonic band gap $\Delta\omega$ for both TE and TM waves (see light gray frequency range).

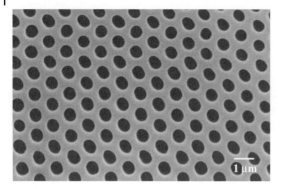

Fig. 6.11 SEM image of the experimental realization of the level-set structure $f^I_{4/3}(x,y)$ by interference lithography. The structure resembles a triangular lattice of air cylinders in a photoresist material. Note that the volume fraction of the solid material in this structure is higher than that from the structure shown in Figure 6.10a.

for electromagnetic waves in the visible range of frequencies. Furthermore, by using interference lithography, the photonic crystals can be fabricated rapidly, economically, and free of defects, and also cover a large area (Figure 6.11).

In summary, in the last sections we introduced one of the most important properties exhibited by two-dimensional photonic crystals, which is the existence of a complete photonic band gap for both TE and TM waves that can forbid the propagation of electromagnetic waves of *any* polarization along *any* direction in the plane of periodicity of the photonic crystal. The importance of such physical property is revealed in Chapter 9, where we study specific applications for these two-dimensional periodic structures. Two-dimensional photonic crystals, however, can never prevent the propagation of waves along the z direction (Figure 6.5) because there is no variation of the dielectric constant along this direction. In order to obtain photonic crystals that forbid the propagation of electromagnetic waves along additional directions in space, we next consider that the nonhomogeneous dielectric constant $\varepsilon = \varepsilon(\mathbf{r})$ describing the photonic crystal varies periodically in three dimensions.

6.4
Three-dimensional Photonic Crystals

Figure 6.12 shows an example of a three-dimensional photonic crystal consisting of spheres arranged on a simple cubic Bravais lattice with lattice constant a, where it is considered that the background material is made of a material with dielectric constant ε_1 and the spheres are made of a material with dielectric constant ε_2. As a result of the geometry of the periodic structure, the dielectric constant of

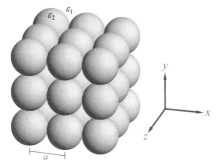

Fig. 6.12 A three-dimensional photonic crystal made of spheres arranged in the simple cubic Bravais lattice. The spheres are made of a material with dielectric constant ε_2, whereas the background material has dielectric constant ε_1. In this case, the dielectric constant of the photonic crystal $\varepsilon = \varepsilon(x, y, z)$ varies periodically in three dimensions.

the photonic crystal $\varepsilon = \varepsilon(\mathbf{r})$ varies periodically along any arbitrary direction in the three-dimensional space.

In this section, we consider three-dimensional photonic crystals that extend infinitely in space. Once again, we are therefore interested in calculating the dispersion relation $\omega = \omega(\mathbf{k})$ in order to examine the presence of photonic band gaps. In contrast to the one- and two-dimensional cases, here we do not place any constraint in the direction of propagation of the electromagnetic waves. That is, we assume that the waves can propagate within the photonic crystal along any arbitrary direction in the three-dimensional space. In fact, we next show that, for certain frequencies, carefully designed three-dimensional photonic crystals can forbid the propagation of electromagnetic waves irrespectively of their propagation direction. In this important case, it is said that the structure possesses a *complete* three-dimensional photonic band gap because for certain frequencies the propagation of electromagnetic waves is forbidden for *all* directions in the three-dimensional space. This significant physical property can be obtained only with three-dimensional photonic crystals and it is due to the fact that the dielectric constant $\varepsilon = \varepsilon(\mathbf{r})$ of the photonic crystal varies periodically in three dimensions.

As in the previous sections, we study the existence of complete photonic band gaps by plotting the dispersion relation $\omega = \omega(\mathbf{k})$ for electromagnetic waves propagating within the three-dimensional photonic crystal. However, we first need to determine the set of wave vectors $\mathbf{k}$ that are required to plot the dispersion relation $\omega = \omega(\mathbf{k})$. Note that, in contrast to one- and two-dimensional photonic crystals, the direction of the wave vector $\mathbf{k}$ is now absolutely arbitrary. In analogy with previous sections, this set of required wave vectors $\mathbf{k}$ is determined by the three-dimensional Brillouin zones. In the following section, we therefore introduce Brillouin zones associated with three-dimensional photonic crystals and establish the wave vectors $\mathbf{k}$ required to plot the dispersion relation $\omega = \omega(\mathbf{k})$.

6.4.1
Reciprocal Lattices and Brillouin Zones in Three Dimensions

We next extend to the three-dimensional case the concepts of reciprocal lattice and Brillouin zone introduced in Section 6.3.1. Readers familiar with these concepts can skip this section and move to Section 6.4.2. Note that reciprocal lattices and Brillouin zones in three dimensions are essential in order to correctly plot the dispersion relation for electromagnetic waves propagating in three-dimensional photonic crystals.

In analogy with the two-dimensional case, any Bravais lattice has associated with it a reciprocal lattice. We start by showing how to obtain the reciprocal lattice and reciprocal lattice vectors $\mathbf{G}$ that correspond to a particular Bravais lattice. Consider an arbitrary Bravais lattice defined by the primitive vectors $\mathbf{a}_1, \mathbf{a}_2$, and $\mathbf{a}_3$ (Chapter 1). The set of reciprocal lattice wave vectors $\mathbf{G}$ associated with the Bravais lattice can be obtained by using the formula

$$\mathbf{G} = m_1 \mathbf{b}_1 + m_2 \mathbf{b}_2 + m_3 \mathbf{b}_3 \tag{6.11}$$

where m_1, m_2, and m_3 are arbitrary integer numbers and $\mathbf{b}_1, \mathbf{b}_2$, and $\mathbf{b}_3$ are the reciprocal lattice primitive vectors, which are given by the formulas

$$\begin{aligned}\mathbf{b}_1 &= 2\pi \frac{\mathbf{a}_2 \times \mathbf{a}_3}{\mathbf{a}_1 \cdot (\mathbf{a}_2 \times \mathbf{a}_3)} \\ \mathbf{b}_2 &= 2\pi \frac{\mathbf{a}_3 \times \mathbf{a}_1}{\mathbf{a}_1 \cdot (\mathbf{a}_2 \times \mathbf{a}_3)} \\ \mathbf{b}_3 &= 2\pi \frac{\mathbf{a}_1 \times \mathbf{a}_2}{\mathbf{a}_1 \cdot (\mathbf{a}_2 \times \mathbf{a}_3)}\end{aligned} \tag{6.12}$$

Equation 6.11 generates a three-dimensional periodic arrangement of reciprocal lattice wave vectors $\mathbf{G}$, which define the corresponding *three-dimensional reciprocal lattice*. As an illustrative example, Table 6.2 shows the real-space primitive vectors $\mathbf{a}_1, \mathbf{a}_2$, and $\mathbf{a}_3$ corresponding to the simple cubic (sc), face-centered-cubic (fcc), and body-centered-cubic (bcc) Bravais lattices, together with the associated reciprocal lattice primitive vectors $\mathbf{b}_1, \mathbf{b}_2$, and $\mathbf{b}_3$, which are calculated by using Equation 6.12.

In addition, Figure 6.13 shows the Bravais lattices, the associated reciprocal lattices, and the corresponding *three-dimensional Brillouin zones*. Note that in the three-dimensional case, the Brillouin zones are bounded by planes halfway between adjacent reciprocal lattice points. Also note the following:

- The reciprocal lattice of a sc Bravais lattice of side a is a sc lattice of side $\frac{2\pi}{a}$.
- The reciprocal lattice of a fcc Bravais lattice of side a is a bcc lattice of side $\frac{4\pi}{a}$.
- The reciprocal lattice of a bcc Bravais lattice of side a is a fcc lattice of side $\frac{4\pi}{a}$.

Table 6.2 Real and reciprocal lattice primitive vectors for the three cubic Bravais lattices

	Primitive vectors	
	Real lattice	Reciprocal lattice
Simple cubic lattice	$\mathbf{a}_1 = a\hat{x}$	$\mathbf{b}_1 = \dfrac{2\pi}{a}\hat{x}$
	$\mathbf{a}_2 = a\hat{y}$	$\mathbf{b}_2 = \dfrac{2\pi}{a}\hat{y}$
	$\mathbf{a}_3 = a\hat{z}$	$\mathbf{b}_3 = \dfrac{2\pi}{a}\hat{z}$
Face-centered-cubic lattice	$\mathbf{a}_1 = \dfrac{a}{2}\hat{x} + \dfrac{a}{2}\hat{y}$	$\mathbf{b}_1 = \dfrac{4\pi}{a}\left(\dfrac{1}{2}\hat{x} + \dfrac{1}{2}\hat{y} - \dfrac{1}{2}\hat{z}\right)$
	$\mathbf{a}_2 = \dfrac{a}{2}\hat{y} + \dfrac{a}{2}\hat{z}$	$\mathbf{b}_2 = \dfrac{4\pi}{a}\left(-\dfrac{1}{2}\hat{x} + \dfrac{1}{2}\hat{y} + \dfrac{1}{2}\hat{z}\right)$
	$\mathbf{a}_3 = \dfrac{a}{2}\hat{x} + \dfrac{a}{2}\hat{z}$	$\mathbf{b}_3 = \dfrac{4\pi}{a}\left(\dfrac{1}{2}\hat{x} - \dfrac{1}{2}\hat{y} + \dfrac{1}{2}\hat{z}\right)$
Body-centered-cubic lattice	$\mathbf{a}_1 = \dfrac{a}{2}\hat{x} + \dfrac{a}{2}\hat{y} - \dfrac{a}{2}\hat{z}$	$\mathbf{b}_1 = \dfrac{4\pi}{a}\left(\dfrac{1}{2}\hat{x} + \dfrac{1}{2}\hat{y}\right)$
	$\mathbf{a}_2 = -\dfrac{a}{2}\hat{x} + \dfrac{a}{2}\hat{y} + \dfrac{a}{2}\hat{z}$	$\mathbf{b}_2 = \dfrac{4\pi}{a}\left(\dfrac{1}{2}\hat{y} + \dfrac{1}{2}\hat{z}\right)$
	$\mathbf{a}_3 = \dfrac{a}{2}\hat{x} - \dfrac{a}{2}\hat{y} + \dfrac{a}{2}\hat{z}$	$\mathbf{b}_3 = \dfrac{4\pi}{a}\left(\dfrac{1}{2}\hat{x} + \dfrac{1}{2}\hat{z}\right)$

Because the dispersion relation $\omega = \omega(\mathbf{k})$ is a periodic function of $\mathbf{k}$, the wave vectors required to plot the dispersion relation are those enclosed within the corresponding three-dimensional Brillouin zones, which are shown as gray regions in Figure 6.13. To consider wave vectors $\mathbf{k}$ outside of the Brillouin zone is unnecessary since the dispersion relation is periodic and the values of $\omega = \omega(\mathbf{k})$ will be repeated. In addition, if the three-dimensional photonic crystal is invariant under certain three-dimensional symmetry operations (e.g. rotations, mirror reflections, and inversions), only part of the three-dimensional Brillouin zone needs to be considered because the values of $\omega = \omega(\mathbf{k})$ in some regions of the Brillouin zone are the same as those in other regions. In general, the larger the number of symmetries present in the photonic crystal, the smaller the region of the Brillouin zone that needs to be considered. Figure 6.14 shows the irreducible Brillouin zone for the three cubic Bravais lattices, which is the smallest region of the Brillouin zone that cannot be reduced by the presence of additional symmetries in the photonic crystal. The irreducible Brillouin zones are defined by the set of wave vectors $\mathbf{k}$ bounded by the polyhedra defined by thick black solid lines with arrows.

For three-dimensional photonic crystals with symmetries such that only the irreducible Brillouin zone needs to be considered, the dispersion relation $\omega = \omega(\mathbf{k})$ is usually plotted for those wave vectors $\mathbf{k}$ whose tips are located in the thick solid black lines and whose origins are located at the Γ point (Figure 6.14). The black arrows

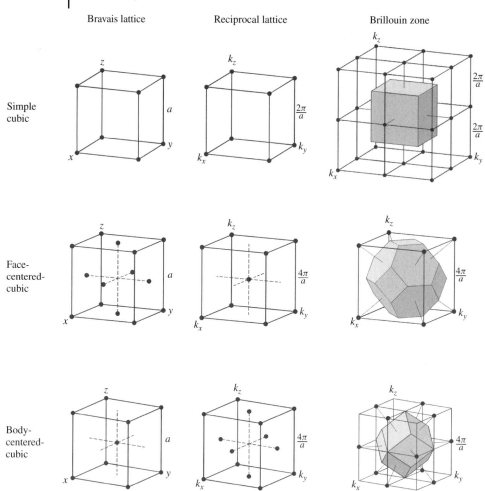

Fig. 6.13 *Left column*: Simple cubic (sc), face-centered-cubic (fcc), and body-centered-cubic (bcc) Bravais lattices. *Center column*: The corresponding reciprocal lattices. A sc Bravais lattice of side a has a sc reciprocal lattice of side $2\pi/a$; a fcc Bravais lattice of side a has a bcc reciprocal lattice of side $4\pi/a$; and a bcc Bravais lattice of side a has a fcc reciprocal lattice of side $4\pi/a$. *Right column*: The corresponding Brillouin zones (gray regions) in reciprocal space. Note that the Brillouin zones are bounded by planes located half way between adjacent reciprocal lattice points. In the case of the Brillouin zone for the bcc Bravais lattice, the origin of the reciprocal lattice is shifted by $(2\pi/a, 0, 0)$ in order to clearly visualize its symmetry.

indicate the standard convention for the direction of the path over which the dispersion relation is plotted.

In summary, in this section we established the wave vectors **k** required to calculate and plot the dispersion relation $\omega = \omega(\mathbf{k})$ for electromagnetic waves propagating

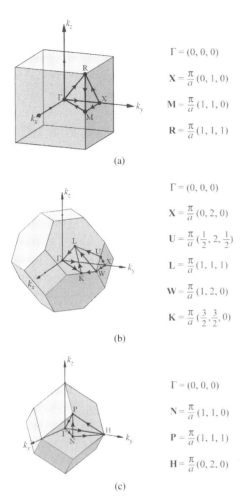

$\Gamma = (0, 0, 0)$

$X = \frac{\pi}{a}(0, 1, 0)$

$M = \frac{\pi}{a}(1, 1, 0)$

$R = \frac{\pi}{a}(1, 1, 1)$

(a)

$\Gamma = (0, 0, 0)$

$X = \frac{\pi}{a}(0, 2, 0)$

$U = \frac{\pi}{a}(\frac{1}{2}, 2, \frac{1}{2})$

$L = \frac{\pi}{a}(1, 1, 1)$

$W = \frac{\pi}{a}(1, 2, 0)$

$K = \frac{\pi}{a}(\frac{3}{2}, \frac{3}{2}, 0)$

(b)

$\Gamma = (0, 0, 0)$

$N = \frac{\pi}{a}(1, 1, 0)$

$P = \frac{\pi}{a}(1, 1, 1)$

$H = \frac{\pi}{a}(0, 2, 0)$

(c)

Fig. 6.14 (a–c) The irreducible Brillouin zones for the simple cubic, face-centered-cubic, and body-centered-cubic Bravais lattices, respectively. The vertices of the irreducible Brillouin zones are denoted by the letters Γ, X, M, R, etc., and the origin is denoted by the point $\Gamma = (0, 0, 0)$. The dispersion relation $\omega = \omega(\mathbf{k})$ for three-dimensional photonic crystals is plotted for wave vectors $\mathbf{k}$ along the paths defined by the edges of the irreducible Brillouin zones (thick black solid lines with arrows). The arrows indicate the standard convention for the direction in which the dispersion relation is plotted.

within three-dimensional photonic crystals (Figure 6.14). As in the two-dimensional case, to establish this set of required wave vectors $\mathbf{k}$ one needs to determine the Bravais lattice, the reciprocal lattice, and the Brillouin zone associated with the particular three-dimensional photonic crystal. We next investigate the existence of complete photonic band gaps in three dimensions by plotting the dispersion relation $\omega = \omega(\mathbf{k})$ corresponding to three-dimensional photonic crystals.

6.4.2
Band Diagrams and Photonic Band Gaps in Three Dimensions

Now that we have established which set of wave vectors **k** are needed to plot the dispersion relation $\omega = \omega(\mathbf{k})$ in three dimensions, we can explore for the existence of complete photonic band gaps in three-dimensional photonic crystals. Note that in the three-dimensional case, the presence of a *complete* photonic band gap means that the periodic structure does not allow the propagation of electromagnetic waves with certain frequencies *irrespective of their propagation direction in the three-dimensional space*. That is, in whatever direction the wave intends to propagate, the photonic crystal will not allow its propagation. This is a significant physical result with enormous consequences, some of which are introduced in Chapter 9.

We could, in principle, consider three-dimensional photonic crystals consisting of a periodic arrangement of dielectric spheres in a different dielectric background. In fact, these types of periodic structures were the first photonic crystals that theoretically showed a complete three-dimensional photonic band gap. Unfortunately, these periodic structures are difficult to fabricate at small length scales. In particular, they are difficult to fabricate at submicron length scales, where the structures show complete photonic band gaps for visible light. For this reason, we concentrate on three-dimensional photonic crystals defined by level-set formulas such as those presented in Chapter 2. These three-dimensional photonic crystals can be fabricated at small length scales by interference lithography as shown in Chapter 4.

We first consider the three-dimensional periodic structure represented by the function (Chapter 2)

$$f_3^{4,\mathrm{II}}(x,y,z) = +\cos\left[\frac{2\pi}{a}(x+y+z)\right] + \cos\left[\frac{2\pi}{a}(x+y-z)\right] \\ + \cos\left[\frac{2\pi}{a}(x-y+z)\right] - \cos\left[\frac{2\pi}{a}(x-y-z)\right] - 1.4 \quad (6.13)$$

Figure 6.15a shows the unit cell of this structure, which is also called the *diamond structure*. The structure is made of solid dielectric ($\varepsilon_1 = 13\varepsilon_0$) material (shown in gray) and air ($\varepsilon_2 = \varepsilon_0$). In this particular case, the volume fraction of the solid material in the periodic structure is 21%.

We calculate and plot the dispersion relation $\omega = \omega(\mathbf{k})$ for electromagnetic waves propagating within this three-dimensional periodic structure in Figure 6.15b. Because the Bravais lattice that corresponds to the diamond structure is the fcc Bravais lattice, the dispersion relation $\omega = \omega(\mathbf{k})$ is plotted for wave vectors **k** along directions in the edges of the Brillouin zone that corresponds to the fcc Bravais lattice (Figure 6.14b).

The graph of the dispersion relation $\omega = \omega(\mathbf{k})$ shows a range of frequencies $\Delta\omega$ (gray region) for which there is no associated wave vector **k**. As previously mentioned, this means that an electromagnetic wave with frequency ω within this frequency range will not be allowed to propagate within the photonic crystal because there is no propagating mode with wave vector **k** that can sustain that

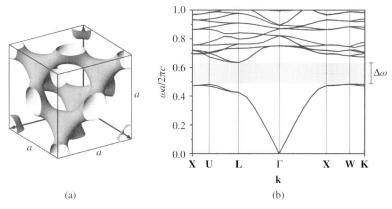

Fig. 6.15 (a) The level-set periodic diamond structure, which is given by the formula $f_3^{4,\text{II}}(x, y, z)$. The dielectric solid material (gray) is silicon ($\varepsilon_1 = 13\varepsilon_0$) and the background material is air ($\varepsilon_2 = \varepsilon_0$). The volume fraction of the solid material is 21%. (b) The dispersion relation $\omega = \omega(\mathbf{k})$ for electromagnetic waves propagating within the diamond structure. The graph shows that electromagnetic waves with frequencies within the range $\omega a/2\pi c = 0.48 - 0.64$ (see gray region) are not allowed to propagate within the structure regardless of their propagation direction. For this frequency range, the structure possesses a complete three-dimensional photonic band gap (see Reference 28).

frequency. Note that, in contrast to the one- and two-dimensional cases, the wave vectors **k** in Figure 6.15b have different directions in three dimensions. As a result, the periodic diamond structure possesses a *complete* three-dimensional photonic band gap, and electromagnetic waves with frequencies within the gap $\Delta\omega$ are not allowed to propagate within the structure independent of their propagation direction in three dimensions. This result is not trivial, and it took several years of intense efforts to find a structure having a complete three-dimensional photonic band gap since the original idea of the existence of such a photonic band gap was proposed (see Reference 11).

Figure 6.15 also shows that the complete photonic band gap for the diamond structure is centered around $\omega a/2\pi c = 0.56$. This means that in order to forbid the propagation of visible electromagnetic waves with frequency $f = \omega/2\pi = 6 \times 10^{14}$ Hz, the lattice constant a of the diamond unit cell must be equal to $a = 0.28\,\mu\text{m}$. This example shows the difficulties associated with the fabrication of three-dimensional photonic crystals that can forbid the propagation of visible light, since the required length scales of the structure are significantly small. Moreover, at visible frequencies silicon is opaque, and other materials with high dielectric constants (but having low absorption) are needed for the solid network.

We would like to remark that the existence and characteristics of complete photonic band gaps depend on the following three key factors:

- The specific geometry of the three-dimensional periodic structure;
- The ratio $\varepsilon_1/\varepsilon_2$ between the dielectric constants of the different dielectric materials; and

- The volume fraction of the high (or low) dielectric material in the structure.

The dependence of photonic band gaps on the specific geometry of the periodic structure is at present not fully understood. In fact, it is not possible to determine *a priori* whether or not a given periodic dielectric structure possesses a complete photonic band gap. For example, since the first diamond photonic crystal with a complete photonic band gap was discovered, researchers have pursued the task of obtaining photonic crystals possessing larger complete band gaps and/or photonic crystals with large and complete band gaps that are more easily fabricated.

The second factor is less complicated and fairly well understood. As previously mentioned, the larger the ratio $\varepsilon_1/\varepsilon_2$ between the dielectric constants of the different dielectric materials forming the periodic structure, the larger the complete photonic band gaps. That is, given a periodic structure with a complete photonic band gap, if the dielectric contrast between the dielectric materials is increased, in general the frequency range $\Delta\omega$ for which electromagnetic waves are not allowed to propagate increases as well.

Finally, the third factor can be examined by calculating the frequency range $\Delta\omega$ of the complete photonic band gaps for different volume fractions of the solid material in the periodic structure. By plotting the frequency range $\Delta\omega$ corresponding to the complete photonic band gap as a function of the volume fraction of the solid material, we obtain what is called a *photonic band gap map*. For example, Figure 6.16 shows the photonic band gap maps for the diamond structure, where we choose two different dielectric contrasts between the solid material and air in order to show the dependence of the complete photonic band gaps with respect to the ratio $\varepsilon_1/\varepsilon_2$ between the dielectric constants of the materials. The dielectric contrasts for the photonic band gap maps in Figures 16a and b are, respectively, $\varepsilon_{\text{solid}}/\varepsilon_{\text{air}} = 13/1$ and $6/1$.

6.4.3
Photonic Band Gaps in Three-dimensional Simple Periodic Structures

In the last section, we introduced the level-set diamond structure as an example of a photonic crystal possessing a complete three-dimensional photonic band gap for electromagnetic waves. However, this structure is not the only one presenting such an important physical property. In fact, many of the three-dimensional simple periodic structures introduced in Chapter 2 also possess complete photonic band gaps in three dimensions.

In this section, we study the complete three-dimensional photonic band gaps exhibited by these simple periodic structures with the purpose of developing a large set of complete photonic band gap structures that can be fabricated at small length scales by interference lithography.

For the calculations that follow, we assume that the solid dielectric material forming the simple periodic structures is silicon ($\varepsilon_1 = 13\varepsilon_0$) and that the background material is air ($\varepsilon_2 = \varepsilon_0$). Before calculating complete photonic band gaps for simple periodic structures, we would like to have a magnitude that allows us to compare

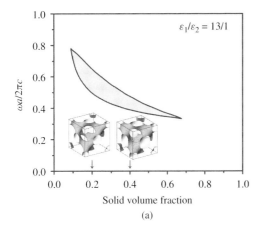

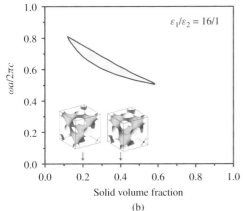

Fig. 6.16 Photonic band gap maps for the three-dimensional simple periodic structure $f_3^{4,\text{II}}(x,y,z)$ or diamond structure. The frequency range $\Delta\omega$ of the complete three-dimensional photonic band gap is shown as a function of the volume fraction of the solid (gray) dielectric material. The dielectric ratio between the solid dielectric material and air is (a) $\varepsilon_{\text{solid}}/\varepsilon_{\text{air}} = 13/1$ and (b) $\varepsilon_{\text{solid}}/\varepsilon_{\text{air}} = 6/1$. The insets show the simple periodic diamond structures at solid volume fractions $f_S = 0.20$ and $f_S = 0.40$, respectively.

complete photonic band gaps between different structures. For example, we can measure the "size" of a complete photonic band gap by the width of the frequency range $\Delta\omega$ for which electromagnetic waves are not allowed to propagate within the structure. However, owing to the scaling property explained in Section 6.2.1, this magnitude depends on the lattice constant a of the structure and it is therefore not a universal quantity. For example, if the lattice constant a of the structure is increased by a factor r, all frequencies in the graph of the dispersion relation scale down by a factor of r, and the width $\Delta\omega$ of the photonic band gap decreases to $\Delta\omega/r$. In order to measure the "size" of a photonic band gap independently of the scale of the structure, we define the gap/midgap frequency ratio $\Delta\omega/\omega_0$, where $\Delta\omega = \omega_U - \omega_L$ is the difference between the upper and lower frequencies ω_U and

6 Photonic Crystals

ω_L of the gap, and $\omega_0 = (\omega_U + \omega_L)/2$ is the middle frequency of the gap. Note that in this case, if the lattice constant a of the structure is increased/reduced, all frequencies are correspondingly reduced/increased, but the gap/midgap frequency ratio $\Delta\omega/\omega_0$ remains constant.

Next, we display complete three-dimensional photonic band gap maps for simple periodic structures, noting that the set of wave vectors $\mathbf{k}$ used to obtain such maps depends on the specific symmetries present in the corresponding periodic structures. For example, in some cases, the set of wave vectors $\mathbf{k}$ required to obtain the dispersion relations is not only within the irreducible Brillouin zone.

6.4.3.1 Simple Cubic Photonic Crystals

The three-dimensional simple periodic structure represented by the formula

$$f_1(x,y,z) = +\cos\left(\frac{2\pi x}{a}\right) + \cos\left(\frac{2\pi y}{a}\right) + \cos\left(\frac{2\pi z}{a}\right) + t \tag{6.14}$$

possesses a complete photonic band gap for volume fractions of solid dielectric material (silicon) ranging from $f_S = 0.22$ to $f_S = 0.43$ (Figure 6.17), where different solid volume fractions are obtained by varying the parameter t. For this structure, the largest complete photonic band gap ($\Delta\omega/\omega_0 = 13\%$) is reached at a solid volume fraction $f_S \sim 0.26$ ($t = -0.85$). The inset in Figure 6.17 shows the unit cell of the structure that presents the largest complete photonic band gap.

6.4.3.2 Face-centered-cubic Photonic Crystals

In the case of three-dimensional simple periodic structures with fcc Bravais lattices, we find two level-set structures that have complete three-dimensional photonic band

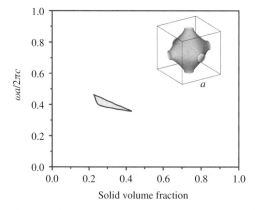

Fig. 6.17 Photonic band gap map for the simple periodic structure $f_1(x,y,z)$. The range of nondimensional frequencies $\omega a/2\pi c$ corresponding to the complete three-dimensional photonic band gap is plotted as a function of the solid volume fraction in the structure. An electromagnetic wave with frequency within the gray region is not allowed to propagate within the structure independently of its propagation direction in three dimensions. The ratio between the dielectric constants of the solid material and air is $\varepsilon_{solid}/\varepsilon_{air} = 13/1$.

gaps (Figure 6.18). They are given by the formulas

$$f_3^{4,II}(x,y,z) = +\cos\left[\frac{2\pi}{a}(x+y+z)\right] + \cos\left[\frac{2\pi}{a}(x+y-z)\right]$$
$$+ \cos\left[\frac{2\pi}{a}(x-y+z)\right] - \cos\left[\frac{2\pi}{a}(x-y-z)\right] + t \qquad (6.15)$$

and

$$f_3^3(x,y,z) = +\cos\left[\frac{2\pi}{a}(x+y-z)\right] + \cos\left[\frac{2\pi}{a}(x-y+z)\right]$$
$$+ \cos\left[\frac{2\pi}{a}(x-y-z)\right] + t \qquad (6.16)$$

The $f_3^{4,II}(x,y,z)$ structure (or diamond structure) possesses a complete photonic band gap for volume fractions of solid dielectric material ranging from $f_S = 0.09$ to $f_S = 0.67$, and the largest complete photonic band gap ($\Delta\omega/\omega_0 = 27\%$) is reached at a solid volume fraction $f_S \sim 0.21$ ($t = -1.40$). This value is very close to the largest (30%) complete photonic band gap to date, which is obtained with a rod-connected version of the diamond structure (see Reference 17). In the case of the periodic structure $f_3^3(x,y,z)$, the complete photonic band gap extends from $f_S = 0.27$ to $f_S = 0.56$, and the largest gap (6%) occurs at $f_S \sim 0.39$ ($t = -0.40$).

6.4.3.3 Body-centered-cubic Photonic Crystals

In the case of three-dimensional simple periodic structures with bcc Bravais lattices, we find that the following structures exhibit complete three-dimensional photonic band gaps:

$$f_2^{6,III}(x,y,z) = +\sin\left[\frac{2\pi}{a}(x+y)\right] + \sin\left[\frac{2\pi}{a}(x-y)\right] + \sin\left[\frac{2\pi}{a}(x+z)\right]$$
$$+ \sin\left[\frac{2\pi}{a}(x-z)\right] + \sin\left[\frac{2\pi}{a}(y+z)\right] + \sin\left[\frac{2\pi}{a}(y-z)\right] + t \qquad (6.17)$$

$$f_2^{6,IV}(x,y,z) = +\sin\left[\frac{2\pi}{a}(x+y)\right] + \sin\left[\frac{2\pi}{a}(x-y)\right] + \sin\left[\frac{2\pi}{a}(x+z)\right]$$
$$- \sin\left[\frac{2\pi}{a}(x-z)\right] + \sin\left[\frac{2\pi}{a}(y+z)\right] + \sin\left[\frac{2\pi}{a}(y-z)\right] + t \qquad (6.18)$$

$$f_2^{5,IV}(x,y,z) = +\sin\left[\frac{2\pi}{a}(x-y)\right] + \sin\left[\frac{2\pi}{a}(x+z)\right]$$
$$+ \sin\left[\frac{2\pi}{a}(x-z)\right] + \sin\left[\frac{2\pi}{a}(y+z)\right] - \sin\left[\frac{2\pi}{a}(y-z)\right] + t \qquad (6.19)$$

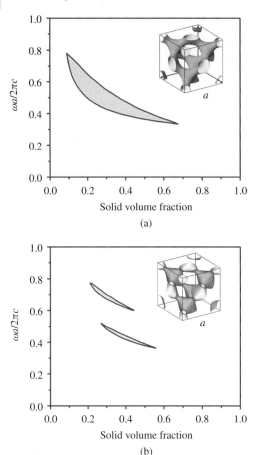

Fig. 6.18 Photonic band gap maps for the simple periodic structures (a) $f_3^{4,\text{II}}$ and (b) f_3^3. The range of nondimensional frequencies $\omega a/2\pi c$ corresponding to complete three-dimensional photonic band gaps is plotted as a function of the solid volume fraction in the structure. The ratio between the dielectric constants of solid material and air is $\varepsilon_{\text{solid}}/\varepsilon_{\text{air}} = 13/1$.

$$f_2^{4,\text{II}}(x, y, z) = +\cos\left[\frac{2\pi}{a}(x+y)\right] + \cos\left[\frac{2\pi}{a}(x-y)\right] + \cos\left[\frac{2\pi}{a}(x+z)\right]$$
$$-\cos\left[\frac{2\pi}{a}(x-z)\right] + t \qquad (6.20)$$

Figure 6.19a shows the corresponding gap map for the $f_2^{6,\text{III}}(x, y, z)$ structure where the complete photonic band gap extends from $f_S = 0.29$ to $f_S = 0.55$ and reaches a maximum $\Delta\omega/\omega_0 = 8\%$ at $f_S \sim 0.39$ ($t = -0.5$). Figure 6.19b shows the gap map for the $f_2^{6,\text{IV}}(x, y, z)$ structure extending from $f_S = 0.05$ to $f_S = 0.60$ and reaching a maximum $\Delta\omega/\omega_0 = 27\%$ at $f_S \sim 0.17$ ($t = -2.0$). The complete photonic

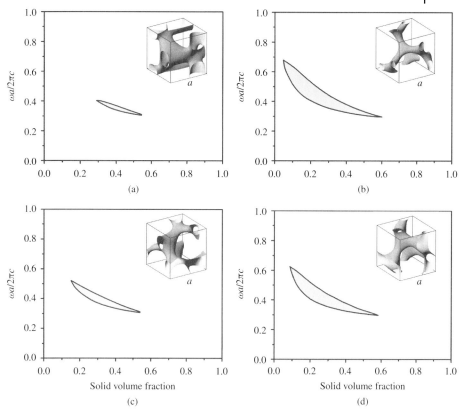

Fig. 6.19 Photonic band gap maps for the simple periodic structures (a) $f_2^{6,III}$, (b) $f_2^{6,IV}$, (c) $f_2^{5,IV}$, and (d) $f_2^{4,II}$. The range of nondimensional frequencies $\omega a/2\pi c$ corresponding to complete three-dimensional photonic band gaps is plotted as a function of the solid volume fraction in the structure. The ratio between the dielectric constants of air and the solid material is $\varepsilon_{solid}/\varepsilon_{air} = 13/1$.

gap map in Figure 6.19c corresponds to the $f_2^{5,IV}(x, y, z)$ structure, where the solid volume fraction range is $f_S = 0.15 - 0.54$ and the maximum gap $\Delta\omega/\omega_0 = 15\%$ is reached at $f_S \sim 0.27$ ($t = -1.2$). Finally, Figure 6.19d shows the photonic gap map for the $f_2^{4,II}(x, y, z)$ structure, where the complete photonic band gap extends from $f_S = 0.09$ to $f_S = 0.58$ and reaches a maximum $\Delta\omega/\omega_0 = 24\%$ at $f_S \sim 0.21$ ($t = -1.4$). The insets show the unit cell of the structures that possess the largest complete photonic band gaps.

Finally, we would like to mention that the periodic structure given by Equation 2.56, which is represented by a combination of simple periodic functions, also presents a complete photonic band gap. This structure is interesting since it is the level-set version (Figure 2.9) of a periodic structure made of air spheres arranged in the fcc Bravais lattice (also called inverse opal) which is a photonic crystal usually fabricated by self-assembly of colloidal particles.

The large number of simple periodic structures (whose formulas are composed of just a few Fourier terms) having complete three-dimensional photonic band gaps for a broad range of solid volume fractions shows that the interference lithography technique is a promising method to fabricate three-dimensional photonic crystals. The hope is that the easy fabrication of photonic crystals at small length scales will enable the miniaturization and integration of optical components and devices that can control the propagation of visible light. Some of these valuable properties of photonic crystals are discussed in Chapter 9.

Further Reading

One-dimensional photonic crystals

Theory

1 Winn, J.N., Fink, Y., Fan, S. and Joannopoulos, J.D. (**1998**) "Omnidirectional reflection from one-dimensional photonic crystals". *Optics Letters*, **23**, 1573–75.
2 Southwell, W.H. (**1999**) "Omnidirectional mirror design with quarter-wave dielectric stacks". *Applied Optics*, **38**, 5464–67.

Experiments

3 Fink, Y., Winn, J.N., Fan, S., Chen, C., Michel, J., Joannopoulos, J.D. and Thomas, E.L. (**1998**) "A dielectric omnidirectional reflector". *Science*, **282**, 1679–82.

Two-dimensional photonic crystals

Theory

4 Plihal, M. and Maradudin, A.A. (**1991**) "Photonic band structure of two-dimensional systems: the triangular lattice". *Physical Review B*, **44**, 8565–71.
5 Meade, R.D., Brommer, K.D., Rappe, A.M. and Joannopoulos, J.D. (**1992**) "Existence of a photonic band gaps in two dimensions". *Applied Physics Letters*, **61**, 495–97.
6 Villeneuve, P.R. and Piche, M. (**1992**) "Photonic band gaps in two-dimensional square and hexagonal lattice". *Physical Review B*, **46**, 4969–72.
7 Villeneuve, P.R. and Piche, M. (**1992**) "Photonic band gaps in two-dimensional square lattices: square and circular rods". *Physical Review B*, **46**, 4973–75.

Experiments

8 Wendt, J.R., Vawter, G.A., Gourley, P.L., Brennan, T.M. and Hammons, B.E. (**1993**) "Nanofabrication of photonic lattice structures in GaAs/AlGaAs". *Journal of Vacuum Science and Technology*, **11**, 2637–40.
9 Inoue, K., Wada, M., Sakoda, K., Yamanaka, A., Hayashi, M. and Haus, J.W. (**1994**) "Fabrication of two-dimensional photonic band-structure with near-infrared band gap". *Japanese Journal of Applied Physics*, **33**, L1463–65.
10 Gruning, U., Lehmann, V., Ottow, S. and Busch, K. (**1996**) "Macroporous silicon with a complete two-dimensional photonic band gap centered at 5 μm". *Applied Physics Letters*, **68**, 747–49.

Three-dimensional photonic crystals

Theory

11 Yablonovitch, E. (**1987**) "Inhibited spontaneous emission in solid-state physics and electronics". *Physical Review Letters*, **58**, 2059–62.
12 John, S. (**1987**) "Strong localization of photons in certain disordered dielectric superlattices". *Physical Review Letters*, **58**, 2486–89.
13 Yablonovitch, E. and Gmitter, T.J. (**1989**) "Photonic band structure: the face-centered-cubic case". *Physical Review Letters*, **63**, 1950–53.
14 Leung, K.M. and Liu, Y.F. (**1990**) "Full wave vector calculation

of photonic band structures in face-centered-cubic dielectric media". *Physical Review Letters*, **65**, 2646–49.
15. Zhang, Z. and Satpathy, S. (**1990**) "Electromagnetic wave propagation in periodic structures: Bloch wave solution of Maxwell's equations". *Physical Review Letters*, **65**, 2650–53.
16. Ho, K.M., Chan, C.T. and Soukoulis, C.M. (**1990**) "Existence of a photonic gap in periodic dielectric structures". *Physical Review Letters*, **65**, 3152–55.
17. Chan, C.T., Ho, K.M. and Soukoulis, C.M. (**1991**) "Photonic band gaps in experimentally realizable periodic dielectric structures". *Europhysics Letters*, **16**, 563–68.
18. Yablonovitch, E. and Gmitter, T.J. (**1991**) "Photonic band structure: the face-centered cubic case employing nonspherical atoms". *Physical Review Letters*, **67**, 2295–98.
19. Sozuer, H.S., Haus, J.W. and Inguva, R. (**1992**) "Photonic bands: convergence problems with the plane-wave method". *Physical Review B*, **45**, 13962–72.
20. Chan, C.T., Datta, S., Ho, K.M. and Soukoulis, C.M. (**1994**) "A-7 structure: a family of photonic crystals". *Physical Review B*, **50**, 1988–91.
21. Ho, K.M., Chan, C.T., Soukoulis, C.M., Biswas, R. and Sigalas, M. (**1994**) "Photonic band gaps in three-dimensions: new layer-by-layer periodic structures". *Solid State Communications*, **89**, 413–16.
22. Fan, S., Villeneuve, P.R., Meade, R.D. and Joannopoulos, J.D. (**1994**) "Design of three-dimensional photonic crystals at submicron length scales". *Applied Physics Letters*, **65**, 1466–68.
23. Leung, K.M. (**1997**) "Diamond like photonic band gap crystal with a sizable band gap". *Physical Review B*, **56**, 3517–19.
24. Joannopoulos, J.D., Villeneuve, P.R. and Fan, S.H. (**1997**) "Photonic crystals: putting a new twist on light". *Nature*, **386**, 143–49.
25. Chutinan, A. and Noda, S. (**1998**) "Spiral three-dimensional photonic-band gap structure". *Physical Review B*, **57**, R2006–8.
26. Martin-Moreno, L., Garcia-Vidal, F.J. and Somoza, A.M. (**1999**) "Self-assembled triply periodic minimal surfaces as molds for photonic band gap materials". *Physical Review Letters*, **83**, 73–75.
27. Johnson, S.G. and Joannopoulos, J.D. (**2000**) "Three-dimensionally periodic dielectric layered structure with omnidirectional photonic band gap". *Applied Physics Letters*, **77**, 3490–92.
28. Maldovan, M., Urbas, A.M., Yufa, N., Carter, W.C. and Thomas, E.L. (**2002**) "Photonic properties of bicontinuous cubic microphases". *Physical Review B*, **65**, 165123.
29. Moroz, A. (**2002**) "Metallo-dielectric diamond and zinc-blende photonic crystal". *Physical Review B*, **66**, 115109.
30. Roundy, D. and Joannopoulos, J.D. (**2003**) "Photonic crystal structure with square symmetry within each layer and a three-dimensional band gap". *Applied Physics Letters*, **82**, 3835–37.
31. Maldovan, M., Thomas, E.L. and Carter, W.C. (**2003**) "A layer-by-layer diamond-like woodpile structure with a large photonic band gap". *Applied Physics Letters*, **84**, 362–64.
32. Ullal, C.K., Maldovan, M., Wohlgemuth, M. and Thomas, E.L. (**2003**) "Triply periodic bicontinuous structures through interference lithography: a level set approach". *Journal of the Optical Society of America A*, **20**, 948–54.
33. Maldovan, M., Ullal, C.K., Carter, W.C. and Thomas, E.L. (**2003**) "Exploring for 3D photonic band gap structures: the 11 fcc groups". *Nature Materials*, **2**, 664–67.
34. Maldovan, M., Carter, W.C. and Thomas, E.L. (**2003**) "Three-dimensional dielectric network structures with large photonic band gaps". *Applied Physics Letters*, **83**, 5172–74.
35. Sharp, D.N., Turberfield, A.J. and Denning, R.G. (**2003**) "Holographic photonic crystals with

36 Maldovan, M. (2004) "A layer-by-layer photonic crystal with a two-layer periodicity". *Applied Physics Letters*, **85**, 911–13.

37 Toader, O. and John, S. (2004) "Photonic band gap architectures for holographic lithography". *Physical Review Letters*, **92**, 043905.

38 Maldovan, M. and Thomas, E.L. (2004) "Diamond structured photonic crystals". *Nature Materials*, **3**, 593–600.

39 Garcia-Adeva, A.J. (2006) "band gap atlas for photonic crystals having the symmetry of the kagome and pyrochlore lattices". *New Journal of Physics*, **8**, 86.

40 Ngo, T.T., Liddell, C.M., Ghebrebrhan, M. and Joannopoulos, J.D. (2006) "Tetrastack: colloidal diamond-inspired structure with omnidirectional photonic band gap for low refractive index". *Applied Physics Letters*, **88**, 241920.

41 Hynninen, A.P., Thijssen, J.H.J., Vermolen, E.C.M., Dijkstra, M. and Van Blaaderen, A. (2007) "Self-assembly route for photonic crystals with a band gap in the visible region". *Nature Materials*, **6**, 202–5.

Experiments

42 Ozbay, E., Abeyta, A., Tuttle, G., Tringides, M., Biswas, R., Chan, C.T., Soukoulis, C.M. and Ho, K.M. (1994) "Measurement of a three-dimensional photonic band gap in a crystal structure made of dielectric rods". *Physical Review B*, **50**, 1945–48.

43 Lin, S.Y., Fleming, J.G., Hetherington, D.L., Smith, B.K., Biswas, R., Ho, K.M., Sigalas, M.M., Zubrzycki, W., Kurtz, S.R. and Bur, J. (1998) "A three-dimensional photonic crystal operating at infrared wavelengths". *Nature*, **394**, 251–53.

44 Wijnhoven, J.E. and Vos, W.L. (1998) "Preparation of photonic crystals made of air spheres in titania". *Science*, **281**, 802–4.

diamond symmetry". *Physical Review B*, **68**, 205102.

45 Fleming, J.G. and Lin, S.Y. (1999) "Three-dimensional photonic crystals with a stop band from 1.35 to 1.95 µm". *Optics Letters*, **24**, 49–51.

46 Cumpston, B.H., Ananthavel, S.P., Barlow, S., Dyer, D.L., Ehrlich, J.E., Erskine, L.L., Heikal, A.A., Kuebler, S.M., Lee, I.-Y.S., McCord-Maughon, D., Qin, J., Rockel, H., Rumi, M., Wu, X.-L., Marder, S.R. and Perry, J.W. (1999) "Two-photon polymerization initiators for three-dimensional optical data storage and microfabrication". *Nature*, **398**, 51–54.

47 Noda, S., Tomoda, K., Yamamoto, N. and Chutinan, A. (2000) "Full three-dimensional photonic band gap crystals at near-infrared wavelengths". *Science*, **289**, 604–6.

48 Campbell, M., Sharp, D.N., Harrison, M.T., Denning, R.G. and Turberfield, A.J. (2000) "Fabrication of photonic crystals for the visible spectrum by holographic lithography". *Nature*, **404**, 53–56.

49 Vlasov, Y.A., Bo, X.Z., Sturm, J.C. and Norris, D.J. (2001) "On-chip natural assembly of silicon photonic band gap crystals". *Nature*, **414**, 289–93.

50 Toader, O. and John, S. (2001) "Proposed square spiral microfabrication architecture for large three-dimensional photonic band gap crystals". *Science*, **292**, 1133–35.

51 Garcia-Santamaria, F., Miyazaki, H.T., Urquia, A., Ibisate, M., Belmonte, M., Shinya, N., Meseguer, F. and Lopez, C. (2002) "Nanorobotic manipulation of microspheres for on-chip diamond architectures". *Advanced Materials*, **14**, 1144–47.

52 Urbas, A.M., Maldovan, M., DeRege, P. and Thomas, E.L. (2002) "Bicontinuous cubic block copolymer photonic crystals". *Advanced Materials*, **14**, 1850–53.

53 Aoki, K., Miyazaki, H.T., Hirayama, H., Inoshita, K., Baba, T., Shinya, N. and Aoyagi, Y. (2002) "Three-dimensional photonic crystals for optical wavelengths assembled by micromanipulation". *Applied Physics Letters*, **81**, 3122–24.

54 Lee, W., Pruzinsky, S.A. and Braun, P.V. (2002) "Multi-photon polymerization of waveguide structure within three-dimensional photonic crystals". *Advanced Materials*, **14**, 271–74.

55 Ullal, C.K., Maldovan, M., Chen, G., Han, Y.J., Yang, S. and Thomas, E.L. (2004) "Photonic crystals through holographic lithography: simple cubic, diamond like and gyroid like structures". *Applied Physics Letters*, **84**, 5434–36.

56 Deubel, M., Von Freymann, G., Wegener, M., Pereira, S., Busch, K. and Soukoulis, C.M. (2004) "Direct laser writing of three-dimensional photonic-crystal templates for telecommunications". *Nature Materials*, **3**, 444–47.

57 Seet, K.K., Mizeikis, V., Matsuo, S., Juodkazis, S. and Misawa, H. (2005) "Three-dimensional spiral-architecture photonic crystals obtained by direct laser writing". *Advanced Materials*, **17**, 541–45.

58 Hermatschweiler, M., Ledermann, A., Ozin, G.A., Wegener, M. and Von Freymanm, G. (2007) "Fabrication of silicon inverse woodpile photonic crystal". *Advanced Functional Materials*, **17**, 2273–77.

59 Garcia-Santamaria, F., Xu, M., Lousse, V., Fan, S.H., Braun, P.V. and Lewis, J.A. (2007) "A germanium inverse woodpile structure with a large photonic band gap". *Advanced Materials*, **19**, 1567–70.

Books

60 Ashcroft, N.W. and Mermin, N.D. (1976) *Solid State Physics*, Saunders, Philadelphia.

61 Kittel, C. (1995) *Introduction to Solid State Physics*, John Wiley & Sons, Ltd, New York.

62 Joannopoulos, J.D., Meade, R. and Winn, J. (1995) *Photonic Crystals*, Princeton Press, Princeton, NJ.

Problems

6.1 An electromagnetic wave with angular frequency ω is incident on the surface of a photonic crystal possessing a complete photonic band gap and the wave continues its propagation within the crystal.

(a) Does the angular frequency ω lie inside or outside the photonic band gap?

(b) Does the frequency ω of the wave change as the wave propagates within the crystal?

(c) Does the wavelength λ of the wave change as the wave propagates within the crystal?

6.2 Consider the graph of the reflectance R versus $\omega a/2\pi c$ shown in Figure 6.3.

(a) If the lattice constant of the one-dimensional photonic crystal is $a = 2\,\mu m$, find the frequencies f of the normal-incident electromagnetic waves that are totally reflected by the crystal.

(b) Repeat the calculations for $a = 10\,\mu m$.

(c) If for a particular frequency ω the value of the reflectance is $R = 0.6$, what percentage of the energy of the incident wave is reflected by the structure and what percentage is transmitted through the structure?

6.3 By using the program in Appendix B

(a) Obtain the graph of the reflectance R as a function of $\omega a/2\pi c$ shown in Figure 6.3, which corresponds to normal-incident electromagnetic waves on a one-dimensional photonic crystal embedded in air consisting of 10 dielectric layers with $\varepsilon_1 = 11\varepsilon_0, \varepsilon_2 = 2\varepsilon_0$, and $t_1 = t_2 = 0.5a$.

(b) Decrease the value of ε_2 and show that the photonic band gaps increase.

(c) Show how the photonic band gaps vary with respect to the number of layers.

6.4 Does the graph of the dispersion relation $\omega = \omega(\mathbf{k})$ describe waves propagating inside or outside the photonic crystal? How can we use this graph to study external incident electromagnetic waves on a finite size photonic crystal?

6.5 Reciprocal lattices in two dimensions:
By using Equation 6.8, find the reciprocal lattice primitive vectors $\mathbf{b}_1$ and $\mathbf{b}_2$ for the square and triangular lattices (Table 6.1).

6.6 Reciprocal lattices in three dimensions:
By using Equation 6.12, find the reciprocal lattice primitive vectors $\mathbf{b}_1, \mathbf{b}_2$, and $\mathbf{b}_3$ for the cubic Bravais lattices (Table 6.2).

6.7 Hexagonal lattice in three dimensions:
The real-space unit vectors for the hexagonal Bravais lattice are given by $\mathbf{a}_1 = a\hat{\mathbf{x}}, \mathbf{a}_2 = a/2\,\hat{\mathbf{x}} + \sqrt{3}a/2\,\hat{\mathbf{y}}$, and $\mathbf{a}_3 = c\hat{\mathbf{z}}$.

(a) Draw the lattice.

(b) Find the corresponding reciprocal lattice primitive vectors $\mathbf{b}_1, \mathbf{b}_2$, and $\mathbf{b}_3$ and draw the reciprocal lattice.

(c) Find the Brillouin zone and the irreducible Brillouin zone.

6.8 Consider two different three-dimensional photonic crystals, one of which has mirror symmetry with respect to a horizontal plane. In order to plot the dispersion relation $\omega = \omega(\mathbf{k})$, which crystal would require the smaller number of wave vectors $\mathbf{k}$?

6.9 Omnidirectional reflection:
Consider two identical finite, one-dimensional periodic dielectric structures made of 10 alternate dielectric layers with $\varepsilon_1 = 21.16$ (tellurium), $\varepsilon_2 = 2.56$ (polystyrene), and $t_1 = 0.33a$. These structures possess a frequency range $\Delta\omega$ for which external incident electromagnetic waves are totally reflected by the structures regardless of the angle of incidence of the waves (i.e. omnidirectional dielectric mirrors). Considering that the structures are separated by a distance d and assuming that the frequency range $\Delta\omega$ corresponds to visible light, what happens to the light coming from a light bulb located in between the structures?

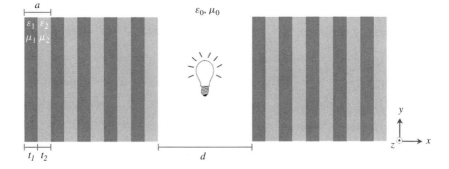

6.10 Omnidirectional reflection:

Imagine now creating a long hollow cylindrical structure composed of concentric annular dielectric layers made of tellurium and polystyrene with air inside. What happens to a light beam smoothly sent into the hollow tube?

6.11 The two-dimensional triangular photonic crystal shown in Figure 6.8 does not allow the propagation of TE and TM waves, with frequencies within the range $\Delta\omega$, irrespectively of the direction of the wave vector **k** in the plane of periodicity of the structure (which is the xy plane). What happens if the component k_z of the wave vector is no longer zero? Can you think of a way to confine the light so that it does not escape along the z direction?

6.12 From the perspective of manufacturing a photonic crystal with a complete three-dimensional photonic band gap, describe different criteria that you would use to determine which of several candidate structures (consider those in Figures 6.16–6.19) would make the best practical photonic crystal.

7
Phononic Crystals

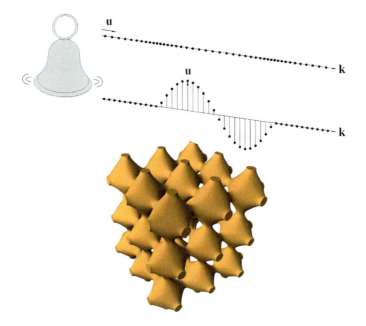

Phononic crystals are periodic structures made of different elastic materials designed to control the propagation of mechanical waves. In this chapter, we introduce the most fundamental property of phononic crystals, that is, the existence of a phononic band gap. Although phononic and photonic crystals have many common characteristics, they also have significant differences because mechanical waves (unlike electromagnetic waves) propagate differently in solid and fluid materials. After introducing some basic theoretical concepts on mechanical waves, we study phononic crystals in one, two, and three dimensions and also show that simple periodic structures such as those presented in Chapter 2 can be usefully employed as phononic crystals.

Periodic Materials and Interference Lithography. M. Maldovan and E. Thomas
Copyright © 2009 WILEY-VCH Verlag GmbH & Co. KGaA, Weinheim
ISBN: 978-3-527-31999-2

7.1
Introduction

Phononic crystals are periodic structures made of two materials with different mechanical properties, which can be used to control mechanical waves. With the purpose of introducing some important basic concepts on the propagation of mechanical waves, it is useful to consider the propagation of these waves in homogeneous materials before studying their propagation in phononic crystals.

One important characteristic of mechanical waves is the fact that there exist different mechanisms for propagation and consequently there exist different types of mechanical waves. The propagation mechanisms depend on the type of material (i.e. solid or fluid) within which the waves propagate. As a consequence, mechanical waves propagating in solid materials are usually called *elastic waves* while those propagating in fluid materials (i.e. gases or liquids) are called *acoustic waves*. Elastic and acoustic waves have different associated characteristics, and we therefore study them separately.

7.1.1
Elastic Waves in Homogeneous Solid Materials

In this section, we consider the propagation of elastic waves in homogeneous *solid* materials. It is important to mention that many solid materials are crystalline. These type of materials are made of a precise periodic arrangement of atoms in space (Figure 7.1a). The atoms in the perfectly ordered crystal are not actually at rest, but are in spontaneous small random oscillation around their equilibrium positions. However, by considering the time average, the atoms can be assumed to be located at their equilibrium positions, which are represented by the black dots in Figure 7.1a.

When an elastic plane wave with wave vector **k** (Chapter 3) propagates within a perfectly ordered crystal, coherent and collective movements of atoms such as those shown in Figures 7.1b and c take place. The amount to which an atom is displaced from its equilibrium position, as a result of the progressive disturbance caused by the elastic plane wave, is represented by the displacement vector **u**(**r**,t). When the atoms move perpendicularly to the direction of propagation of the wave (which is given by the wave vector **k**), the displacement vector **u**(**r**,t) is thus perpendicular to the direction of propagation and the elastic plane wave is called a *transverse* plane wave (Figure 7.1b). On the other hand, when the atoms move along the direction of propagation of the wave, the vector **u**(**r**,t) is parallel to the propagation direction and the elastic plane wave is called a *longitudinal* plane wave (Figure 7.1c).

The description of the propagation of elastic waves within crystalline solids depends on the ratio λ/ℓ between the wavelength λ of the wave and the characteristic distance ℓ between the atoms. In this book, we are interested in wavelengths λ for which $\lambda \gg \ell$ and the solid materials are thus assumed to be continuous. The propagation of transverse and longitudinal elastic plane waves in continuous solid materials is still described by a displacement vector **u**(**r**,t). In this case, however, the vector **u**(**r**,t) represents the instantaneous position of a volume element in the

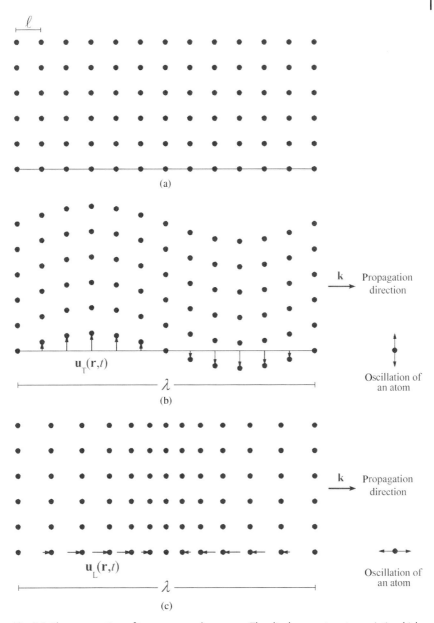

Fig. 7.1 The propagation of transverse and longitudinal elastic plane waves within a two-dimensional crystalline solid material, where it is considered that the waves propagate horizontally with wave vectors **k**. (a) The distribution of atoms (black dots) in the ordered crystalline material. (b) The propagation of a transverse elastic plane wave. The displacement vector $\mathbf{u}_T(\mathbf{r},t)$, which measures the instantaneous displacement of the atoms, is perpendicular to the direction of propagation of the wave. (c) Propagation of a longitudinal elastic plane wave, where the displacement vector $\mathbf{u}_L(\mathbf{r},t)$ is parallel to the direction of propagation of the wave.

solid (which contains many atoms) with respect to its equilibrium position as the wave propagates in the material. As on the atomic level, a transverse elastic plane wave is described by a displacement vector $\mathbf{u}(\mathbf{r},t)$ *perpendicular* to the direction of propagation of the wave, whereas a longitudinal elastic plane wave is described by a vector $\mathbf{u}(\mathbf{r},t)$ *parallel* to the direction of propagation.

An important feature of the propagation of elastic waves within homogeneous solid materials is the fact that transverse and longitudinal elastic plane waves propagate with different velocities and independently of each other. The velocity at which a transverse elastic plane wave propagates in the solid material is denoted by c_T, and the velocity of a longitudinal elastic plane wave is denoted by c_L.

The formulas that describe transverse and longitudinal elastic plane waves propagating with frequency ω within a homogeneous solid material are given by

$$\mathbf{u}_T(\mathbf{r}, t) = \mathrm{Re}(\mathbf{u}_{T0}\, e^{i(\mathbf{k}\cdot\mathbf{r}-\omega t)}) \tag{7.1}$$

$$\mathbf{u}_L(\mathbf{r}, t) = \mathrm{Re}(\mathbf{u}_{L0}\, e^{i(\mathbf{k}\cdot\mathbf{r}-\omega t)}) \tag{7.2}$$

where $\mathbf{u}_T(\mathbf{r},t)$ and $\mathbf{u}_L(\mathbf{r},t)$ are the instantaneous transverse and longitudinal displacement vectors, which are perpendicular and parallel to the wave vector $\mathbf{k}$, respectively (Figures 7.1b and c), Re is an abbreviation for "the real part of", and $\mathbf{u}_{T0}$ and $\mathbf{u}_{L0}$ are the displacement amplitude vectors, which are complex vectors that are constant in space and time.

In addition, the propagation of transverse and longitudinal elastic waves in a homogeneous solid material is governed by the elastic wave equations

$$\nabla^2 \mathbf{u}_T = \frac{1}{c_T^2} \frac{\partial^2 \mathbf{u}_T}{\partial t^2} \tag{7.3}$$

$$\nabla^2 \mathbf{u}_L = \frac{1}{c_L^2} \frac{\partial^2 \mathbf{u}_L}{\partial t^2} \tag{7.4}$$

By replacing the plane waves (7.1 and 7.2) into the wave equations (7.3 and 7.4), we find the relations between the frequency ω and the wave vector $\mathbf{k}$ for transverse and longitudinal elastic plane waves

$$k = |\mathbf{k}| = \frac{\omega}{c_T} \quad \text{for transverse elastic waves} \tag{7.5}$$

$$k = |\mathbf{k}| = \frac{\omega}{c_L} \quad \text{for longitudinal elastic waves} \tag{7.6}$$

Equations 7.5 and 7.6 are the dispersion relations for elastic plane waves propagating within a homogeneous solid material. The velocities c_T and c_L at which transverse and longitudinal elastic waves propagate are determined by the mechanical properties of the underlying solid material. In fact, the velocities c_T and c_L are given by the formulas

$$c_T = \sqrt{\frac{\mu}{\rho}} \qquad c_L = \sqrt{\frac{\lambda + 2\mu}{\rho}} \tag{7.7}$$

where ρ is the density, and λ and μ are the Lame coefficients, which describe the mechanical properties of the homogeneous solid material. The Lame's coefficients are in turn related to more familiar mechanical properties such as Young's modulus E and Poisson's ratio ν (Chapter 8) by the formulas

$$\lambda = \frac{E\nu}{(1+\nu)(1-2\nu)}, \qquad \mu = \frac{E}{2(1+\nu)} \tag{7.8}$$

The parameters ρ, λ, μ, E, and ν describing the mechanical properties of a homogeneous solid material are constant in space and time. In addition, the velocity c_L of a longitudinal elastic plane wave is larger than the velocity c_T of a transverse elastic plane wave.

In brief, in this section we showed that in homogeneous solid materials, mechanical plane waves propagate as elastic plane waves, which are vector waves having in this case either a transverse or a longitudinal character. Furthermore, transverse and longitudinal elastic plane waves propagate independently of each other and at different velocities, which are determined by the mechanical properties of the homogeneous solid material through which the waves propagate.

7.1.2
Acoustic Waves in Homogeneous Fluid Materials

We now consider the propagation of acoustic waves in homogeneous *fluid* materials. The crucial difference between solid and fluid materials is the fact that fluid materials do not support shear deformations and transverse mechanical waves cannot propagate through them. Therefore, fluid materials only allow for the propagation of longitudinal mechanical waves, which are called *acoustic* waves.

It is common practice to describe the propagation of acoustic waves in homogeneous fluid materials in terms of the instantaneous pressure $p = p(\mathbf{r},t)$,[12] which is defined by the formula

$$p(\mathbf{r}, t) = -\lambda \, \nabla \cdot \mathbf{u}(\mathbf{r}, t) \tag{7.9}$$

For example, an acoustic (or pressure) plane wave propagating with frequency ω within a homogeneous fluid material is described by the formula

$$p(\mathbf{r}, t) = \mathrm{Re}(p_0 \, e^{i(\mathbf{k} \cdot \mathbf{r} - \omega t)}) \tag{7.10}$$

where the pressure amplitude p_0 is a complex number constant in space and time.

---

[12] We can use Figure 7.1c to show how a longitudinal wave can be understood as a pressure wave. Note that the number of atoms per unit area at the center of the figure is larger than at both sides. This means that the propagating wave generates regions where the pressure is maximum and others where the pressure is minimum. These periodic regions of maximum and minimum pressure move forward as the longitudinal wave propagates horizontally in the material creating a "pressure wave".

In addition, by using Equation 7.9, we can transform the wave equation (7.4) for longitudinal elastic waves propagating in homogeneous solid materials into the wave equation that governs the propagation of acoustic waves in homogeneous fluid materials

$$\nabla^2 p = \frac{1}{c_0^2} \frac{\partial^2 p}{\partial t^2} \tag{7.11}$$

where $p = p(\mathbf{r},t)$ is the instantaneous pressure in the fluid material, and c_0 is the velocity of the acoustic wave.

By inserting Equation 7.10 into Equation 7.11, we obtain the relation between the frequency ω and the wave vector $\mathbf{k}$ of the acoustic plane wave, which is given by

$$k = |\mathbf{k}| = \frac{\omega}{c_0} \quad \text{for acoustic waves} \tag{7.12}$$

As in the case of solid materials, the velocity c_0 of the acoustic wave is determined by the mechanical properties of the fluid material. Since in the case of fluid materials the Lame's coefficient μ is equal to zero, the velocity of the acoustic wave is given by the formula

$$c_0 = \sqrt{\frac{\lambda}{\rho}} \tag{7.13}$$

In brief, in homogeneous fluid materials, mechanical waves propagate as acoustic (or pressure) waves, which are scalar waves having a longitudinal character. The velocity at which the acoustic waves propagate is determined by the mechanical properties of the fluid material.

7.2
Phononic Crystals

In the last sections, we considered the propagation of mechanical waves in homogeneous materials and showed that mechanical waves propagate differently in solid and fluid materials. We now want to study the propagation of mechanical waves in phononic crystals, which are nonhomogeneous materials made of two different elastic materials. The fact that mechanical waves propagate differently in solid and fluid materials is an important consideration in the study of phononic crystals since these nonhomogeneous materials can be made of a combination of solid–solid, fluid–fluid, or solid–fluid materials. Before studying phononic crystals, we would like to note that many of the theoretical concepts discussed in the following sections are similar to those introduced in Chapter 6 for photonic crystals.

Phononic crystals are periodic structures made of two elastic materials with different mechanical properties (Figure 7.2). The basic property of phononic crystals is that mechanical (either elastic or acoustic) waves, having frequencies within a specific range, are not allowed to propagate within the periodic structure.

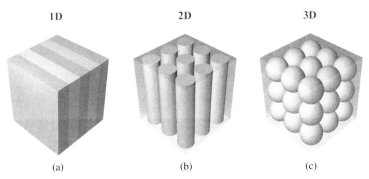

Fig. 7.2 Examples of phononic crystals with periodicities in one, two, and three dimensions. (a) A one-dimensional phononic crystal consisting of elastic layers made of materials with different mechanical properties. (b) A two-dimensional phononic crystal consisting of elastic cylinders in a background elastic material. (c) A three-dimensional phononic crystal consisting of elastic spheres in a background elastic material.

This range of forbidden frequencies is called a *phononic band gap* and it is the central subject of this chapter.

In analogy with the photonic case, the origin of phononic band gaps in periodic elastic structures lies in the multiple scattering of a mechanical wave at the interfaces between materials with different mechanical properties.

Consider, for example, a periodic structure having a phononic band gap. As a consequence of the existence of the phononic band gap, when a mechanical wave with frequency within the phononic band gap is incident on the surface of the phononic crystal, the mechanical wave is totally reflected by the crystal. This is because the wave is not allowed to propagate within the periodic structure. On the other hand, if the mechanical wave is generated *inside* of the phononic crystal, its propagation (or existence) is prohibited.

As in the case of photonic crystals, since the formation of phononic band gaps is based on diffraction, the wavelengths of the mechanical waves that are not allowed to propagate within the phononic crystal are on the order of the spatial periodicity of the structure (e.g. the thickness of the layers or the distance between the cylinders or spheres in Figure 7.2). For example, phononic crystals with periodicities on the order of meters to centimeters will forbid the propagation of mechanical waves with frequencies in the range $(20-20 \times 10^3$ Hz), whereas those with periodicities on the order of microns will forbid the propagation of hypersonic mechanical waves with frequencies in the range $(10^9-10^{12}$ Hz).

Before studying the existence of phononic band gaps in one-, two-, and three-dimensional phononic crystals, we want to mention that, because mechanical waves propagate differently in solid and fluid materials, it is convenient to consider separately phononic crystals made of solid–solid, fluid–fluid, or solid–fluid materials. For example, in the case of solid–solid phononic crystals the mechanical waves propagate (or not) throughout the periodic structure as elastic waves, which are vector waves. In fluid–fluid phononic crystals the mechanical waves propagate (or not) throughout the structure as acoustic waves, which are scalar waves.

On the other hand, in the case of solid–fluid phononic crystals (which is the more complicated case), the mechanical waves propagate as elastic waves in the solid material and as acoustic waves in the fluid material. Therefore, our choice of the materials forming the phononic crystal is critical since this determines the types of mechanical waves propagating within the periodic structure.

7.3
One-dimensional Phononic Crystals

We next consider one-dimensional phononic crystals made of two different *solid* homogeneous materials because this is a reasonable choice for their experimental realization. As we showed in Section 7.1.1, a homogeneous solid material can be characterized by its density ρ, and the transverse and longitudinal velocities c_T and c_L at which elastic waves propagate in the material. Note that by inverting Equations 7.7 and 7.8, a homogeneous solid material can also be characterized by the density ρ and the Lame's coefficients λ and μ or, equivalently, by the density ρ, the Young's modulus E, and the Poisson's ratio ν. This means that the solid materials forming the one-dimensional phononic crystal can be characterized by three different sets of mechanical properties, that is, $\{\rho, c_T, c_L\}$, $\{\rho, \lambda, \mu\}$, or $\{\rho, E, \nu\}$. In this book, we choose to characterize homogeneous solid materials by using the set of mechanical properties $\{\rho, c_T, c_L\}$.

7.3.1
Finite Periodic Structures

The simplest phononic crystal is a one-dimensional periodic structure made of alternate homogeneous solid layers with densities ρ_1 and ρ_2, transverse velocities c_{T1} and c_{T2}, and longitudinal velocities c_{L1} and c_{L2} embedded in a homogeneous solid material characterized by the mechanical parameters ρ_0, c_{T0}, and c_{L0} (Figure 7.3). The thicknesses of the layers along the x direction are given by t_1 and t_2. Because the structure is one dimensional, the homogeneous solid layers extend infinitely in the yz plane and the mechanical properties of the phononic crystal $\rho = \rho(x)$, $c_T = c_T(x)$, and $c_L = c_L(x)$ vary periodically along a single direction in space (x direction in our case).

We consider that the elastic waves are normally incident on the surface of the one-dimensional phononic crystal. Since the one-dimensional phononic crystal is made of two solid materials, mechanical waves propagate throughout the periodic structures as elastic waves. In complete analogy with the electromagnetic case, consider an elastic (either transverse or longitudinal) plane wave that is normally incident on the surface of a phononic crystal having a phononic band gap (Figure 7.3). If the frequency ω of the incident wave is outside of the phononic band gap, part of the wave is reflected while part is transmitted through the structure and emerges on the other side. On the other hand, if the frequency ω of the incident wave is inside of the phononic band gap, the wave is not allowed to propagate

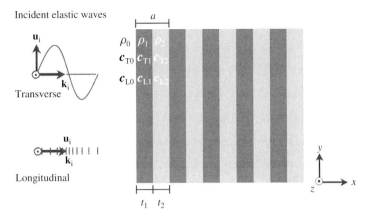

Fig. 7.3 A finite one-dimensional phononic crystal. The periodic structure consists of 10 alternate homogeneous solid layers with densities ρ_1, and ρ_2, transverse and longitudinal velocities c_{T1}, c_{T2}, c_{L1}, and c_{L2}, and thicknesses t_1, and t_2, respectively. The mechanical properties of the phononic crystal $\rho = \rho(x)$, $c_T = c_T(x)$, and $c_L = c_L(x)$ vary periodically along the x direction. Elastic waves (either transverse or longitudinal) are incident perpendicularly to the layers from a homogeneous solid medium characterized by ρ_0, c_{T0}, and c_{L0}.

within the structure and it is therefore completely reflected. In this case, there is no transmitted wave on the other side of the structure.

Because the elastic waves are incident perpendicularly to the layers, the transverse or longitudinal character of the waves remains unchanged. That is, a normally incident transverse elastic wave is reflected by the structure (or propagates within the structure) as a transverse elastic wave and the same is valid for a normally incident longitudinal wave.

The reflectance R of the phononic crystal is defined as the ratio between the intensities of the reflected and incident elastic waves (Chapter 3). As in the case of photonic crystals, the calculation of the reflectance R can help us determine whether or not an elastic wave with frequency ω is allowed to propagate within the structure, which in turn determines the existence of a phononic band gap.

The incident and reflected elastic plane waves are described by the formulas

$$\mathbf{u}_i(\mathbf{r}, t) = \mathrm{Re}(\mathbf{u}_{0i}\, e^{i(\mathbf{k}_i \cdot \mathbf{r} - \omega t)}) \quad \text{incident wave} \tag{7.14}$$

$$\mathbf{u}_r(\mathbf{r}, t) = \mathrm{Re}(\mathbf{u}_{0r}\, e^{i(\mathbf{k}_r \cdot \mathbf{r} - \omega t)}) \quad \text{reflected wave} \tag{7.15}$$

and the reflectance R is thus given by

$$R = \frac{|\mathbf{u}_{0r}|^2}{|\mathbf{u}_{0i}|^2} \tag{7.16}$$

where $\mathbf{u}_{0r}$ and $\mathbf{u}_{0i}$ are the displacement amplitude vectors of the reflected and incident elastic plane waves, respectively.

The reflectance R depends on the structural and mechanical parameters of the phononic crystal and its surroundings, which are given by ρ_0, ρ_1, ρ_2, c_{T0}, c_{T1}, c_{T2},

c_{L0}, c_{L1}, c_{L2}, t_1, t_2, and the number of layers. For a given phononic crystal, the reflectance R also depends on the frequency ω of the normally incident elastic plane wave (Figure 7.4). Because there is no fundamental length scale to the basic physics (Section 6.2.1), it is convenient to plot the reflectance R as a function of the nondimensional variable $\omega a/2\pi c_T$, where c_T is the transverse velocity of one of the materials forming the phononic crystal, and $a = t_1 + t_2$ is the lattice constant of the phononic crystal. This allows us to obtain the reflectance R

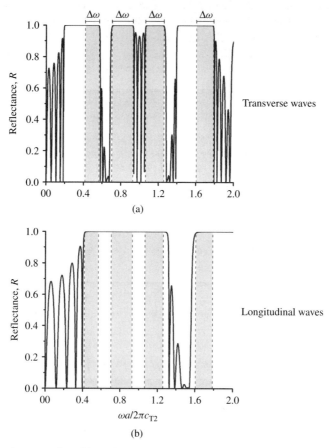

Fig. 7.4 A plot of the reflectance R as a function of the nondimensional frequency $\omega a/2\pi c_{T2}$ for elastic waves that are normally incident on the surface of a one-dimensional phononic crystal. The crystal consists of 10 alternate solid layers made of lead and epoxy; $\rho_1 = 10\,760$ kg m^{-3}, $\rho_2 = 1140$ kg m^{-3}, $c_{T1} = 850$ m s^{-1}, $c_{T2} = 1300$ m s^{-1}, $c_{L1} = 1960$ m s^{-1}, and $c_{L2} = 2770$ m s^{-1} and thicknesses $t_1 = t_2 = 0.5a$. It is assumed that the surrounding material in contact with the structure is nylon. (a) Reflectance for normal incident transverse waves. (b) Reflectance for normal incident longitudinal waves. For frequencies within the ranges $\Delta\omega$ marked in gray, the reflectance R is larger than 0.99 for both transverse and longitudinal incident waves. For these frequency ranges a normally incident elastic wave is totally reflected by the phononic crystal independently whether is transverse or longitudinal.

for phononic crystals with different lattice constants a from a single graph, just as we did for photonic crystals (Section 6.2.1).

Figure 7.4 shows the reflectance R as a function of $\omega a/2\pi c_{T2}$ for normal incident elastic plane waves on a one-dimensional phononic crystal consisting of 10 solid layers with densities $\rho_1 = 10\,760$ kg m^{-3} and $\rho_2 = 1140$ kg m^{-3}, velocities $c_{T1} = 850$ m s^{-1}, $c_{T2} = 1300$ m s^{-1}, $c_{L1} = 1960$ m s^{-1}, and $c_{L2} = 2770$ m s^{-1}, and thicknesses $t_1 = t_2 = 0.5a$. These values correspond to the choice of lead and epoxy for the material layers, respectively. The solid surrounding material is assumed to be nylon, with $\rho_0 = 1110$ kg m^{-3}, $c_{T0} = 1100$ m s^{-1}, and $c_{L0} = 2600$ m s^{-1}. The choice of these materials is based on the fact that this one-dimensional phononic crystal has been realized experimentally (Reference 3). The values of the reflectance R shown in Figure 7.4 were obtained by using the MATLAB code provided in Appendix C.

We can see in Figure 7.4 the existence of frequency ranges $\Delta\omega$ for which the reflectance of normally incident elastic waves on the phononic crystal is nearly equal to 1 for both transverse and longitudinal elastic waves. This means that a normally incident elastic wave with frequency ω within the $\Delta\omega$ ranges is totally reflected (independent of whether is a transverse or a longitudinal wave), and it is therefore not allowed to propagate within the structure. *The frequency ranges $\Delta\omega$ for which elastic waves are not allowed to propagate within the structure are called the phononic band gaps.* In this case, the phononic band gaps $\Delta\omega$ determine the frequency ranges for which both transverse and longitudinal elastic waves are not allowed to propagate within the structure. This means that a general elastic wave (which is made of the sum of transverse and longitudinal elastic waves) with frequency ω within the phononic band gaps $\Delta\omega$ cannot propagate in the structure.

We next examine some particular cases. For example, if the incident wave is a purely transverse elastic wave with frequency ω such as $\omega a/2\pi c_{T2} = 0.3$, Figure 7.4a establishes that the wave will be totally reflected by the phononic crystal because the reflectance for transverse waves at that particular frequency is $R \approx 1$. On the other hand, a purely longitudinal elastic wave with frequency ω such as $\omega a/2\pi c_{T2} = 0.3$ will be allowed to propagate within the structure because $R < 1$ (Figure 7.4b). In this case, it is said that the structure possesses a phononic band gap only for transverse waves. As a result, a general normal incident elastic wave will be allowed to propagate because the structure does not possess a phononic band gap for both transverse and longitudinal waves. (Note that the inverse case occurs for elastic waves with frequencies ω such as $\omega a/2\pi c_{T2} = 0.6$.)

The frequency ranges $\Delta\omega$ of the phononic band gaps depend on the specific structural parameters of the phononic crystal. However, the following statements are valid in general:

- The larger the number of layers, the larger the width of the phononic band gaps.
- The larger the ratio between the densities or the velocities of the elastic waves in the respective layers, the larger the width of the phononic band gaps.

7.3.2
Infinite Periodic Structures

As mentioned in Section 6.2.2, there exists an alternative and powerful approach for the theoretical evaluation of phononic band gaps, which is to assume that the phononic crystal is infinite. That is, *all* space is considered to be filled with the one-dimensional periodic structure. This assumption allows us to use relatively simple numerical techniques to establish the existence of phononic band gaps (especially in the three-dimensional case), and it is also more appropriate because it deals directly with the propagation of waves within the phononic crystal. Note that, by definition, phononic band gaps are frequency ranges $\Delta\omega$ for which mechanical waves are not allowed to propagate *within* the periodic structure.

In this section, we assume that the one-dimensional phononic crystal extends infinitely in space and we are interested in determining whether or not an elastic wave with frequency ω is allowed to propagate within the infinite periodic structure.

We have seen that for an elastic wave propagating within a *homogeneous* solid material, the relation between the frequency ω and the wave vector $\mathbf{k}$ is given by Equation 7.5 or 7.6 depending on whether the wave is transverse or longitudinal. On the other hand, for an elastic wave propagating within a *nonhomogeneous* solid material (such as a solid–solid phononic crystal), the relation between the frequency ω and the wave vector $\mathbf{k}$ is determined by a complicated function $\omega = \omega(\mathbf{k})$. As in the case of photonic crystals, we next show that calculating and plotting the dispersion relation $\omega = \omega(\mathbf{k})$ allows us to explore for the presence of phononic band gaps.

In infinite solid–solid phononic crystals (including one-, two-, and three-dimensional periodic structures), elastic waves do not propagate as plane waves but they propagate as Bloch waves. An elastic Bloch wave, propagating with wave vector $\mathbf{k}$, is represented by the formula

$$\mathbf{u_k}(\mathbf{r}, t) = \mathrm{Re}\left[\mathbf{f_k}(\mathbf{r})\, e^{i(\mathbf{k}\cdot\mathbf{r} - \omega(\mathbf{k})t)}\right] \qquad (7.17)$$

where $\mathbf{f_k}(\mathbf{r})$ is a periodic vector function, with the same spatial period as the phononic crystal, that depends on the particular value of the wave vector $\mathbf{k}$. This periodic vector function satisfies $\mathbf{f_k}(\mathbf{r}) = \mathbf{f_k}(\mathbf{r} + \mathbf{R})$, where $\mathbf{R}$ is the vector that generates the point lattice associated with the phononic crystal (Section 1.2). Importantly, in the more general case, an elastic Bloch wave propagating within a solid–solid phononic crystal is neither transverse nor longitudinal.

It is useful to group the spatial and temporal dependences and write the Bloch wave (7.17) as

7.3 One-dimensional Phononic Crystals

$$\mathbf{u_k}(\mathbf{r}, t) = \mathrm{Re}\left[\mathbf{u_k}(\mathbf{r})\, e^{-i\omega(\mathbf{k})t}\right] \quad \text{where} \quad \mathbf{u_k}(\mathbf{r}) = \mathbf{f_k}(\mathbf{r})\, e^{i\mathbf{k}\cdot\mathbf{r}} \qquad (7.18)$$

The dispersion relation $\omega = \omega(\mathbf{k})$ for elastic waves propagating in infinite solid–solid phononic crystals can now be obtained by numerically solving the elastic wave equation for nonhomogeneous solid materials

$$\nabla \cdot \left(\rho c_T^2 \nabla u_i + \rho c_T^2 \frac{\partial \mathbf{u}}{\partial x_i}\right) + \frac{\partial}{\partial x_i}\left((\rho c_L^2 - 2\rho c_T^2)\nabla \cdot \mathbf{u}\right) = -\rho(\omega(\mathbf{k}))^2 u_i \qquad (7.19)$$

where $\rho = \rho(\mathbf{r})$, $c_T = c_T(\mathbf{r})$, and $c_L = c_L(\mathbf{r})$ are, respectively, the density and the transverse and longitudinal velocities, which describe the mechanical properties of the phononic crystal, and $\mathbf{u} = \mathbf{u_k}(\mathbf{r})$ is the spatial part of the displacement vector of the elastic Bloch wave propagating with wave vector $\mathbf{k}$ within the crystal.

The numerical solution of Equation 7.19 establishes the frequency $\omega = \omega(\mathbf{k})$ that corresponds to a given elastic Bloch wave propagating with wave vector $\mathbf{k}$ within the infinite solid–solid phononic crystal characterized by $\rho(\mathbf{r})$, $c_T(\mathbf{r})$, and $c_L(\mathbf{r})$. Furthermore, the solution of Equation 7.19 also determines the displacement vector $\mathbf{u_k}(\mathbf{r})$, and hence the instantaneous displacement vector $\mathbf{u_k}(\mathbf{r},t)$ of the elastic Bloch wave within the crystal.

In particular, in this section, we consider infinite one-dimensional solid–solid phononic crystals made of alternate homogeneous solid layers, where the elastic waves propagate perpendicular to the layers with wave vectors $\mathbf{k} = k_x \hat{\mathbf{x}}$ (Figure 7.5). In this particular case, the wave equation (7.19) reduces to

$$\frac{\partial}{\partial x}\left(\rho c_L^2 \frac{\partial u_x}{\partial x}\right) = -\rho(\omega(\mathbf{k}))^2 u_x \qquad (7.20a)$$

$$\frac{\partial}{\partial x}\left(\rho c_T^2 \frac{\partial u_y}{\partial x}\right) = -\rho(\omega(\mathbf{k}))^2 u_y \qquad (7.20b)$$

$$\frac{\partial}{\partial x}\left(\rho c_T^2 \frac{\partial u_z}{\partial x}\right) = -\rho(\omega(\mathbf{k}))^2 u_z \qquad (7.20c)$$

where $\rho = \rho(x)$, $c_T = c_T(x)$, and $c_L = c_L(x)$ are the mechanical properties of the one-dimensional phononic crystal, and $u_x = u_x(x)$, $u_y = u_y(x)$, and $u_z = u_z(x)$ are the Cartesian components of the spatial displacement vector $\mathbf{u} = \mathbf{u_k}(x)$. Note that the wave equations (7.20) for the displacement vector components u_x, u_y, and u_z are independent. This means that elastic waves can propagate independently within the one-dimensional phononic crystal as longitudinal elastic waves (which are given by u_x) or transverse elastic waves (which are given by u_y and u_z). Furthermore, the wave equations (7.20b and c) show that transverse elastic waves are doubly degenerate. This is consistent with the symmetry of the one-dimensional phononic crystal (Figure 7.5), where the propagation of a transverse elastic wave with wave vector $\mathbf{k} = k_x \hat{\mathbf{x}}$ and a displacement component u_y oriented along the y direction is equivalent to the propagation of a transverse elastic wave with wave vector $\mathbf{k} = k_x \hat{\mathbf{x}}$ and a displacement component u_z along the z

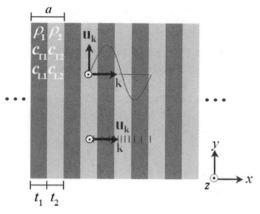

Fig. 7.5 An infinite one-dimensional solid–solid phononic crystal. The periodic structure consists of an infinite arrangement of alternate homogeneous solid layers with densities ρ_1 and ρ_2, transverse and longitudinal velocities c_{T1}, c_{T2}, c_{L1}, and c_{L2}, and thicknesses t_1 and t_2. In infinite phononic crystals, the presence of phononic band gaps is studied by calculating and plotting the dispersion relation $\omega = \omega(\mathbf{k})$ for elastic waves propagating *within* the phononic crystal.

direction (Note: this statement is valid only for waves propagating perpendicularly to the layers).

We numerically solve the wave equations (7.20) and show in Figure 7.6 the dispersion relations $\omega = \omega(\mathbf{k})$ for transverse and longitudinal elastic waves propagating with wave vectors $\mathbf{k} = k_x\hat{\mathbf{x}}$ within an infinite one-dimensional solid–solid phononic crystal made of lead and epoxy layers (see the mechanical parameters for lead and epoxy in Section 7.3.1). The solid layers have thicknesses $t_1 = t_2 = 0.5a$, where $a = t_1 + t_2$ is the lattice constant of the phononic crystal (Figure 7.5). Because transverse and longitudinal elastic waves propagate independently within the crystal, we plot the dispersion relations $\omega = \omega(\mathbf{k})$ for transverse and longitudinal waves in separate graphs. Nondimensional frequencies $\omega a / 2\pi c_{T2}$ are plotted as a function of k_x, where c_{T2} is the transverse velocity of elastic waves in epoxy. In this case, the dispersion relations are plotted over the one-dimensional Brillouin zone, which extends from $k_x = -\pi/a$ to $k_x = \pi/a$.

From the graphs showing the dispersion relations, we can observe the existence of frequency ranges $\Delta\omega$ on the vertical axis (see gray regions) for which there is no associated wave vector $\mathbf{k}$ on the horizontal axis independent of whether the elastic wave is transverse or longitudinal (Figure 7.6). This means that a general elastic wave (made of the sum of a transverse and a longitudinal wave) with frequency within these $\Delta\omega$ ranges is not allowed to propagate within the phononic crystal. These frequency ranges $\Delta\omega$ are, therefore, *phononic band gaps* for both transverse and longitudinal elastic waves.

We can also notice that for frequencies ω such as $\omega a / 2\pi c_{T2} = 0.3$, the dispersion relations show that there is no associated wave vector $\mathbf{k}$ for transverse elastic waves, but there are associated wave vectors for longitudinal elastic waves. This means that a normally incident transverse elastic wave with frequency ω is not allowed

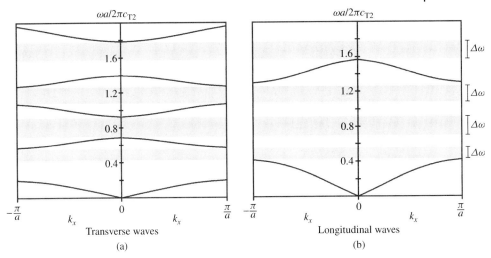

Fig. 7.6 The dispersion relation in one dimension. The graphs show the nondimensional frequencies $\omega a/2\pi c_{T2}$ versus the wave vector k_x for elastic waves propagating within an infinite one-dimensional solid–solid phononic crystal. The phononic crystal consists of alternate lead and epoxy layers with thicknesses $t_1 = t_2 = 0.5a$, where $a = t_1 + t_2$ is the lattice constant of the phononic crystal. It is assumed that the elastic waves propagate perpendicularly to the layers with wave vectors $\mathbf{k} = k_x \hat{\mathbf{x}}$. The quantity c_{T2} represents the transverse velocity of elastic waves in the epoxy material. (a) The dispersion relation for transverse elastic waves. (b) The dispersion relation for longitudinal elastic waves. The frequency gaps $\Delta\omega$ (shown in gray) represent frequency ranges for which both transverse and longitudinal elastic waves are not allowed to propagate within the crystal.

to propagate in the crystal, but a longitudinal elastic wave with the same frequency is allowed to propagate (and it will propagate with the associated wave vector $\mathbf{k}$). In this case, the structure possesses a phononic band gap only for transverse elastic waves. Note that the inverse case occurs for frequencies ω such as $\omega a/2\pi c_{T2} = 1.0$. In these particular cases, the phononic crystal does not possess a phononic band gap for both transverse and longitudinal elastic waves, and thus a general elastic wave is allowed to propagate within the structure.

It is interesting to note the agreement between the phononic band gaps $\Delta\omega$ corresponding to the infinite periodic structure (Figure 7.6) and those corresponding to the finite periodic structure made of only 10 layers (Figure 7.4). Although the frequency ranges $\Delta\omega$ for the finite structure are smaller than those for the infinite structure, the ranges $\Delta\omega$ are comparable. However, it is important to mention that the number of layers in the finite structure required to obtain phononic band gaps similar to those corresponding to the infinite structure depends on the mechanical properties $\rho = \rho(x)$, $c_T = c_T(x)$, and $c_L = c_L(x)$ of the phononic crystal. The larger the contrast between the mechanical properties of the layers, the smaller the number of required layers.

The discussion in Section 6.2.3 about finite versus infinite periodic structures and the oblique incidence of waves is also valid in the case of phononic crystals.

In addition, one-dimensional phononic crystals can never prevent the propagation of mechanical waves parallel to the layers (e.g. along the y and z directions in Figure 7.5) because there is no variation in the mechanical properties $\rho = \rho(x)$, $c_T = c_T(x)$, and $c_L = c_L(x)$ of the phononic crystal along any direction in the yz plane. To have phononic crystals that forbid the propagation of mechanical waves along additional directions in space, we next consider that the mechanical properties $\rho = \rho(\mathbf{r})$, $c_T = c_T(\mathbf{r})$, and $c_L = c_L(\mathbf{r})$ of the phononic crystal vary periodically in two dimensions.

7.4
Two-dimensional Phononic Crystals

As previously mentioned, there exist different types of mechanical waves and different propagation mechanisms depending on whether the two materials forming the phononic crystal are solid–solid, fluid–fluid, or solid–fluid materials. Although two-dimensional phononic crystals made of solid–solid and fluid–fluid materials have been demonstrated to have phononic band gaps, in this section we concentrate on two-dimensional phononic crystals made of solid and fluid materials. This choice is based on the fact that solid–fluid phononic crystals are more accessible in terms of their experimental realization. Imagine the difficulty to form a periodic structure comprising two liquids with different densities. On the other hand, as we showed in the previous chapters, two-dimensional periodic structures made of solid–air materials can be relatively easily created by interference lithography.

Despite the fact that two-dimensional solid–fluid phononic crystals are easy to fabricate, the study of phononic band gaps in these type of phononic crystals is difficult. This is because mechanical waves propagate as elastic waves (which are vector waves) in the solid material and as acoustic waves (which are scalar waves) in the fluid material. As a result, one of the most widely used numerical techniques to obtain the dispersion relation (i.e. the plane-wave method) fails, producing unphysical horizontal frequency bands. Therefore, to apply simple numerical techniques to study the existence of phononic band gaps in solid–fluid phononic crystals, some approximations must be made. For example, in the next two sections, we consider phononic crystals made of vacuum cylinders in a solid background and phononic crystals made of solid cylinders in air.

7.4.1
Vacuum Cylinders in a Solid Background

In this section, we study two-dimensional phononic crystals made of vacuum cylinders in a solid background arranged on the square and triangular lattices (see insets in Figure 7.7). We consider that the vacuum cylinders are aligned along the z direction and that the phononic crystal extends infinitely in the xy plane. The mechanical properties of vacuum are given by $\rho_2 = c_{T2} = c_{L2} = 0$, whereas the mechanical properties of the solid background material are given by ρ_1, c_{T1}, and

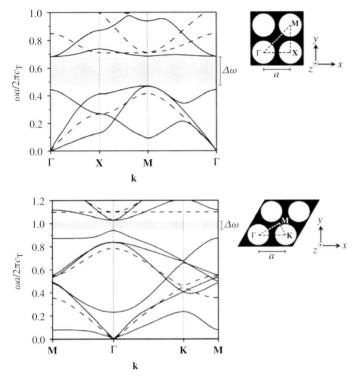

Fig. 7.7 The dispersion relation in two dimensions. The graphs show the nondimensional frequencies $\omega a/2\pi c_T$ versus the wave vector **k** for out-of-plane (dashed lines) and in-plane (solid lines) elastic waves propagating within infinite two-dimensional solid–vacuum phononic crystals. The phononic crystals consist of vacuum cylinders with radius $r = 0.45a$ arranged in the square and triangular lattices, respectively. The background is a solid material with density $\rho = 1190$ kg m^{-3}, and velocities $c_T = 1800$ m s^{-1} and $c_L = 3100$ m s^{-1}. It is assumed that the elastic waves propagate with wave vectors **k** in the xy plane. The square and triangular phononic crystals possess phononic band gaps $\Delta\omega$ for which out-of-plane *and* in-plane elastic waves are not allowed to propagate within the structure irrespectively of their propagation direction in the xy plane (see gray frequency ranges).

c_{L1}. This means that the mechanical properties of the phononic crystal $\rho = \rho(x,y)$, $c_T = c_T(x,y)$, and $c_L = c_L(x,y)$ vary periodically in the xy plane.

Because the two-dimensional phononic crystal is assumed to extend infinitely in the xy plane, we are interested in calculating the dispersion relation $\omega = \omega(\mathbf{k})$ for mechanical waves propagating within the phononic crystal to investigate the presence of phononic band gaps. Here, we assume that the mechanical waves propagate with wave vectors **k** in the xy plane and we want to find whether a two-dimensional phononic crystal can forbid the propagation of mechanical waves independent of their propagation direction in the xy plane.

In solid–vacuum phononic crystals, mechanical waves propagate only through the solid background material because vacuum does not allow the propagation of mechanical waves. This means that in solid–vacuum phononic crystals, mechanical waves must propagate as elastic waves. To calculate and plot the dispersion relation $\omega = \omega(\mathbf{k})$, we therefore consider the wave equation for elastic waves (7.19), which in the two-dimensional case transforms into

$$\frac{\partial}{\partial x}\left(\rho c_L^2 \frac{\partial u_x}{\partial x} + (\rho c_L^2 - 2\rho c_T^2)\frac{\partial u_y}{\partial y}\right) + \frac{\partial}{\partial y}\left(\rho c_T^2 \frac{\partial u_x}{\partial y} + \rho c_T^2 \frac{\partial u_y}{\partial x}\right) = -\rho(\omega(\mathbf{k}))^2 u_x$$
(7.21a)

$$\frac{\partial}{\partial y}\left(\rho c_L^2 \frac{\partial u_y}{\partial y} + (\rho c_L^2 - 2\rho c_T^2)\frac{\partial u_x}{\partial x}\right) + \frac{\partial}{\partial x}\left(\rho c_T^2 \frac{\partial u_y}{\partial x} + \rho c_T^2 \frac{\partial u_x}{\partial y}\right) = -\rho(\omega(\mathbf{k}))^2 u_y$$
(7.21b)

$$\frac{\partial}{\partial x}\left(\rho c_T^2 \frac{\partial u_z}{\partial x}\right) + \frac{\partial}{\partial y}\left(\rho c_T^2 \frac{\partial u_z}{\partial y}\right) = -\rho(\omega(\mathbf{k}))^2 u_z$$
(7.21c)

where $\rho = \rho(x,y)$, $c_T = c_T(x,y)$, and $c_L = c_L(x,y)$ are the mechanical properties of the two-dimensional phononic crystal, and $u_x = u_x(x, y)$, $u_y = u_y(x,y)$, and $u_z = u_z(x,y)$ are the Cartesian components of the spatial displacement vector $\mathbf{u} = \mathbf{u_k}(x,y)$.

Equations 7.21 indicate that elastic waves propagating within a two-dimensional phononic crystal can be separated into two distinct types of waves. For example, Equation 7.21c describes the propagation of an independent transverse elastic wave, which is given by the component u_z of the displacement vector. Note that this elastic wave is transverse because the component u_z of the displacement vector (which oscillates along the z direction) is always perpendicular to the wave vector $\mathbf{k}$ (which is in the xy plane). In addition, this transverse elastic wave propagates independently within the phononic crystal because the components u_x and u_y of the displacement vector do not appear in Equation 7.21c. Independent transverse elastic waves are usually called *out-of-plane* waves because the displacement vector oscillates along the z direction.

On the other hand, Equations 7.21a and b describe the propagation of a mixed elastic wave (neither transverse nor longitudinal) represented by the coupled displacement components u_x and u_y. This elastic wave is mixed because the coupled components u_x and u_y determine that the displacement vector in the xy plane $\mathbf{u}_{//} = (u_x, u_y)$ oscillate neither perpendicular nor parallel to the wave vector $\mathbf{k}$. In fact, the displacement vector $\mathbf{u}_{//}$ may rotate in the xy plane as the wave propagates within the crystal. Because the displacement vector is in the plane of periodicity xy, mixed elastic waves are usually called *in-plane* waves.

We calculate and plot the dispersion relations $\omega = \omega(\mathbf{k})$ for out-of plane (transverse) and in-plane (mixed) elastic waves in Figure 7.7. To obtain the dispersion relations, we separately solve the wave equations (7.21c) and (7.21a and b), respectively. The set of wave vectors $\mathbf{k}$ required to plot the dispersion relations $\omega = \omega(\mathbf{k})$ was established in Chapter 6, where we explained the concepts of reciprocal lattice and Brillouin zone. We encourage readers not familiar with these concepts to read Section 6.3.1.

7.4 Two-dimensional Phononic Crystals

We consider that the background solid material in our two-dimensional solid–vacuum phononic crystal is made of a strongly cross-linked epoxy for which $\rho = 1190$ kg m^{-3}, $c_T = 1800$ m s^{-1}, and $c_L = 3100$ m s^{-1}. The radius of the cylinders is assumed to be $r = 0.45a$.

Note that the band diagrams in Figure 7.7, which correspond to out-of-plane and in-plane elastic waves propagating in the square and triangular phononic crystals, respectively, show a frequency range $\Delta\omega$ for which an elastic wave with frequency ω has no associated wave vector **k**. This means that, for frequencies within these $\Delta\omega$ ranges, neither out-of-plane nor in-plane elastic waves are allowed to propagate within the phononic crystals independent of the direction of propagation of the waves in the xy plane. The frequency range $\Delta\omega$ represents a *complete* phononic band gap for both out-of-plane and in-plane elastic waves, where the word complete means that the respective waves are not allowed to propagate along any direction in the xy plane (see Section 6.3.2 for details).

Note that by considering two-dimensional phononic crystals, we are able to prevent (for certain frequencies) the propagation of *any* elastic wave (either out-of-plane or in-plane) along *any* direction in the plane of periodicity of the phononic crystal (i.e. the xy plane). Some intriguing physical effects arising from this property are discussed in Chapter 9.

It is interesting to examine how the complete phononic band gaps shown in Figure 7.7 change with respect to the radius of the cylinders. Note that since we want the solid material to stay continuous, the upper limit for the radius of the cylinders in these solid–vacuum phononic crystals is $r = 0.5a$. Figure 7.8 shows separately the complete phononic band gaps for out-of-plane (labeled OUT) and in-plane (labeled IN) elastic waves as a function of the radius of the cylinders. Because the

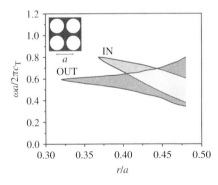
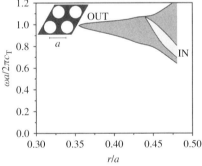

Fig. 7.8 Phononic band gap maps for vacuum cylinders in a solid background material arranged in the square and triangular lattices. The graphs show the range of nondimensional frequencies $\omega a/2\pi c_T$ corresponding to complete phononic band gaps for out-of-plane (OUT) and in-plane (IN) elastic waves as a function of r/a, where r is the radius of the cylinders. When these gaps overlap, the phononic crystal possesses a complete phononic band gap for both transverse and mixed elastic waves (see light gray regions). The solid background material is a strongly cross-linked epoxy with a density $\rho = 1190$ kg m^{-3} and velocities $c_T = 1800$ m s^{-1} and $c_L = 3100$ m s^{-1}.

cylinders are made of vacuum, the complete phononic band gaps for out-of-plane elastic waves do not explicitly depend on the density ρ of the solid material and will exist at the same nondimensional frequencies for *any* solid background material. On the other hand, the complete phononic band gaps for in-plane elastic waves only depend on the c_L/c_T ratio of the solid background material, which is equal to 1.72 in the case of a strongly cross-linked epoxy.

In general, for isotropic homogeneous solid materials, we have $(c_L/c_T)^2 = (2 - 2\nu)/(1 - 2\nu)$, where ν is the Poisson ratio of the solid (Equations 7.7 and 7.8). This means that solid materials with Poisson ratios ν in the typical range of solids 0.250–0.405 have c_L/c_T ratios in the range 1.73–2.50. Importantly, our calculations show that the complete phononic band gaps for in-plane elastic waves do not strongly vary when the homogeneous solid background material of the solid–vacuum phononic crystal has c_L/c_T ratios within this range.

In the case of the square lattice, the complete phononic band gaps for out-of-plane and in-plane elastic waves overlap when the radius of the cylinders is $r > 0.41a$, whereas in the case of the triangular lattice, the complete gaps overlap for $r > 0.44a$. This means that, for these radius values, the square and triangular phononic crystals possess a complete phononic band gap for both out-of-plane (transverse) and in-plane (mixed) elastic waves and a general elastic wave with frequency within these ranges cannot propagate within the phononic crystal independent of its propagation direction in the plane of periodicity.

7.4.2
Solid Cylinders in Air

We next consider two-dimensional solid–fluid phononic crystals made of disconnected solid cylinders in air arranged in the square and triangular lattices (see insets in Figure 7.9). As in the previous section, the cylinders are aligned along the z direction and the phononic crystal extends infinitely in the xy plane. Therefore, the mechanical properties $\rho = \rho(x,y)$, $c_T = c_T(x,y)$, and $c_L = c_T(x,y)$ of the phononic crystal vary periodically in the xy plane.

It is interesting to examine the differences between these solid–fluid phononic crystals and the previously studied solid–vacuum phononic crystals. In the solid–vacuum phononic crystals, the vacuum regions (i.e. cylinders) are disconnected, whereas the solid region (i.e. the background material) is continuous. As a result, mechanical waves propagate as elastic waves within these solid–vacuum phononic crystals. On the other hand, in the solid–fluid phononic crystals shown in Figure 7.9, the solid regions (i.e. cylinders) are disconnected, whereas the fluid region (i.e. air) is continuous. In these solid–fluid phononic crystals, it is assumed that mechanical waves propagate as acoustic (or pressure) waves. This approximation is reasonable because, in these phononic crystals, it can be considered that the mechanical waves propagate mainly through the air region (which is a fluid) due to the rigidity of the cylinders.

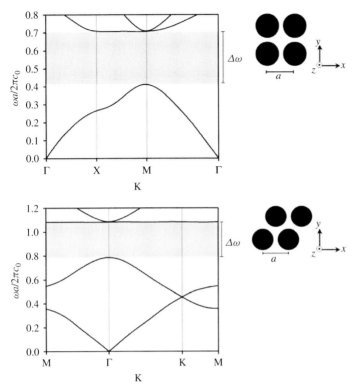

Fig. 7.9 The dispersion relation in two dimensions. The graphs show the nondimensional frequency $\omega a/2\pi c_0$ versus the wave vector **k** for acoustic (pressure) waves propagating within infinite two-dimensional solid–air phononic crystals. The phononic crystals consist of solid cylinders with $r = 0.45a$ arranged in the square and triangular lattices. The background material is assumed to be air, for which the speed of sound is $c_0 = 343$ m s^{-1}. It is assumed that the acoustic waves propagate with wave vectors **k** in the xy plane. These square and triangular phononic crystals possess phononic band gaps $\Delta \omega$ for which acoustic waves are not allowed to propagate within the structure irrespectively of the direction of propagation of the waves in the xy plane (see gray frequency ranges).

Acoustic (or pressure) waves propagate within these solid–fluid phononic crystals as Bloch waves, which are described by the formula

$$p_\mathbf{k}(\mathbf{r}, t) = \mathrm{Re}\left[f_\mathbf{k}(\mathbf{r})\, e^{i(\mathbf{k}\cdot\mathbf{r} - \omega(\mathbf{k})t)} \right] \tag{7.22}$$

where $f_\mathbf{k}(\mathbf{r})$ is a periodic scalar function, with the same spatial period as the underlying phononic crystal, that depends on the particular value of the wave vector **k**. Once again, it is useful to group the spatial and temporal dependences and write the Bloch wave (7.22) as

$$p_\mathbf{k}(\mathbf{r}, t) = \mathrm{Re}\left[p_\mathbf{k}(\mathbf{r})\, e^{-i\omega(\mathbf{k})t} \right] \quad \text{where} \quad p_\mathbf{k}(\mathbf{r}) = f_\mathbf{k}(\mathbf{r})\, e^{i\mathbf{k}\cdot\mathbf{r}} \tag{7.23}$$

7 Phononic Crystals

The propagation of these acoustic Bloch waves within the phononic crystal is governed by the acoustic wave equation

$$\nabla \cdot \left(\frac{1}{\rho}\nabla p\right) = -\frac{1}{\rho c_L^2}(\omega(\mathbf{k}))^2 p \tag{7.24}$$

where $\rho = \rho(\mathbf{r})$ is the density, $c_L = c_L(\mathbf{r})$ is the longitudinal velocity of acoustic waves within the phononic crystal, and $p = p_\mathbf{k}(\mathbf{r})$ is the spatial part of the pressure $p_\mathbf{k}(\mathbf{r},t)$ of an acoustic Bloch wave propagating with wave vector $\mathbf{k}$ within the phononic crystal.

In particular, we consider two-dimensional phononic crystals and Equation 7.24 transforms into

$$\frac{\partial}{\partial x}\left(\frac{1}{\rho}\frac{\partial p}{\partial x}\right) + \frac{\partial}{\partial y}\left(\frac{1}{\rho}\frac{\partial p}{\partial y}\right) = -\frac{1}{\rho c_L^2}(\omega(\mathbf{k}))^2 p \tag{7.25}$$

where $\rho = \rho(x,y)$ and $c_L = c_L(x,y)$ are the mechanical properties of the two-dimensional phononic crystal, and p is the spatial pressure $p_\mathbf{k}(x,y)$.

We numerically solve the wave equation (7.25) and show in Figure 7.9 the dispersion relations $\omega = \omega(\mathbf{k})$ for acoustic waves propagating within phononic crystals made of solid cylinders ($r = 0.45a$) in air arranged in the square and triangular lattices, respectively. The acoustic waves are assumed to propagate with wave vectors $\mathbf{k}$ in the xy plane and the mechanical properties of the solid cylinders are given by $\rho_2 = 2580$ kg m^{-3} and $c_{L2} = 5960$ m s^{-1}, whereas the mechanical properties of air (20° C, atm) are $\rho_1 = 1.29$ kg m^{-3} and $c_{L1} = 343$ m s^{-1}. The longitudinal velocity c_{L1} is usually noted as the speed of sound in air c_0.

We can see in Figure 7.9 the existence of frequency ranges $\Delta\omega$ (see gray regions) for which acoustic waves are not allowed to propagate within the square and triangular phononic crystals independent of their propagation direction in the xy plane. These frequency ranges $\Delta\omega$ are *complete* phononic band gaps for acoustic waves. Note that the phononic band gap for acoustic waves in solid–air phononic crystals (Figure 7.9) are larger than those for elastic waves in solid–vacuum phononic crystals (Figure 7.7). Significantly, by recalling the results from the previous section, we can conclude that depending on the choice of the constituent materials, two-dimensional phononic crystals can present phononic band gaps either for elastic waves (which propagate in solid materials as transverse or mixed waves) or for acoustic waves (which propagate in fluid materials as longitudinal waves).

Figure 7.10 shows the complete phononic band gaps $\Delta\omega$ for acoustic waves as a function of the radius of the cylinders. It is important to remark that because of the large difference between the densities of typical solid materials and air, these *universal* gap maps fairly describe the complete acoustic phononic band gaps corresponding to two-dimensional phononic crystals made of disconnected solid cylinders (made of *any* material) in air.[13]

---

[13] If the air region is filled with a different fluid material such as a typical liquid, the density contrast between the cylinders and the fluid decreases substantially and the phononic band gaps may narrow.

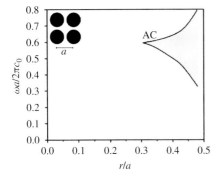 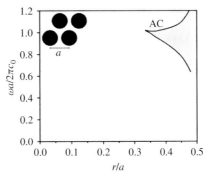

Fig. 7.10 Phononic band gap maps for solid cylinders in air arranged in the square and triangular lattices. The graphs show the range of nondimensional frequencies $\omega a/2\pi c_0$ corresponding to complete phononic band gaps for acoustic (AC) waves as a function of r/a, where r is the radius of the cylinders and $c_0 = 343$ m s^{-1} is the speed of sound in air. For these frequency ranges, the phononic crystals do not allowed the propagation of acoustic waves irrespectively of their propagation direction in the plane of periodicity of the phononic crystal.

7.4.3
Phononic Band Gaps in Two-dimensional Simple Periodic Structures

To conclude the study of phononic band gaps in two-dimensional phononic crystals, we show an example of a solid–vacuum phononic crystal that can be fabricated through interference lithography and possesses a complete phononic band gap for out-of-plane (transverse) and in-plane (mixed) elastic waves. The periodic structure is represented by the analytical formula

$$f(x,y) = \cos\left[\frac{2\pi x}{a}\right] + \cos\left[\frac{2\pi y}{a}\right] + 0.4\cos\left[\frac{2\pi}{a}(x+y)\right]$$
$$+ 0.4\cos\left[\frac{2\pi}{a}(x-y)\right] + 0.7 \quad (7.26)$$

This phononic crystal is shown in Figure 7.11a, where white and gray regions are assumed to be vacuum and solid material, respectively. Owing to the strong similarity between this structure and the structure made of vacuum cylinders in a solid background arranged in the square lattice, the structure is expected to possess a phononic band gap for elastic waves propagating along any direction in the xy plane (Figure 7.8). Figure 7.11b shows the dispersion relation $\omega = \omega(\mathbf{k})$ showing the presence of such a complete phononic band gap.

This example shows that simple periodic structures, which are defined by a few trigonometric terms, together with the interference lithography technique present an interesting method to fabricate two-dimensional phononic crystals. The experimental realization of a phononic crystal representing an arrangement of cylinders in the square lattice is shown in Figure 7.12.

In summary, in this section, we showed that specific two-dimensional phononic crystals made of two different elastic materials can forbid the propagation of

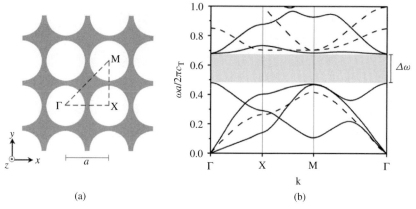

Fig. 7.11 (a) The phononic crystal defined by Equation 7.26, where white and gray regions are vacuum and solid material, respectively. This periodic structure can be fabricated by interference lithography. (b) The dispersion relation in two dimensions. The graph shows the nondimensional frequencies $\omega a/2\pi c_T$ versus the wave vector **k** for out-of-plane (dashed lines) and in-plane (solid lines) elastic waves propagating within the phononic crystal with wave vectors **k** in the xy plane. The solid material has density $\rho = 1190$ kg m^{-3} and velocities $c_T = 1800$ m s^{-1} and $c_L = 3100$ m s^{-1}. The phononic crystal possesses a complete phononic band gap $\Delta\omega$ for both out-of-plane and in-plane elastic waves (see gray frequency ranges).

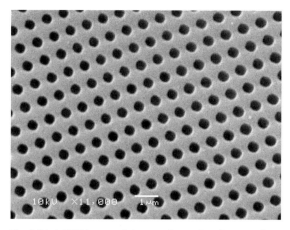

Fig. 7.12 A SEM image of the experimental realization of an interference lithography structure resembling a square lattice of air cylinders in a photoresist material.

mechanical waves, with specific frequencies, along *any* direction in the plane of periodicity. In particular, phononic crystals made of vacuum cylinders in a solid background possess complete phononic band gaps for *elastic* waves, while solid cylinders in air possess complete phononic band gaps for *acoustic* waves. The disadvantage of two-dimensional phononic crystals is the fact that they can never prevent the propagation of mechanical waves along the z direction because there is no variation of the mechanical properties of the phononic crystal along this direction. In order to obtain phononic crystals that forbid the propagation of mechanical waves along additional directions in space, we next consider that the mechanical properties $\rho = \rho(\mathbf{r})$, $c_T = c_T(\mathbf{r})$, and $c_L = c_L(\mathbf{r})$ of the phononic crystal vary periodically in three dimensions.

7.5
Three-dimensional Phononic Crystals

Before considering phononic band gaps in three dimensions, once again we pay attention to the type of materials comprising the phononic crystal. We next discuss some issues associated with solid–solid, solid–fluid, and fluid–fluid phononic crystals in three dimensions.

Three-dimensional fluid–fluid phononic crystals: This type of structure requires a three-dimensional distribution of two different liquids, which have to have a periodic arrangement in space. Needless to say, these periodic structures are very difficult to obtain experimentally. Even though complete acoustic phononic band gaps in fluid–fluid phononic crystals have been demonstrated by means of theoretical calculations, the difficulties associated with their experimental realization strongly limits the applications of these structures.

Three-dimensional solid–fluid phononic crystals: Note that in order to have a three-dimensional self-supporting solid–fluid phononic crystal, the solid material must be continuous throughout the periodic structure. For example, a three-dimensional phononic crystal made of disconnected air spheres in a solid background is a self-supporting structure. But a phononic crystal made of disconnected solid spheres in a fluid background is not a self-supporting structure and collapses because of gravity. A third situation occurs when the solid and the fluid materials are both continuous (i.e. bicontinuous structures). These structures are certainly self-supported. We have seen in previous chapters that the fabrication of this type of structure by interference lithography is relatively easy. However, the theoretical examination of phononic band gaps by means of the dispersion relation is complicated because, as previously discussed, mechanical waves propagate as elastic waves in the solid material and as acoustic waves in the fluid material. The study of the existence of phononic band gaps in three-dimensional solid–fluid periodic structures is thus not extensive due to the fact that relatively complex numerical techniques must be used. In the future, it would be very interesting to examine the presence of phononic band gaps in these types of periodic structures that can present large contrast between the mechanical properties of the constituent materials.

Three-dimensional solid–solid phononic crystals: These types of structures are not easy to fabricate but some specific architectures have been experimentally realized and tested. Note that in this case the structures are always self-supported since they are made of two different solid materials. The advantage of solid–solid phononic crystals is that simple numerical techniques can be used to investigate the existence of phononic band gaps. In the next section, we focus on this solid–solid case and

7.5.1
Solid Spheres in a Solid Background Material

Figure 7.13a shows an example of a three-dimensional solid–solid phononic crystal made of solid spheres in a solid background material (not shown for clarity) arranged in the face-centered-cubic (fcc) Bravais lattice. The mechanical properties of the solid material forming the spheres are given by the density ρ_1 and the velocities c_{T1} and c_{L1}, whereas the solid background material has density ρ_2 and velocities c_{T2} and c_{L2}. Thus, the mechanical parameters of the phononic crystal $\rho = \rho(x,y,z)$, $c_T = c_T(x,y,z)$, and $c_L = c_L(x,y,z)$ vary periodically in three dimensions.

In three-dimensional solid–solid phononic crystals, mechanical waves propagate as elastic waves and their propagation and dispersion relation is governed by the wave equation (7.19). In this case, elastic waves are neither transverse nor longitudinal because the displacement vector components u_x, u_y, and u_z are, in general, coupled to each other.

We now show two representative examples of solid–solid phononic crystals that can be found in the literature (see References 27 and 30). In our first example, we consider that the spheres in the phononic crystal shown in Figure 7.13a are

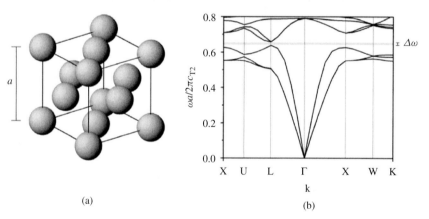

Fig. 7.13 (a) A three-dimensional phononic crystal made of gold spheres ($r = 0.18a$) arranged in the face-centered-cubic Bravais lattice. The solid background material is silicon, which is not shown for clarity. The volume fraction of gold within the structure is 10%. The density and transverse and longitudinal velocities for gold and silicon are, respectively, $\rho_1 = 19\,500$ kg m^{-3}, $c_{T1} = 1240$ m s^{-1}, $c_{L1} = 3360$ m s^{-1}, $\rho_2 = 2330$ kg m^{-3}, $c_{T2} = 5359$ m s^{-1}, and $c_{L2} = 8950$ m s^{-1}. (b) The dispersion relation $\omega = \omega(\mathbf{k})$ for elastic waves propagating within the structure. The graph shows a frequency range $\Delta\omega$ for which elastic waves are not allowed to propagate within the structure independently of the direction of propagation in the three-dimensional space. The structure possesses a *complete* three-dimensional phononic band gap.

made of gold and that the background material is made of silicon. Gold has density $\rho_1 = 19\,500\,\text{kg m}^{-3}$ and transverse and longitudinal velocities $c_{T1} = 1240\,\text{m s}^{-1}$ and $c_{L1} = 3360\,\text{m s}^{-1}$, whereas silicon has density $\rho_2 = 2330\,\text{kg m}^{-3}$ and velocities $c_{T2} = 5359\,\text{m s}^{-1}$ and $c_{L2} = 8950\,\text{m s}^{-1}$. The radius of the spheres is assumed to be $r = 0.18a$.

Figure 7.13b shows the dispersion relation $\omega = \omega(\mathbf{k})$ for elastic waves propagating within this phononic crystal, where nondimensional frequencies $\omega a/2\pi c_{T2}$ are plotted as a function of the wave vector $\mathbf{k}$ along directions in the edges of the Brillouin zone that corresponds to the fcc Bravais lattice (Section 6.4). We can see in Figure 7.13b the existence of a range of frequencies $\Delta\omega$ for which there is no associated wave vector $\mathbf{k}$. Note that, in contrast to the one- and two-dimensional cases, the wave vectors $\mathbf{k}$ in Figure 7.13b have different directions in three dimensions. As a result, elastic waves with frequencies within the range $\Delta\omega$ are not allowed to propagate within the structure irrespectively of their propagation direction in three dimensions. In this case, it is said that the phononic crystal possesses a *complete* three-dimensional phononic band gap. This remarkable physical property can only be achieved by considering phononic crystals where the mechanical properties $\rho = \rho(x,y,z)$, $c_T = c_T(x,y,z)$, and $c_L = c_L(x,y,z)$ vary periodically in three dimensions.

As a second example, we consider that the spheres are made of lead and the background material is made of epoxy. Lead has density $\rho_1 = 11\,357\,\text{kg m}^{-3}$ and transverse and longitudinal velocities $c_{T1} = 860\,\text{m s}^{-1}$ and $c_{L1} = 2158\,\text{m s}^{-1}$, whereas epoxy has density $\rho_2 = 1180\,\text{kg m}^{-3}$ and velocities $c_{T2} = 1160\,\text{m s}^{-1}$ and $c_{L2} = 2540\,\text{m s}^{-1}$. In this case, the radius of the spheres is assumed to be $r = 0.25a$. Figure 7.14b shows the dispersion relation $\omega = \omega(\mathbf{k})$ for elastic waves propagating within the structure, where we can observe the existence of a larger complete three-dimensional phononic band gap.

Note that the complete three-dimensional phononic band gap shown in Figure 7.14b is centered around $\omega a/2\pi c_{T2} = 1.03$. This means that, for example, in order to forbid the propagation of elastic waves with frequency $f = 100$ MHz, the lattice constant a of the periodic structure must be equal to $a = 11.95\,\mu\text{m}$.

As in the case of photonic crystals, the existence and characteristics of complete three-dimensional phononic band gaps depend on the following three key factors:
- The specific geometry of the periodic structure;
- The ratio between the mechanical properties of the different materials; and
- The relative volume fraction of the materials forming the structure.

To date, the first factor is not fully understood and it is not possible to determine *a priori* if a specific periodic structure possesses a complete three-dimensional phononic band gap. The second factor is less complicated and it can be investigated by considering structures with different materials. As previously mentioned, in general, the larger the ratio between equivalent mechanical properties of the constituent materials, the larger the complete phononic band gap. The third factor can systematically be examined by calculating phononic band gaps for different

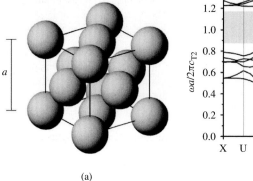

Fig. 7.14 (a) A three-dimensional phononic crystal made of lead spheres ($r = 0.25a$) arranged in the face-centered-cubic Bravais lattice. The solid background material is epoxy, which is not shown for clarity. The volume fraction of lead within the structure is 26%. The density and transverse and longitudinal velocities for lead and epoxy are, respectively, $\rho_1 = 11\,357$ kg m^{-3}, $c_{T1} = 860$ m s^{-1}, $c_{L1} = 2158$ m s^{-1}, $\rho_2 = 1180$ kg m^{-3}, $c_{T2} = 1160$ m s^{-1}, and $c_{L2} = 2540$ m s^{-1}. (b) The dispersion relation $\omega = \omega(\mathbf{k})$ for elastic waves propagating within the structure. The graph shows a large frequency range $\Delta\omega$ for which elastic waves are not allowed to propagate within the structure independently of the direction of propagation. The structure possesses a complete three-dimensional phononic band gap.

volume fractions of the constituent solid materials and plotting the resultant gaps as phononic band gap maps. However, in contrast with the photonics case, the number of material parameters $\{\rho_1, c_{T1}, c_{L1}, \rho_2, c_{T2}, \text{and } c_{L2}\}$ in the case of phononic crystals is relatively large and the systematic study of complete phononic band gaps in three dimensions is a lot of work!

In summary, we introduced in this chapter the concept of phononic band gaps in one, two, and three dimensions and presented a large number of periodic structures having complete phononic band gaps. As we show in Chapter 9, the existence of complete phononic band gaps in periodic structures allows us to control the propagation of mechanical waves. The basic idea behind phononic crystals is that they provide means to control mechanical waves in the same way that photonic crystals control electromagnetic waves. The hope is that the easy fabrication of phononic crystals at small length scales will enable the miniaturization and integration of acoustical components and devices (such as filters, mirrors, waveguides, etc.) that can control the propagation of sound.

Further Reading

One-dimensional phononic crystals

Theory

1 Bria, D., Djafari-Rouhani, B., Bousfia, A., Boudouti, E.H. and Nougaoui, A. (2001) "Absolute acoustic band gap in coupled multilayer structures". *Europhysics Letters*, **55**, 841–46.
2 Bria, D. and Djafari-Rouhani, B. (2002) "Omnidirectional elastic band gap in finite lamellar structures". *Physical Review E*, **66**, 056609.

Experiments

3. Manzanares-Martinez, B., Sanchez-Dehesa, J., Hakansson, A., Cervera, F. and Ramos-Mendieta, F. (**2004**) "Experimental evidence of omnidirectional elastic band gap in finite one-dimensional phononic systems". *Applied Physics Letters*, **85**, 154–56.

Two-dimensional phononic crystals

Solid-solid crystals: Theory

4. Sigalas, M. and Economou, E.N. (**1993**) "Band structure of elastic waves in two dimensional systems". *Solid State Communications*, **86**, 141–43.
5. Kushwaha, M.S., Halevi, P., Dobrzynski, L. and Djafari-Rouhani, B. (**1993**) "Acoustic band structure of periodic elastic composites". *Physical Review Letters*, **71**, 2022–25.
6. Kushwaha, M.S., Halevi, P., Martinez, G., Dobrzynski, L. and Djafari-Rouhani, B. (**1994**) "Theory of acoustic band structure of periodic elastic composites". *Physical Review B*, **49**, 2313–22.
7. Vasseur, J.O., Djafari-Rouhani, B., Dobrzynski, L., Kushwaha, M.S. and Halevi, P. (**1994**) "Complete acoustic band gaps in periodic fiber reinforced composite materials: the carbon/epoxy composite and some metallic systems". *Journal of Physics: Condensed Matter*, **6**, 8759–70.
8. Sigalas, M.M. and Economou, E.N. (**1994**) "Elastic waves in plates with periodically placed inclusions". *Journal of Applied Physics*, **75**, 2845–50.
9. Sigalas, M.M. (**1997**) "Elastic wave band gaps and defect states in two-dimensional composites". *Journal of the Acoustical Society of America*, **101**, 1256–61.

Solid-solid crystals: Experiments

10. Vasseur, J.O., Deymier, P.A., Frantziskonis, G., Hong, G., Djafari-Rouhani, B. and Dobrzynski, L. (**1998**) "Experimental evidence for the existence of absolute acoustic band gaps in two-dimensional periodic composite media". *Journal of Physics: Condensed Matter*, **10**, 6051–64.
11. Vasseur, J.O., Deymier, P.A., Chenni, B., Djafari-Rouhani, B., Dobrzynski, L. and Prevost, D. (**2001**) "Experimental and theoretical evidence for the existence of absolute acoustic band gaps in two-dimensional solid phononic crystals". *Physical Review Letters*, **86**, 3012–15.

Vacuum-solid crystals: Theory

12. Langlet, P., Hladky-Hennion, A.C. and Decarpingy, J.N. (**1995**) "Analysis of the propagation of plane acoustic waves in passive periodic materials using the finite element method". *Journal of the Acoustical Society of America*, **98**, 2792–800.
13. Tanaka, Y., Tomoyasu, Y. and Tamura, S. (**2000**) "Band structure of acoustic waves in phononic lattices: two-dimensional composites with large acoustic mismatch". *Physical Review B*, **62**, 7387–92.
14. Zalipaev, V.V., Movchan, A.B., Poulton, C.G. and McPhedran, R.C. (**2002**) "Elastic waves and homogenization in oblique periodic structures". *Proceedings of the Royal Society of London, Series A*, **458**, 1887–912.
15. Laude, V., Wilm, M., Benchabane, S. and Khelif, A. (**2005**) "Full band gap for surface acoustic waves in a piezoelectric phononic crystal". *Physical Review E*, **71**, 036607.

Fluid-solid crystals: Experiments

16. Montero de Espinosa, F.R., Jimenez, E. and Torres, M. (**1998**) "Ultrasonic band gap in a periodic two-dimensional composite". *Physical Review Letters*, **6**, 1208–211.
17. Garcia-Pablos, D., Sigalas, M., Montero de Espinosa, F.R., Torres, M., Kafesaki, M. and Garcia, N. (**2000**) "Theory and experiments on elastic band gaps". *Physical Review Letters*, **84**, 4349–52.
18. Gorishnyy, T., Ullal, C.K., Maldovan, M., Fytas, G. and Thomas, E.L.

(2005) "Hypersonic phononic crystals". *Physical Review Letters*, **94**, 115501.

Solid-air crystals: Theory

19   Sigalas, M.M. and Economou, E.N. (1996) "Attenuation of multiple-scattered sound". *Europhysics Letters*, **36**, 241–46.
20   Kushwaha, M.S. (1997) "Stop bands for periodic metallic rods: sculptures that can filter noise". *Applied Physics Letters*, **70**, 3218–20.
21   Sigalas, M.M. (1998) "Defect states of acoustic waves in a two-dimensional lattice of solid cylinders". *Journal of Applied Physics*, **84**, 3026–30.

Solid-air crystals: Experiments

22   Martinez-Sala, R., Sancho, J., Sanchez, J.V., Gomez, V., Llinares, J. and Meseguer, F. (1995) "Sound attenuation by sculpture". *Nature*, **378**, 241.
23   Sanchez-Perez, J.V., Caballero, D., Martinez-Sala, R., Rubio, C., Sanchez-Dehesa, J., Meseguer, F., Llinares, J. and Galvez, F. (1998) "Sound attenuation by a two-dimensional array of rigid cylinders". *Physical Review Letters*, **80**, 5325–28.
24   Robertson, W.M. and Rudy, J.F. (1998) "Measurement of acoustic stop bands in two-dimensional periodic scattering arrays". *Journal of the Acoustical Society of America*, **104**, 694–99.
25   Rubio, C., Caballero, D., Sanchez-Perez, J.V., Martinez-Sala, R., Sanchez-Dehesa, J., Meseguer, F. and Cervera, F. (1999) "The existence of full gaps and deaf bands in two-dimensional sonic crystals". *Journal of Lightwave Technology*, **17**, 2202–7.

Three-dimensional phononic crystals

Solid-solid crystals: Theory

26   Sigalas, M.M. and Economou, E.N. (1992) "Elastic and acoustic wave band structure". *Journal of Sound and Vibration*, **158**, 377–82.
27   Economou, E.N. and Sigalas, M.M. (1994) "Stop bands for elastic waves in periodic composite materials". *Journal of the Acoustical Society of America*, **95**, 1734–40.
28   Kafesaki, M., Sigalas, M.M. and Economou, E.N. (1995) "Elastic wave band gaps in 3-D periodic polymer matrix composites". *Solid State Communications*, **96**, 285–89.
29   Sigalas, M.M. and Garcia, N. (2000) "Theoretical study of three-dimensional elastic band gaps with the finite-difference time-domain method". *Journal of Applied Physics*, **87**, 3122–25.
30   Psarobas, I.E., Stefanou, N. and Modinos, A. (2000) "Phononic crystals with planar defects". *Physical Review B*, **62**, 5536–40.
31   Sainidou, R., Stefanou, N. and Modinos, A. (2002) "Formation of absolute frequency gaps in three-dimensional solid phononic crystals". *Physical Review B*, **66**, 212301.
32   Sainidou, R., Stefanou, N., Psarobas, I.E. and Modinos, A. (2005) "A layer-multiple-scattering method for phononic crystals and heterostructures of such". *Computer Physics Communications*, **166**, 197–240.

Solid-solid crystals: Experiments

33   Liu, Z., Zhang, X., Mao, Y., Zhu, Y.Y., Yang, Z., Chan, C.T. and Sheng, P. (2000) "Locally resonant sonic materials". *Science*, **289**, 1734–36.

Solid-fluid crystals: Theory

34   Kushwaha, M.S., Djafari-Rouhani, B., Dobrzynski, L. and Vasseur, J.O. (1998) "Sonic stop-bands for cubic arrays of rigid inclusions in air". *European Physical Journal B*, **3**, 155–61.
35   Kafesaki, M. and Economou, E.N. (1999) "Multiple-scattering theory for three-dimensional periodic acoustic composites". *Physical Review B*, **60**, 11993–2001.

36 Psarobas, I.E., Modinos, A., Sainidou, R. and Stefanou, N. (**2002**) "Acoustic properties of colloidal crystals". *Physical Review B*, **65**, 064307.

Solid-fluid crystals: Experiments

37 Penciu, R.S., Kriegs, H., Petekidis, G., Fytas, G. and Economou, E.N. (**2003**) "Phonons in colloidal systems". *Journal of Chemical Physics*, **118**, 5224–40.

38 Cheng, W., Wang, J.J., Jonas, U., Fytas, G. and Stefanou, N. (**2006**) "Observation and tuning of hypersonic band gaps in colloidal crystals". *Nature Materials*, **5**, 830–36.

Problems

7.1 How many mechanical parameters are needed to characterize a homogeneous solid material?

How many mechanical parameters are necessary to characterize a homogeneous fluid material?

7.2 Describe the differences between elastic waves and acoustic waves?

7.3 Can transverse waves propagate in solids, fluids, and vacuum?
Can longitudinal waves propagate in solids, fluids, and vacuum?

7.4 Consider the graph of the reflectance R versus $\omega a/2\pi c_{T2}$ shown in Figure 7.4.

(a) If the lattice constant of the one-dimensional phononic crystal is $a = 10\mu m$, find the frequencies f of the normally incident elastic waves that are totally reflected by the crystal.

(b) If a normal incident elastic wave has a frequency ω such as $\omega a/2\pi c_{T2} = 0.6$, is the wave totally reflected by the phononic crystal?

7.5 By using the program in Appendix C:

(a) Obtain the graph of the reflectance R as a function of $\omega a/2\pi c_{T2}$ shown in Figure 7.4.

(b) Increase the value of ρ_2 and show that the phononic band gaps decrease.

(c) Show how the phononic band gaps vary with respect to the number of layers.

7.6 Consider a two-dimensional phononic crystal made of solid cylinders in air arranged in the square lattice. The radius of the cylinders is $r = 0.4a$, where $a = 100\ \mu m$ is the distance between the cylinders. Find the range of frequencies f for which the phononic crystal forbids the propagation of acoustic waves.

8
Periodic Cellular Solids

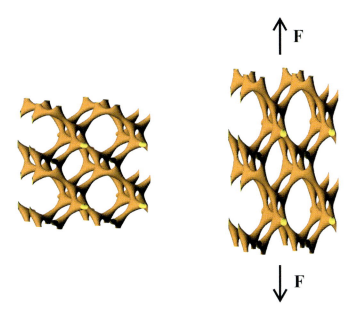

Cellular solids are structures made of a solid material and air. Because the effective physical properties of cellular solids differ from those of the constituent solid material, these structures have a large number of uses that include mechanical, thermal, and acoustical applications. In particular, cellular solids hold the promise of creating lightweight, stiff-and-strong structures. In this chapter, we study the mechanical properties of periodic cellular solids. We first provide a basic review on the elastic properties of materials and present periodic cellular solids that can be fabricated by interference lithography and have useful elastic mechanical properties. Subsequently, we introduce considerations on the plastic deformation and energy absorption of this type of structures. The fabrication of periodic cellular solids at small length scales is important because it can extend the mechanical functionalities of these materials owing to length-scale-dependent mechanical behavior.

8.1
Introduction

Cellular solids are complex structures made of a combination of solid material and air. The word *cellular* reflects the fact that these structures consist of small compartments or cells (which may have different sizes and shapes) that are arranged in space to form the corresponding structure (Figure 8.1). The importance of cellular solids is that they can have improved effective mechanical properties over those corresponding to the constituent solid material. Cellular solids can be classified into two categories: *nonperiodic* and *periodic*. Nonperiodic cellular solids include natural materials such as bone, cork, and wood, and man-made materials such as metallic, ceramic, and polymeric foams, where the arrangement of the cells forming the structure lacks organization or regularity in space (Figure 8.1a). On the other hand, periodic cellular solids include truss architectures used in open mesh plates as well as the simple periodic structures presented in Chapter 2. In this case, the structures are formed by an arrangement of cells or objects that repeat regularly in space (Figure 8.1b).

Cellular solids (both nonperiodic and periodic) are also classified into closed-cell and open-cell cellular solids. In closed-cell cellular solids, the air regions that correspond to each cell are *not* connected with each other. For example, Figure 8.2a shows a closed-cell periodic cellular solid made of nonconnected air spheres in the simple cubic lattice. On the other hand, in open-cell cellular solids, the air regions in the structure are connected with each other and the cellular solid is bicontinuous. Figure 8.2b shows an example of an open-cell periodic cellular solid made of connected air spheres in a simple cubic lattice, where the air and solid regions in the structure are both self-connected.

When a solid material is transformed into a solid–air cellular solid, the physical properties of the material structure can be considerably widened. For example, the

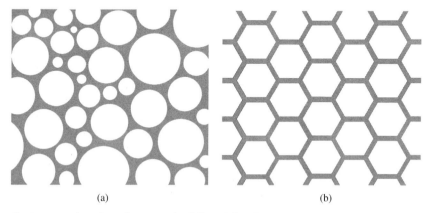

Fig. 8.1 Examples of two-dimensional cellular solids, where the gray regions represent solid material and the white regions are air. (a) A nonperiodic cellular solid. (b) A periodic cellular solid.

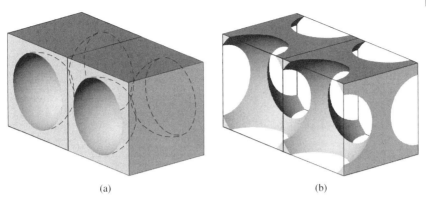

Fig. 8.2 Examples of three-dimensional periodic cellular solids. (a) Two unit cells of a closed-cell periodic cellular solid. (b) Two unit cells of an open-cell periodic cellular solid.

effective stiffness, strength, thermal conductivity, or electrical resistivity of cellular solids can significantly differ from those corresponding to the homogeneous constituent solid material. As a result, cellular solids are useful materials that can be used to design lightweight structures, acoustic filters, thermal insulators, and so on.

A common technique to fabricate three-dimensional nonperiodic cellular solids is by foaming different types of solid materials (e.g. metals, ceramics, and polymers). The foams are obtained by introducing gas bubbles into a liquid material and then solidifying the mixture. As a result of this process, open-cell foams, closed-cell foams, or partially open partially closed foams can be obtained. The mechanical properties of open-cell and closed-cell foams have been extensively studied in the literature and they are commonly used as a reference with which other materials can be compared.

In this chapter, we study elastic and plastic mechanical properties of periodic cellular solids and note that this type of cellular solids can be fabricated at small length scales by interference lithography. The effective linear elastic mechanical properties of periodic cellular solids depend on different factors, which include the mechanical properties of the solid material from which the structure is made, the specific geometry of the structure, and the proportion between the solid material and air in the structure (e.g. the volume fraction of the solid material in the structure). Therefore, it is important to find the dependence of the effective mechanical properties of such cellular solids on the density and microstructure in order to understand how such properties can be optimized for specific applications. In the last part of the chapter, we briefly study the plastic deformation of microframe structures and demonstrate that below a critical length scale, materials that are brittle in bulk form can exhibit considerable plasticity and hence energy absorption.

Before studying the mechanical properties of periodic cellular solids, we introduce basic concepts on mechanical properties of materials such as Hooke's law, Young's modulus, Poisson's ratio, shear modulus, and bulk modulus.

8.2
One-dimensional Hooke's Law

When a force is applied to a solid body, the body alters its original shape. Depending on the intensity of the applied force, the body may or may not return to its original shape after the removal of the force. If the applied force is sufficiently small, the solid body in fact returns to its original shape after the removal of the force and it is said that the solid is in the *elastic* regime. In this case, the relationship between the force applied on the solid body and the resultant deformation of the body is determined by the theory of the elasticity.

One of the most fundamental laws in the theory of elasticity is *Hooke's law*, which states that the amount by which a solid is deformed is proportional to the applied force. For example, consider a bar made of an isotropic solid material subject to an applied force F along the axis of the bar (Figure 8.3). The tensile stress σ acting on the bar is defined as the ratio between the applied force F and the cross-sectional area A of the bar:

$$\sigma = F/A \tag{8.1}$$

Upon the application of the force F, the bar deforms from its original length l_0 to the final length l_1. The longitudinal strain ε of the bar is defined as the relative increase in length, which is given by

$$\varepsilon = (l_1 - l_0)/l_0 \tag{8.2}$$

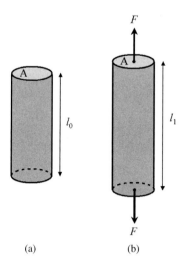

Fig. 8.3 (a) A solid bar of length l_0 and cross-sectional area A. (b) Upon the application of a force F, the solid bar deforms to its new length l_1. Hooke's law states that the stress $\sigma = F/A$ and the strain $\varepsilon = (l_1 - l_0)/l_0$ are related by $\varepsilon = s\sigma$, or equivalently $\sigma = c\varepsilon$, where s and c are the compliance and stiffness elastic constants of the solid material forming the bar.

If the applied force is small, Hooke's law states that the amount of strain ε experienced by the bar is proportional to the applied stress σ. That is,

$$\varepsilon = s\sigma \quad \text{or} \quad \sigma = c\varepsilon \tag{8.3}$$

where s is the *compliance* and c is the *stiffness constant* of the isotropic solid material. (Note: the stiffness constant c is also known as the Young's modulus and it is alternatively denoted by the letter E.)

In the one-dimensional isotropic solid case, the applied stress σ and the resultant strain ε are considered to be scalar numbers. However, in the next sections, we show that general states of stress and strain on solid bodies are described by second-rank tensors σ_{ij} and ε_{ij}, respectively. In addition, these stress and strain tensors are related to each other by the so-called generalized Hooke's law.

8.3
The Stress Tensor

In this section, we introduce the concept of stress tensor, which allows us to describe the forces per unit area acting on volume elements located within a solid body that is subject to applied external forces.

Consider a cubic volume element located at the point P within a solid body that is in a general state of stress (e.g. a solid body with external forces acting on it). As a result of the state of stress, the faces of the cubic volume element are subject to forces that are exerted by the solid material surrounding the cube. These forces are denoted by $\mathbf{F}_1$, $\mathbf{F}_2$, and $\mathbf{F}_3$ in Figure 8.4.

Because these forces are proportional to the area of the faces of the cubic volume element, it is convenient to define the *traction vectors* $\mathbf{T}_i (i = 1, 2, 3)$, which are

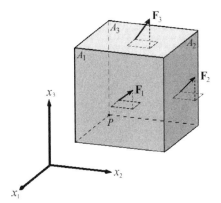

Fig. 8.4 Forces acting on the faces of a cubic volume element located within a solid body that is under a general state of stress.

given by

$$T_1 = \frac{F_1}{A_1}, \quad T_2 = \frac{F_2}{A_2}, \quad T_3 = \frac{F_3}{A_3} \tag{8.4}$$

That is, the traction vector T_i represents the force per unit area acting on the face of the cube normal to the direction x_i. The three traction vectors T_i can in turn be resolved along the orthogonal axes x_1, x_2, and x_3 to generate the nine components $\sigma_{ij}(i,j = 1, 2, 3)$ of the *stress tensor* (Figure 8.5). For example, the traction T_1 acting on the face of the cube normal to the direction x_1 can be resolved into the three stress components σ_{11}, σ_{12}, and σ_{13}, the traction T_2 into the stress components σ_{21}, σ_{22}, and σ_{23}, and the traction T_3 into the stress components σ_{31}, σ_{32}, and σ_{33}. Note the convention used in the definition of the stress components in Figure 8.5. The component j of the traction vector T_i generates the stress component σ_{ij}. That is

$$\sigma_{ij} \text{ is the } j\text{th component of the traction vector } T_i \tag{8.5}$$

Therefore, the components σ_{ij} of the stress tensor represent forces per unit area and are defined by two directions: the i direction, which indicates the face of the cube on which the stress is acting, and the j direction, which indicates the direction along which the stress is acting (Figure 8.5).

The stress components σ_{ij} that are perpendicular to the faces of the cube (i.e. σ_{11}, σ_{22}, and σ_{33}) are called *normal stresses*, whereas the stress components that are tangential to the faces of the cube (i.e. σ_{12}, σ_{13}, σ_{21}, σ_{23}, σ_{31}, and σ_{32}) are called *shear stresses*. Normal stresses tend to deform the solid and change its volume, whereas shear stresses tend to deform the solid without changing its volume. In this book, we adopt the following sign convention for normal stresses: positive values for σ_{11},

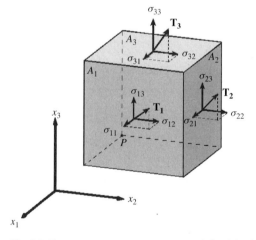

Fig. 8.5 The stress tensor components σ_{ij} defined by the j-components of the traction vectors T_i acting on the faces of a cubic volume element located within the solid body.

σ_{22}, and σ_{33} correspond to stresses that stretch the cubic volume element, whereas negative values correspond to stresses that compress the volume element.

The nine components of stress can be organized in a 3×3 matrix representing the stress tensor σ_{ij}. This matrix completely describes the state of stress of the solid body at the point P and is given by

$$\sigma_{ij} = \begin{pmatrix} \sigma_{11} & \sigma_{12} & \sigma_{13} \\ \sigma_{21} & \sigma_{22} & \sigma_{23} \\ \sigma_{31} & \sigma_{32} & \sigma_{33} \end{pmatrix} \tag{8.6}$$

We note that the components of the stress tensor in Equation 8.6 are not independent. Because it is assumed that the cubic volume element in Figure 8.5 is in equilibrium, the net moments about the x_1, x_2, and x_3 axes must be equal to zero and therefore we must have $\sigma_{23} = \sigma_{32}$, $\sigma_{13} = \sigma_{31}$, and $\sigma_{12} = \sigma_{21}$. These relationships can be written in compact notation as

$$\sigma_{ij} = \sigma_{ji} \tag{8.7}$$

Equation 8.7 determines that the stress tensor σ_{ij} is *symmetric* and that only six of its nine components are independent.

In summary, the stress tensor given by Equation 8.6 completely characterizes the state of stress of the solid body at the point P. That is, it completely defines the forces per unit area acting on the faces of a cubic volume element located at the point P within the solid body (Figure 8.5).

8.4
The Strain Tensor

We next study the effects that the applied stresses σ_{ij} cause on the cubic volume element. For example, when the cubic volume element in Figure 8.5 is subject to a state of stress σ_{ij} given by the stress tensor (Equation 8.6), the cubic element may in general *translate* and *deform*. That is,

$$\text{Effects of applied stresses} \longrightarrow \text{translation} + \text{deformation} \tag{8.8}$$

A translation in space means that the cubic volume element moves from its original location to a final location as a rigid solid; that is, all points within the cubic element move along parallel paths and the distance between any pair of points within the cubic element does not change. As a consequence, the cubic volume element moves in space retaining its cubic shape. On the other hand, a deformation generates a change of shape of the cubic volume element (i.e. the volume element is no longer cubic) and the distance between any pair of points within the cubic element does not remain fixed. From this point forward, we assume that the deformation of the cubic element is small. As we show in the next sections, a small deformation can be separated as the sum of a *strain* and a *rotation*.

$$\text{Deformation} = \text{strain} + \text{rotation} \tag{8.9}$$

A rotation in space means that the cubic volume element turns around some specific axis as a rigid solid; that is, all points within the cubic element move along circular paths around the axis and the distance between any pair of points within the cubic element does not change. Therefore, as the cubic volume element rotates in space it retains its cubic shape. On the other hand, *a strain of the cubic volume element generates the actual change in shape of the cubic element*, and as we show in the next sections, its relation with the applied stresses σ_{ij} determines the mechanical properties of the solid material.

$$\text{Relation between stress and strain}$$
$$\longrightarrow \text{mechanical properties of the material} \tag{8.10}$$

In other words, as a result of applied stresses, the cubic volume element in Figure 8.5 undergoes a translation and a deformation. The deformation of the element can be described as a strain of the original element plus a rotation of the element as a whole. Translations and rotations of the element do not change the shape of the element, whereas the strain of the element generates its actual change in shape. Importantly, the relation between the applied stresses on the element and the resultant strain of the element determines the mechanical properties of the solid material. In the next sections, we study the above considerations and introduce the mathematical scheme to define the deformation, strain, and rotation tensors, which allows us to analytically describe the effects of the applied stresses σ_{ij} on a volume element located within a solid body.

Consider an arbitrary point P located at the position $\mathbf{r} = (x_1, x_2, x_3)$ within a solid body (Figure 8.6). When the solid body is subject to a state of stress, the point P moves to the point P' located at $\mathbf{r}' = (x_1', x_2', x_3')$. The displacement of the point P is defined by the vector $\mathbf{u} = \mathbf{r}' - \mathbf{r}$, which determines the location of the point P' with respect to its original position P. In general, the amount of displacement of the point P depends on the original position of the point within the solid body. That is, the displacement vector $\mathbf{u} = \mathbf{u}(\mathbf{r})$ is a function of the position vector $\mathbf{r}$. Analogously, consider a point Q located near the point P at the position $\mathbf{r} + d\mathbf{r}$. After the stresses are applied, the displacement of the point Q is determined by the vector $\mathbf{u} + d\mathbf{u}$ (Figure 8.6). The difference in displacements between the points P and Q is given by the vector $d\mathbf{u}$, which is an important quantity because it measures the amount of deformation around the point P. For example, note that if $d\mathbf{u}$ is equal to zero, the displacements of the points P and Q are the same and there is no deformation of the solid at the point P since the line segment PQ only translates in space.

The displacement vector $\mathbf{u} = \mathbf{u}(\mathbf{r})$ can be resolved along the Cartesian coordinate axis as

$$\mathbf{u} = u_1 \hat{\mathbf{x}}_1 + u_2 \hat{\mathbf{x}}_2 + u_3 \hat{\mathbf{x}}_3 \tag{8.11}$$

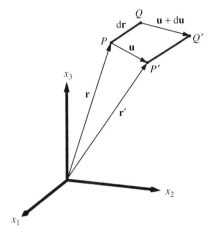

Fig. 8.6 The application of stresses to a solid body causes the points P and Q located within the solid body move to the points P' and Q', respectively. The displacement vector **u** corresponding to the point P is given by $\mathbf{u} = \mathbf{r'} - \mathbf{r}$, whereas the displacement vector corresponding to the point Q is given by $\mathbf{u} + d\mathbf{u}$. The quantity $d\mathbf{u}$ measures the amount of local deformation around the point P in the solid body.

This means that $d\mathbf{u} = d\mathbf{u}(\mathbf{r})$ is given by

$$d\mathbf{u} = du_1 \hat{\mathbf{x}}_1 + du_2 \hat{\mathbf{x}}_2 + du_3 \hat{\mathbf{x}}_3 \tag{8.12}$$

where

$$du_1 = \frac{\partial u_1}{\partial x_1} dx_1 + \frac{\partial u_1}{\partial x_2} dx_2 + \frac{\partial u_1}{\partial x_3} dx_3$$
$$du_2 = \frac{\partial u_2}{\partial x_1} dx_1 + \frac{\partial u_2}{\partial x_2} dx_2 + \frac{\partial u_2}{\partial x_3} dx_3 \tag{8.13}$$
$$du_3 = \frac{\partial u_3}{\partial x_1} dx_1 + \frac{\partial u_3}{\partial x_2} dx_2 + \frac{\partial u_3}{\partial x_3} dx_3$$

or, collectively,

$$du_i = \frac{\partial u_i}{\partial x_j} dx_j \quad (i,j = 1, 2, 3) \tag{8.14}$$

Therefore, after the stresses are applied, the displacement vector **u** given by Equation 8.11 determines the translation of the point P, whereas the vector $d\mathbf{u}$ given by Equation 8.12 measures the local deformation of the solid around the point P. Because we are interested in studying the mechanical properties of the solid material (which are determined by the resultant strains), we do not consider the translation of the point P, and we therefore focus only on the vector $d\mathbf{u}$, which represents the local deformation around the point P. This means that we need to consider the partial derivatives $\partial u_i / \partial x_j$ of the components of the displacement vector and not the displacement vector itself (Equations 8.12–8.14).

The partial derivatives $\partial u_i/\partial x_j$ in Equation 8.13 can be organized in a 3×3 matrix representing the *deformation tensor* e_{ij}:

$$e_{ij} = \begin{pmatrix} e_{11} & e_{12} & e_{13} \\ e_{21} & e_{22} & e_{23} \\ e_{31} & e_{32} & e_{33} \end{pmatrix} = \begin{pmatrix} \dfrac{\partial u_1}{\partial x_1} & \dfrac{\partial u_1}{\partial x_2} & \dfrac{\partial u_1}{\partial x_3} \\ \dfrac{\partial u_2}{\partial x_1} & \dfrac{\partial u_2}{\partial x_2} & \dfrac{\partial u_2}{\partial x_3} \\ \dfrac{\partial u_3}{\partial x_1} & \dfrac{\partial u_3}{\partial x_2} & \dfrac{\partial u_3}{\partial x_3} \end{pmatrix} \qquad (8.15)$$

which can also be written in compact form as

$$e_{ij} = \frac{\partial u_i}{\partial x_j} \qquad (8.16)$$

The deformation tensor (Equation 8.15), which measures the amount of deformation around the point P, completely describes how the cubic element in Figure 8.5 deforms after the stresses σ_{ij} are applied. We show this effect in the next sections, where we define the physical meaning of the deformation tensor components e_{ij}.

Note that Equation 8.16 can equivalently be written as

$$e_{ij} = \frac{1}{2}\left(\frac{\partial u_i}{\partial x_j} + \frac{\partial u_j}{\partial x_i}\right) + \frac{1}{2}\left(\frac{\partial u_i}{\partial x_j} - \frac{\partial u_j}{\partial x_i}\right) \qquad (8.17)$$

The *strain tensor* ε_{ij} is defined as

$$\varepsilon_{ij} = \frac{1}{2}\left(\frac{\partial u_i}{\partial x_j} + \frac{\partial u_j}{\partial x_i}\right) \qquad (8.18)$$

whereas the *rotation tensor* ϖ_{ij} is defined as

$$\varpi_{ij} = \frac{1}{2}\left(\frac{\partial u_i}{\partial x_j} - \frac{\partial u_j}{\partial x_i}\right) \qquad (8.19)$$

This means that the deformation tensor e_{ij} given by Equation 8.17 can be written as the sum of the strain and rotation tensors ε_{ij} and ϖ_{ij}. That is,

$$e_{ij} = \varepsilon_{ij} + \varpi_{ij} \qquad (8.20)$$

Therefore, a small deformation can be regarded as a strain plus a rotation. The reason we separate the deformation tensor e_{ij} into the strain tensor ε_{ij} and rotation tensor ϖ_{ij} is explained in the next sections where we study the physical significance of these tensors. Note that the strain tensor ε_{ij} is symmetric (i.e. $\varepsilon_{ij} = \varepsilon_{ji}$), whereas the rotation tensor ϖ_{ij} is antisymmetric (i.e. $\varpi_{ij} = -\varpi_{ji}$).

For simplicity, to study the physical meaning of the deformation, strain, and rotation tensors, we next consider that the element located within the solid body is two dimensional. The results can subsequently be extended to the more general three-dimensional case.

8.4.1
Expansion

Consider a two-dimensional rectangular element of sides dx_1 and dx_2 located at the point P within a solid body (Figure 8.7). Assume also that after the stresses are applied on the solid body, the rectangular element is displaced by $\mathbf{u} = (u_1, u_2)$ from the point P to the point P' and undergoes an expansion along the two coordinate axes as shown in Figure 8.7.

As previously mentioned, we are not interested in the translation of the element as a whole and we only consider the resultant deformation. Figure 8.7 shows that the rectangular element increases its length by $(\partial u_1/\partial x_1)\, dx_1$ along the x_1 direction and by $(\partial u_2/\partial x_2)\, dx_2$ along the x_2 direction. The strains of the element along the x_1 and x_2 directions are defined as the relative increase in length and are thus given by the ratio between the increase in length and the original length of the element (Section 8.2). Since the original lengths of the element are dx_1 and dx_2, the strains of the element along the x_1 and x_2 directions are given by $\partial u_1/\partial x_1$ and $\partial u_2/\partial x_2$, respectively.

Now, according to Equations 8.16 and 8.18 we have

$$\varepsilon_{11} = e_{11} = \frac{\partial u_1}{\partial x_1}, \quad \varepsilon_{22} = e_{22} = \frac{\partial u_2}{\partial x_2} \tag{8.21}$$

This means that, *the diagonal components ε_{11} and ε_{22} of the strain tensor (or equivalently the diagonal components e_{11} and e_{22} of the deformation tensor) measure the relative increase in length of the element along the x_1 and x_2 directions.*

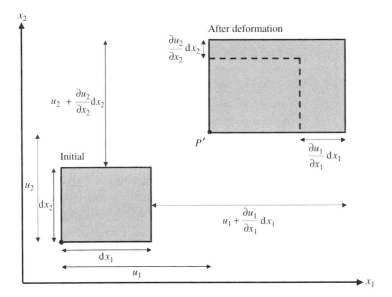

Fig. 8.7 Expansion of a two-dimensional rectangular element within a solid body.

8.4.2
General Deformation

Consider once more a two-dimensional rectangular element of sides dx_1 and dx_2 located at the point P within a solid body (Figure 8.8). Also, assume that after the stresses are applied on the solid body, the rectangular element is displaced by $\mathbf{u} = (u_1, u_2)$ from the point P to the point P' and deforms as shown in Figure 8.8.

We can see in Figure 8.8 that the angles $d\alpha$ and $d\beta$ satisfy

$$\tan d\alpha = \frac{\dfrac{\partial u_1}{\partial x_2} dx_2}{dx_2 + \dfrac{\partial u_2}{\partial x_2} dx_2}, \quad \tan d\beta = \frac{\dfrac{\partial u_2}{\partial x_1} dx_1}{dx_1 + \dfrac{\partial u_1}{\partial x_1} dx_1} \tag{8.22}$$

Because the resultant deformation is assumed to be small, we have $\partial u_i/\partial x_j \ll 1$. In addition, we can assume that $\tan d\alpha \approx d\alpha$ and $\tan d\beta \approx d\beta$. As a result, Equation 8.22 can be written as

$$d\alpha = \frac{\partial u_1}{\partial x_2} = e_{12}, \quad d\beta = \frac{\partial u_2}{\partial x_1} = e_{21} \tag{8.23}$$

This means that *the nondiagonal components e_{12} and e_{21} of the deformation tensor measure respectively the angles $d\alpha$ and $d\beta$ shown in Figure* 8.8.

It is important to note, however, that the actual change in shape of the rectangular element is measured by the angle $d\gamma$, which is equal to $d\alpha + d\beta$ (Figure 8.8).

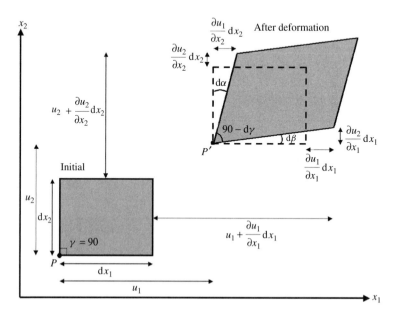

Fig. 8.8 Arbitrary deformation of a two-dimensional rectangular element within a solid body.

By considering Equations 8.23 and 8.18, we have

$$dy = d\alpha + d\beta = \frac{\partial u_1}{\partial x_2} + \frac{\partial u_2}{\partial x_1} = 2\varepsilon_{12} = 2\varepsilon_{21} \qquad (8.24)$$

Therefore, *the nondiagonal components $\varepsilon_{12} = \varepsilon_{21}$ of the symmetric strain tensor measure the change in shape of the element given by dy*. In particular, these tensor components are the average of the angles $d\alpha$ and $d\beta$ that correspond to the general deformation shown in Figure 8.8. (Note: the diagonal components ε_{11} and ε_{22} of the strain tensor have the same physical significance as in the case of pure expansion, which is the measurement of the relative increase in length of the element along the x_1 and x_2 directions.)

In summary, by studying both an expansion and a general deformation of a rectangular element, we demonstrated the physical significance of the components of the deformation tensor e_{ij} and the strain tensor ε_{ij}. In the next section, we introduce the significance of the rotation tensor ϖ_{ij} and show how a general deformation e_{ij} can be separated as the sum of a strain ε_{ij} and a rotation ϖ_{ij}.

8.4.3
Resolving a General Deformation as Strain Plus Rotation

Consider that after the stresses are applied on a solid body, a rectangular element within the solid body deforms as shown in Figure 8.9a.

As previously shown, the deformation experienced by the rectangular element in Figure 8.9a is described by the deformation tensor e_{ij}, where the diagonal components e_{11} and e_{22} determine the relative increase in length of the element along the x_1 and x_2 directions, and the nondiagonal components e_{12} and e_{21} determine the angles $d\alpha$ and $d\beta$, respectively. That is,

$$e_{ij} = \begin{pmatrix} e_{11} & e_{12} \\ e_{21} & e_{22} \end{pmatrix} = \begin{pmatrix} e_{11} & d\alpha \\ d\beta & e_{22} \end{pmatrix} \qquad (8.25)$$

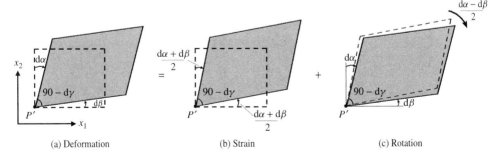

(a) Deformation (b) Strain (c) Rotation

Fig. 8.9 (a) A general deformation of a two-dimensional rectangular element within a solid separated into a strain (b) plus a rotation (c).

The deformation of the element, however, can be understood as the superposition of two transformations as shown in Figure 8.9. The first transformation is a strain of the element represented by the strain tensor ε_{ij} (Figure 8.9b), where the diagonal components $\varepsilon_{11} = e_{11}$ and $\varepsilon_{22} = e_{22}$ determine the relative increase in length of the element along the x_1 and x_2 directions, and the nondiagonal symmetric components $\varepsilon_{12} = \varepsilon_{21}$ measure half the angle $d\gamma = d\alpha + d\beta$, which determines the change in shape of the element. That is, the strain of the element is represented by the strain tensor

$$\varepsilon_{ij} = \begin{pmatrix} \varepsilon_{11} & \varepsilon_{12} \\ \varepsilon_{21} & \varepsilon_{22} \end{pmatrix} = \begin{pmatrix} e_{11} & \dfrac{d\alpha + d\beta}{2} \\ \dfrac{d\alpha + d\beta}{2} & e_{22} \end{pmatrix} \quad (8.26)$$

The second transformation is a rotation of the element as a rigid body by an angle $(d\alpha - d\beta)/2$ (Figure 8.9c). This transformation is represented by the rotation tensor ϖ_{ij}. Note that because the element rotates as a rigid body, there is no expansion of the element and the diagonal components ϖ_{11}, ϖ_{22} of the rotation tensor are equal to zero. On the other hand, the nondiagonal antisymmetric components $\varpi_{12} = -\varpi_{21}$ of the rotation tensor measure the angle of rotation $(d\alpha - d\beta)/2$. As a result, the rotation of the element is described by the rotation tensor

$$\varpi_{ij} = \begin{pmatrix} \varpi_{11} & \varpi_{12} \\ \varpi_{21} & \varpi_{22} \end{pmatrix} = \begin{pmatrix} 0 & \dfrac{d\alpha - d\beta}{2} \\ -\dfrac{d\alpha - d\beta}{2} & 0 \end{pmatrix} \quad (8.27)$$

Equations 8.25–8.27 together with the scheme shown in Figure 8.9 show that $e_{ij} = \varepsilon_{ij} + \varpi_{ij}$ and that an arbitrary deformation of the rectangular element can be considered as the sum of a strain and a rotation.

The significance of separating the deformation tensor into the strain and rotation tensors is to obtain a physical magnitude that only describes the actual change in shape of the element and does not take into account the rotation of the element as a whole. This physical magnitude is the strain tensor. As previously mentioned, the mechanical properties of solid materials are determined by the relation between the applied stress tensor and the resultant strain tensor. In the next sections, we therefore focus only on the stress and strain tensors σ_{ij} and ε_{ij}. Note that in the three-dimensional case, the strain tensor ε_{ij} can be written as

$$\varepsilon_{ij} = \begin{pmatrix} \varepsilon_{11} & \varepsilon_{12} & \varepsilon_{13} \\ \varepsilon_{21} & \varepsilon_{22} & \varepsilon_{23} \\ \varepsilon_{31} & \varepsilon_{32} & \varepsilon_{33} \end{pmatrix}$$

$$= \begin{pmatrix} \dfrac{\partial u_1}{\partial x_1} & \dfrac{1}{2}\left(\dfrac{\partial u_1}{\partial x_2} + \dfrac{\partial u_2}{\partial x_1}\right) & \dfrac{1}{2}\left(\dfrac{\partial u_1}{\partial x_3} + \dfrac{\partial u_3}{\partial x_1}\right) \\ \dfrac{1}{2}\left(\dfrac{\partial u_1}{\partial x_2} + \dfrac{\partial u_2}{\partial x_1}\right) & \dfrac{\partial u_2}{\partial x_2} & \dfrac{1}{2}\left(\dfrac{\partial u_2}{\partial x_3} + \dfrac{\partial u_3}{\partial x_2}\right) \\ \dfrac{1}{2}\left(\dfrac{\partial u_1}{\partial x_3} + \dfrac{\partial u_3}{\partial x_1}\right) & \dfrac{1}{2}\left(\dfrac{\partial u_2}{\partial x_3} + \dfrac{\partial u_3}{\partial x_2}\right) & \dfrac{\partial u_3}{\partial x_3} \end{pmatrix} \quad (8.28)$$

8.5
Stress–Strain Relationship: The Generalized Hooke's Law

In Section 8.1, we studied the relationship between stress and strain in the case of a one-dimensional solid bar and showed that this relationship is given by the one-dimensional Hooke's law (Equation 8.3). In this section, we generalize these results for the more significant three-dimensional case and study the relationship between a general state of stress on a solid body given by the stress tensor σ_{ij} and the resultant strains given by the strain tensor ε_{ij}.

The generalized Hooke's law states that, if the applied stresses on the solid body are small enough, each component of the resultant strain tensor ε_{ij} is linearly dependent on all the components of the applied stress tensor σ_{ij}. That is,

$$\varepsilon_{ij} = s_{ijkl}\, \sigma_{kl} \quad \text{or} \quad \sigma_{ij} = c_{ijkl}\, \varepsilon_{kl} \quad (i,j,k,l = 1,2,3) \tag{8.29}$$

where s_{ijkl} and c_{ijkl} are the *compliances* and *stiffness constants* of the solid material.

For example, the strain tensor components $\varepsilon_{11}, \varepsilon_{12}$, and ε_{13} are related to the nine stress components by

$$\varepsilon_{11} = s_{1111}\sigma_{11} + s_{1112}\sigma_{12} + s_{1113}\sigma_{13} + s_{1121}\sigma_{21} + s_{1122}\sigma_{22} + s_{1123}\sigma_{23}$$
$$+ s_{1131}\sigma_{31} + s_{1132}\sigma_{32} + s_{1133}\sigma_{33} \tag{8.30a}$$

$$\varepsilon_{12} = s_{1211}\sigma_{11} + s_{1212}\sigma_{12} + s_{1213}\sigma_{13} + s_{1221}\sigma_{21} + s_{1222}\sigma_{22} + s_{1223}\sigma_{23}$$
$$+ s_{1231}\sigma_{31} + s_{1232}\sigma_{32} + s_{1233}\sigma_{33} \tag{8.30b}$$

$$\varepsilon_{13} = s_{1311}\sigma_{11} + s_{1312}\sigma_{12} + s_{1313}\sigma_{13} + s_{1321}\sigma_{21} + s_{1322}\sigma_{22} + s_{1323}\sigma_{23}$$
$$+ s_{1331}\sigma_{31} + s_{1332}\sigma_{32} + s_{1333}\sigma_{33} \tag{8.30c}$$

Similarly, the stress tensor components σ_{11}, σ_{12}, and σ_{13} are related to the nine strain components by

$$\sigma_{11} = c_{1111}\varepsilon_{11} + c_{1112}\varepsilon_{12} + c_{1113}\varepsilon_{13} + c_{1121}\varepsilon_{21} + c_{1122}\varepsilon_{22} + c_{1123}\varepsilon_{23}$$
$$+ c_{1131}\varepsilon_{31} + c_{1132}\varepsilon_{32} + c_{1133}\varepsilon_{33} \tag{8.31a}$$

$$\sigma_{12} = c_{1211}\varepsilon_{11} + c_{1212}\varepsilon_{12} + c_{1213}\varepsilon_{13} + c_{1221}\varepsilon_{21} + c_{1222}\varepsilon_{22} + c_{1223}\varepsilon_{23}$$
$$+ c_{1231}\varepsilon_{31} + c_{1232}\varepsilon_{32} + c_{1233}\varepsilon_{33} \tag{8.31b}$$

$$\sigma_{13} = c_{1311}\varepsilon_{11} + c_{1312}\varepsilon_{12} + c_{1313}\varepsilon_{13} + c_{1321}\varepsilon_{21} + c_{1322}\varepsilon_{22} + c_{1323}\varepsilon_{23}$$
$$+ c_{1331}\varepsilon_{31} + c_{1332}\varepsilon_{32} + c_{1333}\varepsilon_{33} \tag{8.31c}$$

The compliances and stiffness constants s_{ijkl} and c_{ijkl} form fourth-rank tensors and are intrinsic physical properties of the solid material. Since there are nine Equations 8.30 and nine Equations 8.31, we have 81 s_{ijkl} and 81 c_{ijkl} constants. However, the symmetry of the stress and strain tensors (i.e. $\sigma_{ij} = \sigma_{ji}$ and $\varepsilon_{ij} = \varepsilon_{ji}$) determines that

$$s_{ijkl} = s_{jikl}, \quad s_{ijkl} = s_{ijlk} \quad \text{and} \quad c_{ijkl} = c_{jikl}, \quad c_{ijkl} = c_{ijlk} \tag{8.32}$$

As a result, the number of independent s_{ijkl} (or c_{ijkl}) constants, which characterize a general solid material, is reduced from 81 to 36.

8.6
The Generalized Hooke's Law in Matrix Notation

By taking advantage of the fact that the stress tensor σ_{ij} and the strain tensor ε_{ij} are symmetric, we can rewrite Equation 8.29 in a simple matrix form.

For example, because the stress tensor σ_{ij} is symmetric (i.e. $\sigma_{ij} = \sigma_{ji}$), only six of its nine components are independent. Hence, we choose to write the stress tensor σ_{ij} in a compact form as a six-component column vector

$$\sigma_{ij} = \begin{pmatrix} \sigma_{11} & \sigma_{12} & \sigma_{13} \\ \sigma_{12} & \sigma_{22} & \sigma_{23} \\ \sigma_{13} & \sigma_{23} & \sigma_{33} \end{pmatrix} = \begin{pmatrix} \sigma_1 & \sigma_6 & \sigma_5 \\ \sigma_6 & \sigma_2 & \sigma_4 \\ \sigma_5 & \sigma_4 & \sigma_3 \end{pmatrix}$$

$$\longrightarrow \sigma_m = \begin{pmatrix} \sigma_1 \\ \sigma_2 \\ \sigma_3 \\ \sigma_4 \\ \sigma_5 \\ \sigma_6 \end{pmatrix} \quad (m = 1, \ldots, 6) \tag{8.33}$$

Similarly, we choose to write the symmetric strain tensor ε_{ij} as a six-component column vector as

$$\varepsilon_{ij} = \begin{pmatrix} \varepsilon_{11} & \varepsilon_{12} & \varepsilon_{13} \\ \varepsilon_{12} & \varepsilon_{22} & \varepsilon_{23} \\ \varepsilon_{13} & \varepsilon_{23} & \varepsilon_{33} \end{pmatrix} = \begin{pmatrix} \varepsilon_1 & \tfrac{1}{2}\varepsilon_6 & \tfrac{1}{2}\varepsilon_5 \\ \tfrac{1}{2}\varepsilon_6 & \varepsilon_2 & \tfrac{1}{2}\varepsilon_4 \\ \tfrac{1}{2}\varepsilon_5 & \tfrac{1}{2}\varepsilon_4 & \varepsilon_3 \end{pmatrix}$$

$$\longrightarrow \varepsilon_n = \begin{pmatrix} \varepsilon_1 \\ \varepsilon_2 \\ \varepsilon_3 \\ \varepsilon_4 \\ \varepsilon_5 \\ \varepsilon_6 \end{pmatrix} \quad (n = 1, \ldots, 6) \tag{8.34}$$

where the introduction of the 1/2 factor in the case of the nondiagonal components is for mathematical convenience as shown next.

In addition, by considering the 36 independent compliance constants s_{ijkl} and the 36 independent stiffness constants c_{ijkl} of the solid material, we define the matrices s_{ij} and c_{ij} as

$$s_{ij} = \begin{pmatrix} s_{11} & s_{12} & s_{13} & s_{14} & s_{15} & s_{16} \\ s_{21} & s_{22} & s_{23} & s_{24} & s_{25} & s_{26} \\ s_{31} & s_{32} & s_{33} & s_{34} & s_{35} & s_{36} \\ s_{41} & s_{42} & s_{43} & s_{44} & s_{45} & s_{46} \\ s_{51} & s_{52} & s_{53} & s_{54} & s_{55} & s_{56} \\ s_{61} & s_{62} & s_{63} & s_{64} & s_{65} & s_{66} \end{pmatrix}$$

8.6 The Generalized Hooke's Law in Matrix Notation

$$= \begin{pmatrix} s_{1111} & s_{1122} & s_{1133} & 2s_{1123} & 2s_{1113} & 2s_{1112} \\ s_{2211} & s_{2222} & s_{2233} & 2s_{2223} & 2s_{2213} & 2s_{2212} \\ s_{3311} & s_{3322} & s_{3333} & 2s_{3323} & 2s_{3313} & 2s_{3312} \\ 2s_{2311} & 2s_{2322} & 2s_{2333} & 4s_{2323} & 4s_{2313} & 4s_{2312} \\ 2s_{1311} & 2s_{1322} & 2s_{1333} & 4s_{1323} & 4s_{1313} & 4s_{1312} \\ 2s_{1211} & 2s_{1222} & 2s_{1233} & 4s_{1223} & 4s_{1213} & 4s_{1212} \end{pmatrix} \quad (8.35)$$

$$c_{ij} = \begin{pmatrix} c_{11} & c_{12} & c_{13} & c_{14} & c_{15} & c_{16} \\ c_{21} & c_{22} & c_{23} & c_{24} & c_{25} & c_{26} \\ c_{31} & c_{32} & c_{33} & c_{34} & c_{35} & c_{36} \\ c_{41} & c_{42} & c_{43} & c_{44} & c_{45} & c_{46} \\ c_{51} & c_{52} & c_{53} & c_{54} & c_{55} & c_{56} \\ c_{61} & c_{62} & c_{63} & c_{64} & c_{65} & c_{66} \end{pmatrix}$$

$$= \begin{pmatrix} c_{1111} & c_{1122} & c_{1133} & c_{1123} & c_{1113} & c_{1112} \\ c_{2211} & c_{2222} & c_{2233} & c_{2223} & c_{2213} & c_{2212} \\ c_{3311} & c_{3322} & c_{3333} & c_{3323} & c_{3313} & c_{3312} \\ c_{2311} & c_{2322} & c_{2333} & c_{2323} & c_{2313} & c_{2312} \\ c_{1311} & c_{1322} & c_{1333} & c_{1323} & c_{1313} & c_{1312} \\ c_{1211} & c_{1222} & c_{1233} & c_{1223} & c_{1213} & c_{1212} \end{pmatrix} \quad (8.36)$$

where the following index abbreviation is used

Matrix notation	1	2	3	4	5	6
Tensor notation	11	22	33	23	13	12

Importantly, note that in the case of the compliance matrix s_{ij} some coefficients are multiplied by a factor of 2 or 4.

Now, by considering all the above definitions, we can write the generalized Hooke's law (Equation 8.29) in a simple matrix form as

$$\begin{pmatrix} \varepsilon_1 \\ \varepsilon_2 \\ \varepsilon_3 \\ \varepsilon_4 \\ \varepsilon_5 \\ \varepsilon_6 \end{pmatrix} = \begin{pmatrix} s_{11} & s_{12} & s_{13} & s_{14} & s_{15} & s_{16} \\ s_{21} & s_{22} & s_{23} & s_{24} & s_{25} & s_{26} \\ s_{31} & s_{32} & s_{33} & s_{34} & s_{35} & s_{36} \\ s_{41} & s_{42} & s_{43} & s_{44} & s_{45} & s_{46} \\ s_{51} & s_{52} & s_{53} & s_{54} & s_{55} & s_{56} \\ s_{61} & s_{62} & s_{63} & s_{64} & s_{65} & s_{66} \end{pmatrix} \begin{pmatrix} \sigma_1 \\ \sigma_2 \\ \sigma_3 \\ \sigma_4 \\ \sigma_5 \\ \sigma_6 \end{pmatrix} \quad \text{or}$$

$$\begin{pmatrix} \sigma_1 \\ \sigma_2 \\ \sigma_3 \\ \sigma_4 \\ \sigma_5 \\ \sigma_6 \end{pmatrix} = \begin{pmatrix} c_{11} & c_{12} & c_{13} & c_{14} & c_{15} & c_{16} \\ c_{21} & c_{22} & c_{23} & c_{24} & c_{25} & c_{26} \\ c_{31} & c_{32} & c_{33} & c_{34} & c_{35} & c_{36} \\ c_{41} & c_{42} & c_{43} & c_{44} & c_{45} & c_{46} \\ c_{51} & c_{52} & c_{53} & c_{54} & c_{55} & c_{56} \\ c_{61} & c_{62} & c_{63} & c_{64} & c_{65} & c_{66} \end{pmatrix} \begin{pmatrix} \varepsilon_1 \\ \varepsilon_2 \\ \varepsilon_3 \\ \varepsilon_4 \\ \varepsilon_5 \\ \varepsilon_6 \end{pmatrix} \quad (8.37)$$

In contrast to the tensor equation (8.29), Equation 8.37 establishes the relation between the applied stresses σ_m on the solid body and the resultant strains ε_n in a simple matrix form.

(Note: by calculating the matrix product in Equation 8.37 and replacing the corresponding values for σ_m, ε_n, s_{ij}, and c_{ij} given by Equations 8.33–8.36 one can easily arrive to Equations 8.30 and 8.31. Therefore, Equation 8.37 is considered to be the generalized Hooke's law in matrix form.)

8.7
The Elastic Constants of Cubic Crystals

The compliance and stiffness matrices s_{ij} and c_{ij} in Equation 8.37 have additional constraints owing to the requirement that the energy stored in an elastic solid material must be a continuous function of the strain. In fact, this determines that the compliance and stiffness matrices s_{ij} and c_{ij} are symmetric (i.e. $s_{ij} = s_{ji}$ and $c_{ij} = c_{ji}$). In addition, the symmetry of the crystalline solid material further reduces the number of independent s_{ij} and c_{ij} components [22]. For example, in the case where the solid material is a crystalline solid with cubic symmetry (which are the type of materials studied in this book) the compliance and stiffness matrices s_{ij} and c_{ij} have the form

$$s_{ij} = \begin{pmatrix} s_{11} & s_{12} & s_{12} & 0 & 0 & 0 \\ s_{12} & s_{11} & s_{12} & 0 & 0 & 0 \\ s_{12} & s_{12} & s_{11} & 0 & 0 & 0 \\ 0 & 0 & 0 & s_{44} & 0 & 0 \\ 0 & 0 & 0 & 0 & s_{44} & 0 \\ 0 & 0 & 0 & 0 & 0 & s_{44} \end{pmatrix}$$

$$c_{ij} = \begin{pmatrix} c_{11} & c_{12} & c_{12} & 0 & 0 & 0 \\ c_{12} & c_{11} & c_{12} & 0 & 0 & 0 \\ c_{12} & c_{12} & c_{11} & 0 & 0 & 0 \\ 0 & 0 & 0 & c_{44} & 0 & 0 \\ 0 & 0 & 0 & 0 & c_{44} & 0 \\ 0 & 0 & 0 & 0 & 0 & c_{44} \end{pmatrix} \quad (8.38)$$

Upon inspection of Equation 8.38, we see that in the case of crystalline solid materials with cubic symmetry the number of independent constants s_{ij} (or c_{ij}) is reduced to 3. The compliance constants s_{11}, s_{12}, and s_{44} (or equivalently the stiffness constants c_{11}, c_{12}, and c_{44}) are called the *elastic constants* of the cubic crystal and their relationship is given by the formulas

$$c_{11} = \frac{s_{11} + s_{12}}{(s_{11} - s_{12})(s_{11} + 2s_{12})}, \quad c_{12} = \frac{-s_{12}}{(s_{11} - s_{12})(s_{11} + 2s_{12})}, \quad c_{44} = \frac{1}{s_{44}} \quad (8.39)$$

By inserting Equation 8.38 into Equation 8.37, we can obtain the generalized Hooke's law for crystalline solid materials with cubic symmetry, which is given by

$$\begin{aligned} \varepsilon_1 &= s_{11}\sigma_1 + s_{12}\sigma_2 + s_{12}\sigma_3, & \sigma_1 &= c_{11}\varepsilon_1 + c_{12}\varepsilon_2 + c_{12}\varepsilon_3 \\ \varepsilon_2 &= s_{12}\sigma_1 + s_{11}\sigma_2 + s_{12}\sigma_3, & \sigma_2 &= c_{12}\varepsilon_1 + c_{11}\varepsilon_2 + c_{12}\varepsilon_3 \\ \varepsilon_3 &= s_{12}\sigma_1 + s_{12}\sigma_2 + s_{11}\sigma_3, & \sigma_3 &= c_{12}\varepsilon_1 + c_{12}\varepsilon_2 + c_{11}\varepsilon_3 \end{aligned} \quad (8.40)$$

$$\varepsilon_4 = s_{44}\sigma_4, \qquad \sigma_4 = c_{44}\varepsilon_4$$
$$\varepsilon_5 = s_{44}\sigma_5, \qquad \sigma_5 = c_{44}\varepsilon_5$$
$$\varepsilon_6 = s_{44}\sigma_6, \qquad \sigma_6 = c_{44}\varepsilon_6$$

That is, Equations 8.40 determine the relationship between the applied stresses and the resultant strains for solid materials with cubic crystal symmetry. Note that in order to establish such an important relationship, only three scalar constants (e.g. s_{11}, s_{12}, and s_{12}) are required. (Note: the form of the matrices s_{ij} and c_{ij} for solid materials with other crystal symmetries can be found in [22].)

8.7.1
Young's Modulus and Poisson's Ratio

We now study the physical significance of the three cubic elastic constants s_{11}, s_{12}, and s_{44}. Indeed, these constants are related to three independent intrinsic mechanical properties of solid crystals known as the *Young's modulus, Poisson's ratio*, and *shear modulus*. In particular, we focus in this section on the Young's modulus and the Poisson's ratio and establish the relationship of the cubic elastic constants with these two mechanical properties.

Consider a cubic element within a crystalline solid material with cubic symmetry, which is deformed by an applied tensile stress σ_3 along the x_3 direction (Figure 8.10). Initially, the cubic element (which is shown in dashed lines) has sides of length

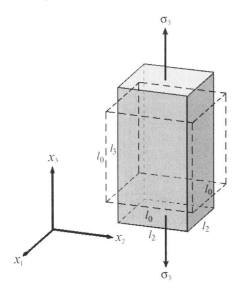

Fig. 8.10 A cubic element within a crystalline solid material with cubic symmetry is deformed by an applied stress σ_3 along the x_3 direction. The resultant longitudinal strain ε_3 experienced by the element is given by $\varepsilon_3 = (l_3 - l_0)/l_0$, whereas the resultant transversal strains ε_1 and ε_2 are given by $\varepsilon_1 = \varepsilon_2 = (l_2 - l_0)/l_0$. Note that according to Figure 8.5 and Equation 8.33, we have $\sigma_3 = \sigma_{33}$.

l_0. After the stress σ_3 is applied, the cubic element deforms into a rectangular parallelepiped of sides l_2, l_2, and l_3 (which is shown in solid lines).

According to Section 8.2, the longitudinal strain ε_3 along the x_3 direction is given by

$$\varepsilon_3 = \frac{(l_3 - l_0)}{l_0} \tag{8.41}$$

whereas the transversal strains ε_1 and ε_2 along the x_1 and x_2 directions are given by

$$\varepsilon_1 = \varepsilon_2 = \frac{(l_2 - l_0)}{l_0} \tag{8.42}$$

Note that because σ_3 is a tensile stress, we have $l_3 > l_0$ and $l_2 < l_0$. As a result, the longitudinal strain ε_3 is positive whereas the transversal strains ε_1 and ε_2 are negative. (Note: the opposite situation occurs when σ_3 is a compressive stress.)

The Young's modulus E is defined as the ratio between the applied stress σ_3 and the corresponding longitudinal strain ε_3, and it is an intrinsic mechanical property of the solid material. We thus have

$$\text{Young's modulus } E = \frac{\sigma_3}{\varepsilon_3} \tag{8.43}$$

Equation 8.43 indicates that the Young's modulus E measures the resistance of the crystalline solid to strain along the direction of the applied stress. This property is also called *stiffness*.

The Poisson's ratio ν is defined as the ratio between the transversal and longitudinal strains, and it is also an intrinsic mechanical property of the solid material.

$$\text{Poisson's ratio } \nu = -\frac{\varepsilon_1}{\varepsilon_3} = -\frac{\varepsilon_2}{\varepsilon_3} \tag{8.44}$$

Because the transversal and longitudinal strains have different signs, the Poisson's ratio formula contains a minus sign in order to obtain positive Poisson's ratios for normal solid materials. Indeed, Equation 8.44 indicates that the Poisson's ratio ν measures the relative value of the transversal strains with respect to the longitudinal strain.

By using the generalized Hooke's law equations (8.40), we can obtain the transversal and longitudinal strains ε_1, ε_2, and ε_3 generated by the applied stress σ_3. Since we know *a priori* the value of the stress component tensor σ_3, we use the equations on the left-hand side of Equation 8.40 and replace $\sigma_i = 0$ ($i \neq 3$) to obtain

$$\begin{aligned}\varepsilon_1 &= s_{12}\sigma_3 \\ \varepsilon_2 &= s_{12}\sigma_3 \\ \varepsilon_3 &= s_{11}\sigma_3 \\ \varepsilon_4 &= 0 \\ \varepsilon_5 &= 0 \\ \varepsilon_6 &= 0\end{aligned} \tag{8.45}$$

Equation 8.45 indicates that $\sigma_3/\varepsilon_3 = 1/s_{11}$, and $\varepsilon_2/\varepsilon_3 = s_{12}/s_{11}$. By considering Equations 8.43 and 8.44, we have

$$s_{11} = \frac{1}{E}, \quad s_{12} = -\frac{\nu}{E} \tag{8.46}$$

This means that the elastic constant s_{11} represents the reciprocal of the Young's modulus, whereas the elastic constant s_{12} represents the Poisson's ratio divided by the Young's modulus. Therefore, the elastic constants s_{11} and s_{12} represent intrinsic mechanical properties of the solid material.

In the next section, we find the physical significance of the remaining elastic constant s_{44}. To do this, we first need to introduce an additional intrinsic mechanical property of solid materials known as the shear modulus G.

8.7.2
The Shear Modulus

Consider a cubic element that deforms under the applied pure shear stresses τ as shown in Figure 8.11.

The shear modulus G is defined as the ratio between the applied shear stresses τ and the resultant angle γ, and it is an intrinsic mechanical property of the solid material.

$$\text{Shear modulus } G = \frac{\tau}{\gamma} \tag{8.47}$$

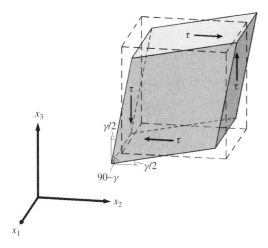

Fig. 8.11 A cubic element within a crystalline solid material with cubic symmetry is deformed by the applied shear stresses τ. The angle γ is directly proportional to the applied stress τ and inversely proportional to the shear modulus G of the solid. Note that according to Figure 8.5 and Equation 8.33, we have $\tau = \sigma_{32} = \sigma_{23} = \sigma_4$.

8 Periodic Cellular Solids

By considering Section 8.4.3, the deformation of the cubic element in Figure 8.11 can be described as a pure strain. For example, by comparing Figures 8.11 and 8.9, we can see that $d\alpha = d\beta = \gamma/2$ and the deformation in Figure 8.11 can therefore be described as (Equations 8.25–8.27)

$$e_{ij} = \varepsilon_{ij} + \varpi_{ij} \tag{8.48}$$

where

$$e_{ij} = \begin{pmatrix} 0 & 0 & 0 \\ 0 & 0 & \gamma/2 \\ 0 & \gamma/2 & 0 \end{pmatrix},$$

$$\varepsilon_{ij} = \begin{pmatrix} 0 & 0 & 0 \\ 0 & 0 & \gamma/2 \\ 0 & \gamma/2 & 0 \end{pmatrix},$$

$$\varpi_{ij} = \begin{pmatrix} 0 & 0 & 0 \\ 0 & 0 & 0 \\ 0 & 0 & 0 \end{pmatrix} \tag{8.49}$$

which corresponds to a deformation that involves only strain of the element and not rotation (i.e. pure strain).

Equation 8.49 determines that the symmetric components $\varepsilon_{23} = \varepsilon_{32}$ of the strain tensor are given by

$$\varepsilon_{23} = \varepsilon_{32} = \frac{\gamma}{2} \tag{8.50}$$

In addition, the generalized Hooke's law equations (8.40) establish that

$$\varepsilon_4 = s_{44}\sigma_4 \tag{8.51}$$

By considering Equations 8.33 and 8.34, the above equation transforms into

$$2\varepsilon_{23} = s_{44}\sigma_{23} \tag{8.52}$$

or, equivalently,

$$\gamma = s_{44}\tau \tag{8.53}$$

Finally, by comparing Equations 8.53 and 8.47, we have

$$s_{44} = \frac{1}{G} \tag{8.54}$$

This means that the cubic elastic constant s_{44} represents the reciprocal of the shear modulus G of the solid material.

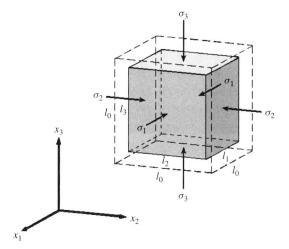

Fig. 8.12 A cubic element within a crystalline solid material with cubic symmetry is deformed by applying normal stresses σ_1, σ_2, and σ_3. The relative change in volume $\Delta V/V_0$, or dilatation, is inversely proportional to the compressibility K of the solid material.

8.7.3
The Bulk Modulus

We next introduce an additional mechanical property of solid materials known as the *bulk modulus*. Consider, for example, the deformation of a cubic element under compressive normal stresses σ_1, σ_2, and σ_3 along the orthogonal directions x_1, x_2, and x_3 as shown in Figure 8.12. Initially, the cubic element has sides of length l_0 (see dashed lines) but after the stresses σ_1, σ_2, and σ_3 are applied, the cubic element deforms into a rectangular parallelepiped of sides l_1, l_2, and l_3 (see solid lines).

The *pressure* on the cubic volume element is defined as the average value of the normal stresses σ_1, σ_2, and σ_3. That is,

$$p = -1/3(\sigma_1 + \sigma_2 + \sigma_3) \tag{8.55}$$

where the minus sign is included in order to obtain positive values for the pressure since compressive stresses are negative by convention.

After the stresses are applied, the cubic volume element decreases its volume by $\Delta V = V_f - V_0$, where V_f is the final volume given by $l_1 l_2 l_3$, and V_0 is the initial volume given by l_0^3 (Figure 8.12). The relative change in volume $\Delta V/V_0$ is called the *dilatation* of the element and is given by

$$\frac{\Delta V}{V_0} = \frac{V_f - V_0}{V_0} = \frac{l_1 l_2 l_3 - l_0^3}{l_0^3} = \frac{l_1 l_2 l_3}{l_0^3} - 1 \tag{8.56}$$

The strains along the x_1, x_2, and x_3 directions are given by

$$\varepsilon_1 = \frac{(l_1 - l_0)}{l_0}, \quad \varepsilon_2 = \frac{(l_2 - l_0)}{l_0}, \quad \varepsilon_3 = \frac{(l_3 - l_0)}{l_0} \tag{8.57}$$

Therefore, we have

$$l_1 = l_0(1 + \varepsilon_1), \quad l_2 = l_0(1 + \varepsilon_2), \quad \text{and} \quad l_3 = l_0(1 + \varepsilon_3) \tag{8.58}$$

and the dilatation of the volume element (Equation 8.56) can be written as

$$\frac{\Delta V}{V_0} = (1 + \varepsilon_1)(1 + \varepsilon_2)(1 + \varepsilon_3) - 1 \approx \varepsilon_1 + \varepsilon_2 + \varepsilon_3 \tag{8.59}$$

The bulk modulus K, which is an intrinsic mechanical property of the solid material, is defined as the ratio between the applied pressure p and the resultant dilatation $\Delta V/V_0$

$$\text{Bulk modulus } K = -\frac{p}{\Delta V/V_0} \tag{8.60}$$

where the minus sign is included because the resultant dilatation $\Delta V/V_0$ is negative under compressive stresses.

Equation 8.60 determines that the bulk modulus K measures the resistance of the solid crystal material to uniform compression. By considering Equations 8.55, 8.59 and 8.40, the bulk modulus can equivalently be written as

$$K = \frac{1}{3}\left(\frac{\sigma_1 + \sigma_2 + \sigma_3}{\varepsilon_1 + \varepsilon_2 + \varepsilon_3}\right) = \frac{1}{3}\left(\frac{\sigma_1 + \sigma_2 + \sigma_3}{(\sigma_1 + \sigma_2 + \sigma_3)(s_{11} + 2s_{12})}\right) = \frac{1}{3(s_{11} + 2s_{12})} \tag{8.61}$$

To summarize, crystalline solid materials with cubic symmetry are characterized by three independent elastic constants (e.g. s_{11}, s_{12}, and s_{44}). These elastic constants are related to intrinsic mechanical properties of the solid material such as the Young's modulus, Poisson's ratio, shear modulus, and bulk modulus by Equations 8.46, 8.54 and 8.61, respectively. As a result, solid materials can be equivalently characterized by their intrinsic mechanical properties. In this book, we choose to characterize solid materials by the following three independent mechanical properties: the Young's modulus E, the shear modulus G, and the bulk modulus K. The relation between these mechanical properties and the cubic elastic constants is summarized below:

$$E = \frac{1}{s_{11}}, \quad G = \frac{1}{s_{44}}, \quad K = \frac{1}{3(s_{11} + 2s_{12})} \tag{8.62}$$

8.8
Topological Design of Periodic Cellular Solids

In the next part of this chapter, we study the effective *elastic* mechanical properties of periodic cellular solids. As previously mentioned, periodic cellular solids are structures made of a combination of solid material and air. A significant characteristic of these structures is the fact that they possess effective mechanical properties that are different from those of the constituent solid material. In fact, the search for periodic cellular solids that are both mechanically robust and weight efficient is extremely important.

In the past, much attention has been focused on the fabrication of these structures at the millimeter scale. In this chapter, we are interested in the prospect of having such complex structures constructed at the submicron scale because these structures would allow us to access novel length-scale-dependent mechanical properties. For example, it was recently found that the strain required to break polymer systems strongly depends on the absolute thickness of the specimen and the local thickness within the microstructure of material. In addition, both the value of this critical thickness and the intrinsic strain at break of a polymer are determined by the relative density of the solid in the structure [4].

To fabricate periodic cellular solids on the submicron scale, we have a variety of techniques. From the perspective of mechanical properties, however, it is desirable to be able to cover large areas (i.e. to obtain large samples) while still having control over the intrinsic geometry of the resultant structures.

Construction-based serial writing techniques allow us to create arbitrary structures but are usually slow and cover small areas. For example, two-photon lithography makes use of the two-photon-absorption phenomena to activate chemical processes within a specific material with accurate spatial resolution and this makes possible the lithographic microfabrication of three-dimensional structures. In addition, concentrated polyelectrolyte inks allow us to directly write three-dimensional periodic structures at micron length scales without the use of masks, and the robotic micromanipulation of particles also demonstrates the feasibility of the assembly of three-dimensional microstructures.

On the other hand, self-assembly fabrication techniques can be rapid and can cover large areas but afford limited control over the geometries of the structures. For example, monodispersed colloidal spheres and block-copolymers (such as styrene and isoprene) spontaneously assemble into specific microstructures and provide simple fabrication techniques, but the number of structures that can be accessed through these techniques is quite restricted.

In this chapter, however, we focus on periodic cellular solids fabricated by interference lithography since this technique allows control over the geometry of the structure while still creating large-area single crystals. In addition, by using this technique, periodic cellular solids can be fabricated at small length scales. The only requirement is that the target periodic structures be bicontinuous in order to allow development of the photoresist.

In order to be able to create periodic cellular solids using interference lithography, we look for periodic structures in the simple cubic (sc), body-centered-cubic (bcc), and face-centered-cubic (fcc) Bravais lattices. One way to achieve this would be to adopt an approach similar to that used in Chapter 2, where the structures are described by the sum of a small number of Fourier terms and correlated to the parameters of the interfering beams so that they can be fabricated by either single or multiple exposures via interference lithography. In this chapter, however, we choose to start our search for lightweight periodic structures accessible via interference lithography by looking at the classic materials science structures; that

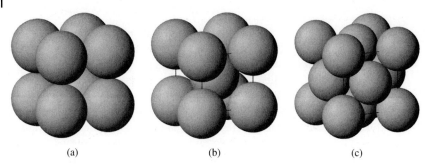

Fig. 8.13 Periodic structures made of spheres arranged in the (a) simple cubic Bravais lattice, (b) body-centered-cubic Bravais lattice, and (c) face-centered-cubic Bravais lattice.

is, structures made of spheres located at the sites of the sc, bcc, and fcc lattices (Figure 8.13).

In principle, we could explore the following two options: structures made of solid spheres in air background and structures made of air spheres in a solid background. In the case of solid spheres in air, in order to obtain self-supported structures, it is necessary that the spheres be connected with each other. This means that these self-supported structures have high volume fractions of solid material and therefore the resultant structures have substantial relative weight. Because we are interested in lightweight periodic cellular solids, the opposite case, which is structures made of air spheres in a solid background, is the appropriate choice. In this inverse case, the structures can reach low volume fractions of solid material while still being self-supported.

We also consider a group of interesting periodic structures that are obtained by examining the geometry of the structures made of air spheres and deriving their rod-connected counterparts. To illustrate this, consider a periodic structure made of air spheres in a solid background arranged in the sc Bravais lattice. Figure 8.14a shows the unit cell of this structure, where the air spheres are located at the corners of the unit cell. Figure 8.14b shows the corresponding rod-connected model, which is obtained by reproducing the geometry of the air–sphere structure by the use of rods. The rod-connected model is therefore a periodic structure with similar geometries to the air–sphere structure but made of connected solid rods. Rod-connected models are also known as *microtruss structures* and are important because they are usually considered as standard models for periodic cellular solids.

In addition, we consider level-set periodic structures that mimic the geometries of the air–sphere and rod-connected models. The advantage of these level-set structures, which are described by trigonometric formulas, is the fact that they can be fabricated at small length scales via interference lithography. For example, Figure 8.14c shows a level-set structure with similar geometries to the air–sphere and rod-connected structures shown in Figures 8.14a and b. Significantly, we next show that such level-set structures can have *better* mechanical properties than the rod-connected model counterparts at a given volume fraction of solid material. In particular, the level-set equation that describes the simple cubic periodic structure

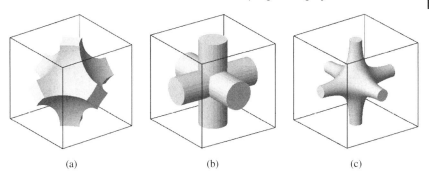

Fig. 8.14 Periodic cellular solids in the simple cubic Bravais lattice. (a) Air spheres in a solid background arranged in the simple cubic Bravais lattice. Note that the air spheres are located at the corners of the cubic unit cell. (b) A corresponding rod-connected model. (c) A corresponding level-set tubular P structure, which is described by Equation 8.63. For each structure, the volume fraction of the solid material is $f = 0.20$, 0.20, and 0.08, respectively.

shown in Figure 8.14c is given by

$$f_{sc}(x, y, z) = \cos\left(\frac{2\pi x}{a}\right) + \cos\left(\frac{2\pi y}{a}\right) + \cos\left(\frac{2\pi z}{a}\right)$$
$$- 0.5 \cos\left(\frac{2\pi x}{a}\right) \cos\left(\frac{2\pi y}{a}\right)$$
$$- 0.5 \cos\left(\frac{2\pi y}{a}\right) \cos\left(\frac{2\pi z}{a}\right) \quad (8.63)$$
$$- 0.5 \cos\left(\frac{2\pi z}{a}\right) \cos\left(\frac{2\pi x}{a}\right) + t$$

where the parameter t determines the volume fraction of the solid material in the structure. By considering different values of the parameter t, Equation 8.63 describes a family of level-set structures with different volume fractions of solid material known as the *tubular P structures*. In particular, the tubular P structure shown in Figure 8.14c has a volume fraction of solid material $f = 0.08$.

Figure 8.15 shows the structures that correspond to the bcc lattice case. The structure made of air spheres is shown on the left, the corresponding rod-connected model at the center, and the level-set structure on the right. In this case, the equation that describes the bcc level-set structure is given by

$$f_{bcc}(x, y, z) = \cos\left(\frac{2\pi x}{a}\right) \cos\left(\frac{2\pi y}{a}\right) + \cos\left(\frac{2\pi y}{a}\right) \cos\left(\frac{2\pi z}{a}\right)$$
$$+ \cos\left(\frac{2\pi z}{a}\right) \cos\left(\frac{2\pi x}{a}\right) + t \quad (8.64)$$

Equation 8.64 describes a family of structures with different volume fractions, which are approximations to a triply periodic minimal surface discovered by

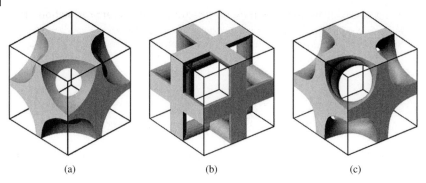

Fig. 8.15 Periodic cellular solids in the body-centered-cubic Bravais lattice. (a) Air spheres in a solid background arranged in the bcc Bravais lattice. Note that the air spheres are located at the corners and at the center of the cubic unit cell. (b) A corresponding rod-connected model. (c) A corresponding level-set I-WP structure, which is described by Equation 8.64. For each structure, the volume fraction of the solid material is $f = 0.20$.

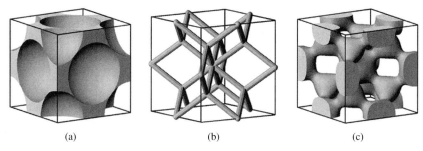

Fig. 8.16 Periodic cellular solids in the face-centered-cubic Bravais lattice. (a) Air spheres in a solid background arranged in the fcc Bravais lattice. Note that the air spheres are located at the corners and at the center of the faces of the cubic unit cell. (b) A corresponding rod-connected model. (c) A corresponding level-set F-RD structure, which is described by the formula 8.65. For each structure, the volume fraction of the solid material is $f = 0.26, 0.04$, and 0.25, respectively.

Schoen, known as the *I-WP*. In particular, the level-set I-WP structure shown in Figure 8.15c has a volume fraction of solid material $f = 0.20$.

Finally, Figure 8.16 shows the structures corresponding to the fcc Bravais lattice. In this case, the equation that describes the fcc level-set structure is given by

$$f_{\text{fcc}}(x, y, z) = 4\cos\left(\frac{2\pi x}{a}\right)\cos\left(\frac{2\pi y}{a}\right)\cos\left(\frac{2\pi z}{a}\right)$$
$$+ \cos\left(\frac{4\pi x}{a}\right)\cos\left(\frac{4\pi y}{a}\right)$$
$$+ \cos\left(\frac{4\pi x}{a}\right)\cos\left(\frac{4\pi z}{a}\right) \quad (8.65)$$
$$+ \cos\left(\frac{4\pi y}{a}\right)\cos\left(\frac{4\pi z}{a}\right) + t$$

Equation 8.65 describes a family of structures that are approximations to a triply periodic minimal surface also discovered by Schoen, known as the *F-RD*. In particular, the level-set F-RD structure shown in Figure 8.16c has a volume fraction of solid material $f = 0.25$.

8.9
Finite Element Program to Calculate Linear Elastic Mechanical Properties

Our aim in the next sections is to study the dependence of the effective elastic mechanical properties of periodic cellular solids on relative density and microstructure. To do this, we calculate the effective linear elastic mechanical properties for the air–sphere structures, the corresponding rod-connected models, and the level-set structures as a function of the relative density of the solid material (or equivalently, as a function of the volume fraction of solid material in the structure). To compute the mechanical properties of periodic cellular solids, we use a freely available FORTRAN 77 finite element code that can be obtained at *ftp://ftp.nist.gov/pub/bfrl/garbocz/FDFEMANUAL/*

The numerical algorithm is based on the following scheme. When a particular structure is subject to applied stresses, the resultant strains are such that the total energy stored in the structure is minimized. This means that the gradient of the total energy with respect to the elastic displacements is equal to zero. In this algorithm, the structure is spatially discretized into finite elements and its total energy is written as a function of the displacements at the nodes of the elements. This expression is then minimized with respect to the nodal displacements by using the conjugate gradient method. Once the displacements that minimize the energy are obtained, they can be used to calculate the effective mechanical properties of the periodic structure. For numerical details about the algorithm, the reader should consult the freely available manual provided in the above link.

8.10
Linear Elastic Mechanical Properties of Periodic Cellular Solids

By using the above-mentioned numerical code, we study the linear elastic mechanical properties of the periodic structures presented in Section 8.8. We consider that the structures are made of a solid material and air. In particular, the solid material forming the solid networks of the structures is itself considered to be *isotropic*.

Isotropic solid materials have compliance and stiffness matrices given by Equation 8.38 but they must satisfy the following additional constraints:

$$s_{44} = 2(s_{11} - s_{12}) \quad \text{and} \quad c_{44} = \frac{1}{2}(c_{11} - c_{12}) \quad (8.66)$$

This means that isotropic solid materials have only two independent elastic constants (e.g. s_{11} and s_{12}). Since the elastic constants are related to the mechanical

Table 8.1 The choice of mechanical properties used to completely describe the isotropic solid material forming the structures and the cubic periodic structures. Other mechanical properties and elastic constants can be calculated by using the formulas provided in the right column

	Mechanical properties	Relationships to other mechanical properties and elastic constants
Solid material (isotropic)	E^s, ν^s	$G^s = \dfrac{E^s}{2(1+\nu^s)}$, $K^s = \dfrac{E^s}{3(1-2\nu^s)}$ $s_{11}^s = \dfrac{1}{E^s}$, $s_{12}^s = -\dfrac{\nu^s}{E^s}$, $s_{44}^s = 2(s_{11}^s - s_{12}^s) = \dfrac{2(1+\nu^s)}{E^s} = \dfrac{1}{G^s}$
Periodic structure (cubic symmetry)	E, G, K	$\nu = \dfrac{3K - E}{6K}$ $s_{11} = \dfrac{1}{E}$, $s_{12} = -\dfrac{\nu}{E}$, $s_{44} = \dfrac{1}{G}$

properties of the material, we choose the Young's modulus E^s and the Poisson's ratio ν^s as the two independent mechanical properties that completely characterize the isotropic solid material forming the periodic structure.

On the other hand, the periodic structures presented in Section 8.8 (which are made of this isotropic solid material and air) have cubic symmetry and they are therefore characterized by three independent elastic constants. In this case, we choose the effective Young's modulus E, shear modulus G, and bulk modulus K as the independent mechanical properties that completely characterize the periodic structures.

Table 8.1 summarizes the set of mechanical properties used to characterize the isotropic solid material forming the structures and the periodic structures.

Before calculating the effective mechanical properties of the periodic structures presented in Section 8.8, we need to establish the values of E^s and ν^s that characterize the isotropic solid material forming the structures. In fact, since we choose to normalize the effective mechanical properties of the periodic structures with respect to the Young's modulus E^s of the isotropic solid material, we need to establish only the value of the Poisson's ratio ν^s. For our calculations, we assume that the Poisson's ratio of the isotropic solid material is $\nu^s = 0.33$, which corresponds to typical values for polymers or aluminum alloys.

In Figure 8.17, we show the effect of solid volume fraction and microstructure on the effective Young's modulus E, shear modulus G, and bulk modulus K of the periodic cellular solids introduced in Section 8.8. As previously mentioned, the calculations are normalized with respect to the Young's modulus E^s of the solid material forming the structure. In the graphs corresponding to the effective Young's modulus E/E^s, lines with filled circles represent the value of E/E^s along the $\langle 100 \rangle$ cubic directions, whereas lines with empty circles represent E/E^s along the $\langle 111 \rangle$ directions. The difference between the values along the $\langle 100 \rangle$ and $\langle 111 \rangle$

cubic directions provides a measurement of the anisotropy of the cubic periodic structure. The thin solid black lines correspond to the relative Young's modulus values E/E^s for open-cell foams, which are proportional to the square of the solid density. Open-cell foams are generally used as a standard comparison for mechanical properties. Since we focus on three-dimensional structures at the submicron scale, for the level-set structures we only consider volume fractions over which the structures are guaranteed to be bicontinuous (i.e. structures where both air and solid regions are continuous) and therefore can be fabricated by interference lithography. For the idealized sphere and rod-connected structures, we consider wider volume fraction ranges in order to establish the trends in the properties. It is important to mention that rod-connected periodic cellular solids have been typically fabricated at about 15% solid volume fraction (at millimeter scales owing to difficulty of assembly).

Figure 8.17 shows that for a wide range of solid volume fractions, at a given volume fraction, the air–sphere architectures possess larger mechanical moduli, the level-set structures present intermediate values, and the rod-connected models have the smaller mechanical moduli. (Note: this trend, however, may not be valid when the solid volume fraction tends to zero.) The reason that the air–sphere and level-set structures have higher stiffness than the rod-connected models is that the solid material connecting the nodes in the first two structures has increasing diameter closer to the nodes, where the bending moments are higher: this is more efficient than having a constant-diameter solid material connecting the nodes of the structure, as in the case of the rod-connected structures. For the sc case, the structures deform by stretching when loaded along the $\langle 100 \rangle$ cubic direction but by bending when loaded along the $\langle 111 \rangle$ direction – this gives rise to the large difference in the Young's modulus in these directions as shown in Figure 8.17. Stretch-dominated moduli vary linearly with solid volume fraction, whereas bending-dominated moduli vary as the square of the solid volume fraction for open-cell foams, so that as the volume fraction decreases the difference in moduli in the two directions increases. On the other hand, the plots of the Young's modulus show that fcc structures are comparatively more isotropic. In some cases, the Young's modulus of the level-set structures (fabricable by interference lithography) compare favorably with the corresponding values for open-cell foams.

As shown in Figure 8.17, the structures made of air spheres have, in general, large mechanical moduli. However, the fabrication of these structures at small length scales as large-area single crystals in the sc and bcc cases is difficult. The construction of structures made of spheres in the fcc lattice is relatively simple in the case that the spheres are in contact with each other (e.g. via colloidal self-assembly), but when the spheres overlap, the control of volume fraction is a further experimental challenge. The rod-connected models, which are commonly considered standard models for periodic cellular solids, have smaller mechanical moduli. In addition, there is no currently available experimental technique that allows us to fabricate these rod-connected structures at small length scales. On the other hand, the proposed level-set structures have in general larger mechanical

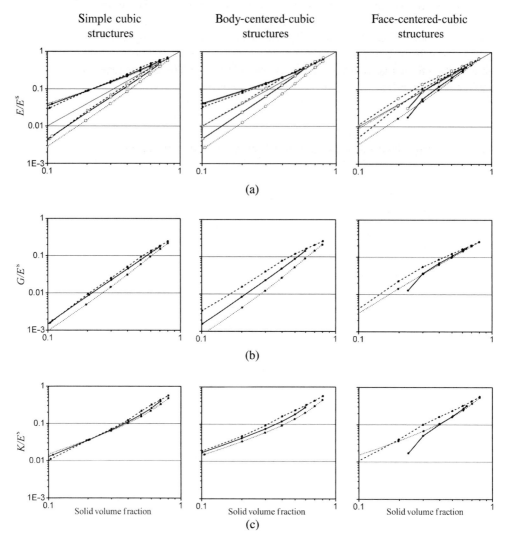

Fig. 8.17 Effective linear elastic mechanical properties as a function of solid material volume fraction for the structures shown in Figures 8.14–8.16. (a) Young's modulus E/E^s, (b) shear modulus G/E^s, and (c) bulk modulus K/E^s for air spheres (dash lines), rod-connected models (dot lines), and level-set structures (solid lines). E^s is the Young's modulus of the isotropic solid material that forms the periodic structures. Lines with filled and empty circles in (a) correspond to the effective Young's modulus E/E^s along the $\langle 100 \rangle$ and $\langle 111 \rangle$ directions, respectively. Thin solid black lines correspond to Young's modulus values E/E^s for open-cell foams.

moduli with respect to their rod-connected counterparts and can be fabricated at small length scales via the interference lithography technique. In conclusion, level-set periodic structures made of the sum of a small number of trigonometric terms open the possibility of having periodic cellular solids at small length scales with different geometries, which can result in the achievement of superior mechanical properties.

8.11
Twelve-connected Stretch-dominated Periodic Cellular Solids via Interference Lithography

When periodic cellular solids deform, they do it either by bending or by stretching of the unit cell components. While most cellular solids deform by bending, those that deform by stretching have a much higher stiffness per unit weight. As a result, there is a search for structures that experience stretch-dominated deformations as opposed to bending-dominated deformations.

Recently there has been much interest in periodic cellular solids that achieve this functionality through the rational design of their geometry. For example, it has been demonstrated that the minimum connectivity between the unit cell components for structured materials to experience stretch-dominated deformations is 6 for two-dimensional cellular solids and 12 in the three-dimensional case [3].

It is therefore interesting to examine a classic periodic structure in materials science, which is twelve-connected. This structure is made of solid spheres in air located at the sites of the fcc Bravais lattice. As shown in Figure 8.18a, each sphere is in contact with 12 neighboring spheres and the structure is therefore twelve-connected. The disadvantage of the structure, however, is the fact that the

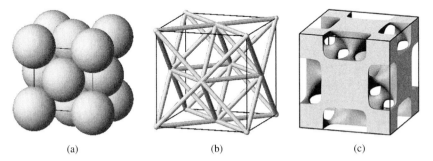

Fig. 8.18 Twelve-connected periodic cellular solids. (a) Solid spheres in air arranged in the fcc Bravais lattice. (b) A corresponding rod-connected model. (c) A corresponding level-set inverse F-RD structure, which is described by the formula 8.65. For each structure, the volume fraction of the solid material is $f = 0.74$, 0.03, and 0.38, respectively.

volume fraction of solid material is large ($f = 0.74$); thus, this prevents the use of this structure as a weigh-efficient periodic cellular solid.

We therefore concentrate on the corresponding rod-connected model (Figure 8.18b). This structure is also twelve-connected and has recently attracted considerable attention because it has been demonstrated to be stretch-dominated both theoretically and experimentally at the millimeter scale [16]. In particular, the stiffness of the structure has been shown to compare favorably with open-cell foams at small volume fractions of solid material and have similar stiffness at high volume fractions. In spite of that, the fabrication of this rod-connected model at small length scales is quite difficult (imagine trying to glue together all the thin rods at submicron length scales).

In order to obtain a lightweight, twelve-connected structure that can be fabricated at small length scales, we consider the level-set analog of the rod-connected model, which is the inverse F-RD structure shown in Figure 8.18c.

We calculate and compare the effective linear elastic mechanical properties of the rod-connected and the level-set structures (Figure 8.19). We do not consider the structure made of solid spheres in air since this architecture is self-supporting only for solid volume fractions larger than 0.74 and, as previously mentioned, is therefore not useful for lightweight structures. Figure 8.19 shows the Young's modulus, shear modulus, and bulk modulus as a function of solid volume fraction for the rod-connected model (dotted lines) and the level-set inverse F-RD structure (solid lines). It can be seen that both structures have comparable linear elastic constants. The main advantage of the level-set inverse F-RD structure, however, is the fact that it can be fabricated at small length scales via interference lithography.

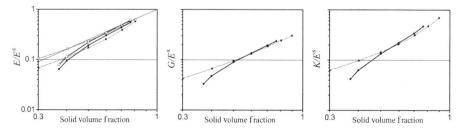

Fig. 8.19 Effective linear elastic mechanical properties as a function of solid volume fraction for the twelve-connected rod-connected model (dotted lines) and the level-set inverse F-RD structure (solid lines). E^s is the Young's modulus of the isotropic solid material that forms the periodic structures. Lines with filled and empty circles correspond to the effective Young's modulus E/E^s along the $\langle 100 \rangle$ and $\langle 111 \rangle$ directions, respectively. Thin solid black lines correspond to Young's modulus values E/E^s for open-cell foams. The graphics show that the rod-connected model and the level-set inverse F-RD structure have comparable linear elastic mechanical response.

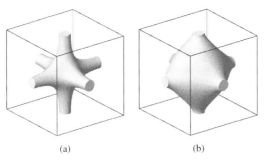

Fig. 8.20 Simple cubic cellular solids. (a) The tubular P structure. (b) The P structure.

8.12
Fabrication of a Simple Cubic Cellular Solid via Interference Lithography

We next consider the experimental realization of a simple cubic cellular solid by interference lithography. In order to do this, we examine the P structure, which is given by Equation 2.33. This structure possesses similar structural characteristics to those of the tubular P structure given by Equation 8.63. This is due to the fact that the formulas describing the structures have commonly retained Fourier terms. Figure 8.20 shows both the tubular P structure and the P structure where the structural similarity is evident.

It is important to mention that the P structure is less difficult to fabricate than the tubular P structure because it requires the superposition of a smaller number of beams when fabricated by interference lithography. The disadvantage of the P structure, however, is that it is self-supported and bicontinuous only for solid volume fractions ranging from 0.21 to 0.79, whereas the tubular P structure can reach

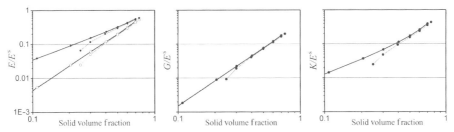

Fig. 8.21 Comparison of effective linear elastic mechanical properties between the tubular P and P level-set structures. Broad and narrow solid lines correspond to the tubular P structure and P structure, respectively. In the Young's modulus plot, symbols (●, ○) correspond to the effective Young's modulus E/E^s along the $\langle 100 \rangle$ and $\langle 111 \rangle$ directions, respectively. Note that the tubular P structure is self-supported and bicontinuous at very low solid volume fractions, whereas the level-set P structure is self-supported and bicontinuous only for volume fractions within the range 0.21–0.79. The advantage of the level-set P structure, however, is its ease of fabrication at submicron length scales.

very low solid volume fractions. Figure 5.5 shows the experimental realization of a P structure through interference lithography by using a SU-8 photoresist material.

We compare in Figure 8.21 the Young's modulus, shear modulus, and bulk modulus as a function of the solid volume fraction for these two cubic structures. It can be seen that owing to their similar structural characteristics, both structures have comparable effective linear elastic mechanical properties at a given volume fraction.

8.13
Plastic Deformation of Microframes

Periodic, bicontinuous polymer/air structures can also have very interesting *large* strain mechanical behavior and are promising candidates for lightweight materials having improved mechanical performance, especially as to do with energy absorption. This is because when a material is structured at and below a critical size, a length-scale dependence can arise in the properties, which is not found in the corresponding homogeneous bulk material. Significant plasticity of glassy

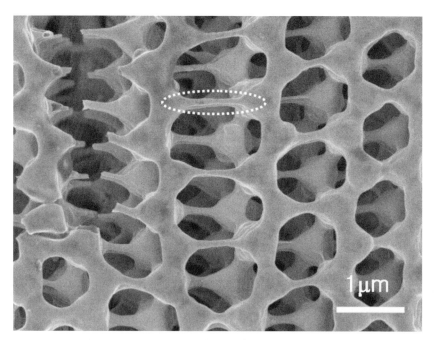

Fig. 8.22 A scanning electron micrograph of a portion of a deformed SU-8 microframe sample exhibiting necking, plastic flow, and fracture of the struts [20]. The circled strut has undergone nearly 300% elongation without fracture. Such microframes hold the promise as lightweight energy absorbing structures.

amorphous polymers, such as the photoresists used to make microframes, requires a combination of two factors: a certain chemical cross-link density and a sample with a thickness or ligament size below a critical value. In this situation, the deformation in the material can approach that of the draw ratio of the entanglement or cross-link network. The required critical thickness (or ligament diameter) for the onset of enhanced plasticity increases with increasing cross-link density (which, however, limits the ultimate strain to break and hence energy absorption) [19].

A bicontinuous SU-8 epoxy microframe fabricated using multibeam interference lithography is shown in Figure 8.22. The right-hand side of the image shows the essentially undeformed unit cells of the original structure. However, the struts to the left-hand side of the image show increasingly large amounts of plastic deformation. By comparing the respective lengths of the undeformed and deformed struts, one can estimate that the most highly deformed struts have undergone approximately 300% deformation. These thin struts exhibit almost 100 times more deformation than a bulk film of the same material that is essentially brittle (<5% deformation). In order to more fully understand the origin of this interesting behavior, simple homogeneous beam-shaped samples made of SU-8 were fabricated by photolithography using a mask. The cross-link density and resultant glass transition temperature, which influence the mechanical properties of SU-8, are sensitive to the fabrication conditions (wavelength of light, exposure dose, baking time, and temperature); thus, it is important to make sure that all beam-shaped samples are processed in a consistent manner.

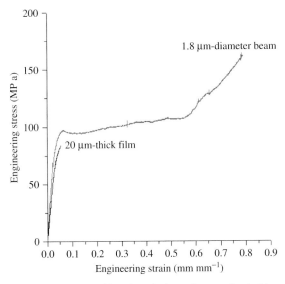

Fig. 8.23 Examples of length-scale-dependent mechanical behavior. The 20-µm-thick epoxy film shows a maximum strain of only about 5%, which is typical of bulk cross-linked SU-8, while the strain to failure of the 1.8 µm diameter beam reaches 80%.

A homogeneous beam-shaped sample with a diameter of 1.8 µm and a homogeneous 20 µm thick film, both made of SU-8, were uniaxially loaded and their constitutive stress-strain behaviour measured. Figure 8.23 shows the superior mechanical behavior of the beam-shaped sample over that of the thick flim. These constitutive stress–strain relations can be used with finite-element modeling to predict the large-strain mechanical behavior of microframes with various defined geometries made by interference lithography (Figure 8.22). The toughness (the area under the stress–strain curve) represents the energy absorbed per unit volume. Importantly, the enhanced modulus, plasticity, strain hardening, and the corresponding larger toughness of the thin beam microframes suggest good prospects for superior miniature microelectromechanical (MEMS) devices as well as lightweight, energy-absorbing structures.

In summary, in this chapter we considered both linear elastic and nonlinear plastic mechanical properties of periodic cellular solids. After reviewing the basic concepts on mechanical properties of materials, we showed that the interference lithography technique can be used to create periodic cellular solids with larger effective elastic mechanical properties than their rod-connected counterparts (or microtruss structures). An essential feature of the holographic fabrication is that the periodic cellular solids can be fabricated at submicron length scales. This opens up the possibility of creating cellular solids on a scale that allows us access to novel length-scale-dependent mechanical properties and multifunctional characteristics.

Further Reading

1 Evans, A.G., Hutchinson, J.W., Fleck, N.A., Ashby, M.F. and Wadley, H.N.G. (**2001**) "The topological design of multifunctional cellular metals". *Progress in Materials Science*, **46**, 309–27.
2 Ashby, M.F. (**2006**) "The properties of foams and lattices". *Philosophical Transactions of the Royal Society A*, **364**, 15–30.
3 Deshpande, V.S., Ashby, M.F. and Fleck, N.A. (**2001**) "Foam topology bending versus stretching dominated architectures". *Acta Materiala*, **49**, 1035–40.
4 Van der Sanden, M.C.M., Meijer, H.E.H. and Lemstra, P.J. (**1993**) "Deformation and toughness of polymeric systems: the concept of a critical thickness". *Polymer*, **34**, 2148–54.
5 Cumpston, B.H., Ananthavel, S.P., Barlow, S., Dyer, D.L., Ehrlich, J.E., Erskine, L.L., Heikal, A.A., Kuebler, S.M., Lee, I-Y.S., McCord-Maughon, D., Qin, J., Rockel, H., Rumi, M., Wu, X.-L., Marder, S.R. and Perry, J.W. (**1999**) "Two-photon polymerization initiators for three-dimensional optical data storage and microfabrication". *Nature*, **398**, 51–54.
6 Gratson, G.M., Xu, M.J. and Lewis, J.A. (**2004**) "Microperiodic structures – direct writing of three-dimensional webs". *Nature*, **428**, 386–86.
7 Miyazaki, H. and Sato, T. (**1997**) "Mechanical assembly of three-dimensional microstructures from fine particles". *Advance Robotics*, **11**, 169–85.
8 Xia, Y.N., Gates, B., Yin, Y.D. and Lu, Y. (**2000**) "Monodispersed colloidal spheres: old materials with new applications". *Advanced Materials*, **12**, 693–713.
9 Thomas, E.L., Alward, D.B., Kinning, D.J., Martin, D.C., Handlin, D.L. and Fetters, L.J. (**1986**) "Ordered bicontinuous double-diamond structure of star block copolymers – a new

equilibrium microdomain morphology". *Macromolecules*, **19**, 2197–202.
10. Mei, D., Cheng, B., Hu, W., Li, Z. and Zhang, D. (**1995**) "Three dimensional ordered patterns by light interference". *Optics Letters*, **20**, 429–31.
11. Campbell, M., Sharp, D.N., Harrison, M.T., Denning, R.G. and Turberfield, A.J. (**2000**) "Fabrication of photonic crystals for the visible spectrum by holographic lithography". *Nature*, **404**, 53–56.
12. Ullal, C.K., Maldovan, M., Chen, G., Han, Y.-J., Yang, S. and Thomas, E.L. (**2004**) "Photonic crystals through holographic lithography: Simple cubic, diamond-like, and gyroid-like structures". *Applied Physics Letters*, **84**, 5434–36.
13. Wohlgemuth, M., Yufa, N., Hoffman, J. and Thomas, E.L. (**2001**) "Triply periodic bicontinuous cubic microdomain morphologies by symmetries". *Macromolecules*, **34**, 6083–89.
14. Schoen, A. (**1970**) *NASA Technical Note TN D-5541*.
15. Garboczi, E.J. and Day, A.R. (**1995**) "An algorithm for computing the effective linear elastic properties of heterogeneous materials: three-dimensional results for composites with equal phase Poisson ratios". *Journal of the Mechanics and Physics of Solids*, **43**, 1349–62.
16. Leunissen, M.E., Christova, C.G., Hynninen, A.P., Royall, C.P., Campbell, A.I., Imhof, A., Dijkstra, M., van Roij, R. and van Blaaderen, A. (**2005**) "Ionic colloidal crystals of oppositely charged particles". *Nature*, **437**, 235–40.
17. Deshpande, V.S., Fleck, N.A. and Ashby, M.F. (**2001**) "Effective properties of the octet-truss lattice material". *Journal of the Mechanics and Physics of Solids*, **49**, 1747–69.
18. Maldovan, M., Ullal, C.K., Jang, J.-H. and Thomas, E.L. (**2007**) "Sub-micrometer scale periodic porous cellular structures: microframes prepared by holographic interference lithography". *Advanced Materials*, **19**, 3809–13.
19. Kramer, E.J. and Berger, L.L. (**1990**) "Fundamental processes of craze growth and fracture". *Advances in Polymer Science*, **91/92**, 1–68.
20. Jang, J.-H., Ullal, C.K., Choi, T., LeMieux, M.C., Tsukruk, V.V. and Thomas, E.L. (**2006**) "3D polymer microframes that exploit length-scale-dependent mechanical behavior". *Advanced Materials*, **18**, 2123–27.

Books

21. Archer, R.R., Cook, N.H., Crandall, S.H., Dahl, N.C., Lardner, T.J., McClintock, F.A., Rabinowicz, E. and Reichenbach, G.S. (**1978**) *An Introduction to the Mechanics of Solids*, McGraw-Hill, New York.
22. Nye, J.F. (**1985**) *Physical Properties of Crystals: Their Representation by Tensors and Matrices*, Oxford University Press, Oxford, UK.
23. Gibson, L.J. and Ashby, M.F. (**1997**) *Cellular Solids*, Cambridge University Press, Cambridge, UK.

9

Further Applications

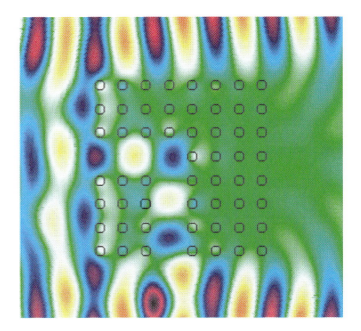

In the previous chapters, we showed that periodic structures can be usefully employed as photonic crystals, phononic crystals, and cellular solids. In this chapter it is our aim to introduce additional potential applications of periodic structures in several research areas. For example, we next show how periodic structures can be used to control the spontaneous emission of light, to localize and guide light, to simultaneously localize light and sound, to create superlenses, to make aberration-free images of objects, to control the flow of heat and electricity, to manage the flow of fluids in microfluidic devices, and to increase the efficiency of thermoelectric devices. From these selected examples, it is amply clear that periodic structures have a multifunctional character and they are of critical importance in technological applications that benefit from the control of important physical phenomena such as light, sound, heat, electricity, or fluid motion.

Periodic Materials and Interference Lithography. M. Maldovan and E. Thomas
Copyright © 2009 WILEY-VCH Verlag GmbH & Co. KGaA, Weinheim
ISBN: 978-3-527-31999-2

9.1
Controlling the Spontaneous Emission of Light

Periodic dielectric structures with complete three-dimensional photonic band gaps (such as those presented in Chapter 6) can be used to control the spontaneous emission of light. The spontaneous emission of light is a physical process in which an atom (or molecule) in a high energy state E_2 decays to a lower energy state E_1 and emits a photon with frequency $\omega = (E_2 - E_1)/\hbar$, where $\hbar$ is Planck's constant. To illustrate this concept, consider the hydrogen atom shown in Figure 9.1a, where the single bounded electron in the atom is in an excited state with energy E_2, which is represented by the electron traveling in a circular orbit of radius r_2 around the nucleus. After a certain period of time, the excited electron spontaneously moves into a lower energy state E_1 represented by the electron traveling in a circular orbit of radius $r_1 < r_2$. As a consequence of this electronic transition, a photon with frequency $\omega = (E_2 - E_1)/\hbar$ (and energy $\hbar\omega = E_2 - E_1$) is emitted (Figure 9.1b). This physical process is called the *spontaneous emission of light* and it is an essential phenomenon in science with a strong impact on technology since the spontaneous emission of light significantly affects the efficiency of semiconductor lasers, heterojunction bipolar transistors, and solar cells.

The spontaneous emission of light was often considered an intrinsic physical phenomenon difficult to control. In recent years, however, it was demonstrated that this phenomenon can, in principle, be inhibited by the use of parallel conducting planes. To illustrate this, consider two perfectly conducting parallel planes extending infinitely in the xz plane and separated by a distance d (Figure 9.2). Because the planes are considered to be perfect conductors, they act as perfect mirrors for electromagnetic waves. For example, if an electromagnetic wave is allowed to propagate in the space between the two planes, the wave will bounce back and forth between the mirrors as it propagates forward along the x axis (Figure 9.2a). Moreover, it can be demonstrated through a mathematical analysis that to propagate between the mirrors, the electromagnetic wave has to have a

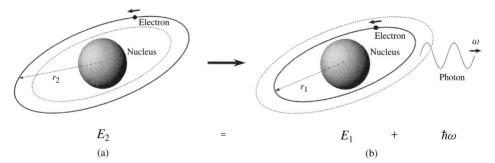

Fig. 9.1 The spontaneous emission of light. (a) An atom with an electron in an excited state of energy E_2. (b) The electron spontaneously decays to a lower energy state E_1 and a photon of frequency $\omega = (E_2 - E_1)/\hbar$ is emitted.

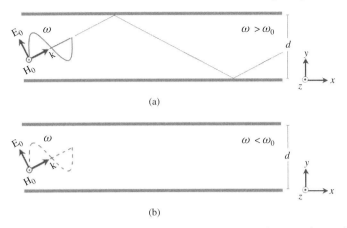

Fig. 9.2 Propagation of electromagnetic waves between two perfectly conducting parallel planes. (a) The frequency ω of the wave is above the cutoff frequency ω_0 of the system, as a result the wave is allowed to propagate between the parallel planes. (b) The frequency ω of the wave is below the cutoff frequency ω_0, and so thus the wave is not allowed to propagate between the parallel planes.

frequency ω above a certain threshold frequency ω_0 (the so-called cutoff frequency). That is, the parallel mirrors allow the propagation of electromagnetic waves with frequencies $\omega > \omega_0$ and forbid the propagation of electromagnetic waves with frequencies $\omega < \omega_0$. The cutoff frequency ω_0 of the system depends on the distance d between the mirrors and can therefore be readily altered to control the frequencies of the waves that are allowed to propagate between the mirrors.

In a recent experiment, to control the spontaneous emission of light, researchers placed excited Rydberg atoms between two conducting parallel plates such as those shown in Figure 9.2 [1]. Rydberg atoms have orbiting electrons in very high energy states that can spontaneously decay to lower energy states by emitting photons with frequencies ω in the microwave region. In particular, in this experiment, the distance between the conducting parallel plates was such that the frequency ω of the photons emitted by the Rydberg atoms was below the cutoff frequency ω_0 of the system (Figure 9.2b).

Now, to decay to a lower energy state, the excited Rydberg atoms must emit photons with frequency ω. These photons cannot propagate between the conducting parallel plates because their frequency ω is below the cutoff frequency ω_0 of the system. As a result, the spontaneous decay of the excited Rydberg atoms (and the resultant emission of photons) is inhibited. In the actual experiment, the lifetime of the Rydberg atoms was measured to increase by a factor of 20.

Imagine now that we want to repeat this experiment, but by using excited atoms that spontaneously decay to lower energy states by emitting photons with frequencies ω in the *visible* region. Even considering the fact that the frequency ω of the photons is below the cutoff frequency ω_0 of the system and that the emitted photons are not allowed to propagate between the parallel plates, it turns out that the spontaneous decay of the excited atoms is *not* inhibited in this case. The reason behind this contradictory result is the fact that, for frequencies in the visible range, the conducting parallel plates absorb photons (something that can be neglected at microwave frequencies). In this case, the spontaneous emission of light is not inhibited because the excited atoms decay to lower energy states emitting photons that are absorbed by the conducting parallel plates.

To forbid the spontaneous emission of light with frequencies in the visible range, we therefore need a system that forbids the propagation of photons with these frequencies and also does not absorb them. Such a system can be provided by periodic dielectric structures possessing complete three-dimensional photonic band gaps [2]. These structures satisfy the two above-mentioned properties. If fabricated at appropriate length scales, the structures can forbid the propagation of electromagnetic waves with frequencies in the visible range. In addition, they do not absorb photons because these structures are made of dielectric materials (which can, in principle, be nearly lossless).

Consider, for example, a periodic dielectric structure possessing a complete three-dimensional photonic band gap $\Delta\omega = \omega_2 - \omega_1$ for electromagnetic waves with frequencies in the visible range, where ω_1 and ω_2 are the lower and upper frequencies of the complete photonic band gap. In addition, suppose that we have an atom with excited electrons that can spontaneously decay to a lower energy state

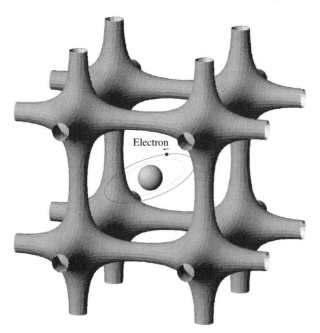

Fig. 9.3 Controlling the spontaneous emission of light. An atom with an excited electron is placed within a periodic dielectric structure having a complete three-dimensional photonic band gap. The electron remains in the excited state and cannot decay to a lower energy state because the complete photonic band gap forbids the emission of photons from the atom. Note: the scale of the atom relative to the periodic structure is greatly magnified.

by emitting photons with frequency ω in the visible range such that $\omega_1 < \omega < \omega_2$. If we now place the atom within the periodic dielectric structure (Figure 9.3), the excited electrons will not be able to decay to the lower energy state because to do that they must emit photons with frequency ω, which are not allowed to propagate within the periodic structure because the frequency ω is within the complete three-dimensional photonic band gap. This shows that *the spontaneous emission of visible light can be inhibited by the use of periodic dielectric structures possessing complete three-dimensional photonic band gaps.*

9.2
Localization of Light: Microcavities and Waveguides

So far in this book, we have considered the common picture of electromagnetic waves propagating progressively from point to point in space and transporting electromagnetic energy. We have also shown that propagating electromagnetic waves with certain frequencies are not allowed to exist within periodic structures possessing complete photonic band gaps (i.e. photonic crystals). To illustrate these two particular cases, consider a periodic structure having a complete photonic band gap. If the frequency of the electromagnetic wave is outside of the photonic band gap, the wave can progressively propagate within the periodic structure. On the other hand, if the frequency of the wave is inside of the photonic band gap, the propagating wave cannot exist within the structure (Chapter 6).

In recent years, however, the occurrence of a novel third situation, which is that the electromagnetic wave exists within the structure but it cannot propagate in any direction, has been demonstrated. In this state, the "nonpropagating" electromagnetic wave is restricted to a particular spatial region within certain structured materials and it is said that the wave is *localized*. In its localized state, the electromagnetic wave forms a standing (or stationary) wave that does not propagate within the structured material, but remains concentrated around a certain region in space.

To show how electromagnetic waves can be localized within certain structured materials, we first need a periodic structure having a complete photonic band gap. Consider, for example, a two-dimensional periodic dielectric structure made of cylinders arranged in the square lattice (Figure 9.4a), where the cylinders are made of air ($\varepsilon = \varepsilon_0$) and the background material is made of silicon ($\varepsilon = 13\varepsilon_0$). The radius of the cylinders is $r = 0.48a$, where a is the lattice constant of the square lattice. This two-dimensional photonic crystal possesses a complete photonic band gap $\Delta\omega$ for TE electromagnetic waves propagating in the xy plane with frequencies within the range $\omega a/2\pi c = 0.232 - 0.290$ (see the gray region in Figure 9.4b), where c is the speed of light in vacuum. Because of the existence of the complete photonic band gap, the perfectly periodic structure has a binary response for electromagnetic waves within the structure. It allows the propagation of TE electromagnetic waves with frequencies outside of the photonic band gap, but it prevents the existence of TE waves with frequencies within the photonic band gap.

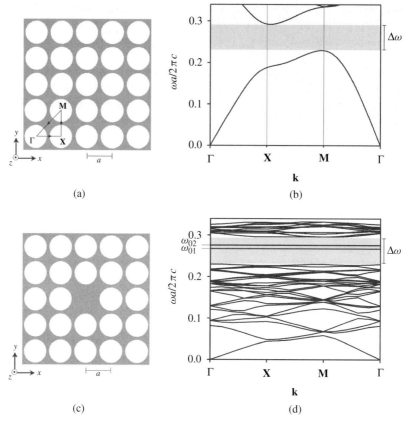

Fig. 9.4 Localization of light. (a) A perfectly periodic dielectric structure made of air ($\varepsilon = \varepsilon_0$) cylinders (white) in a silicon ($\varepsilon = 13\varepsilon_0$) background (gray). The radius of the air cylinders is $r = 0.48a$. (b) The band diagram corresponding to the perfectly periodic structure shows the existence of a complete TE photonic band gap for frequencies in the range $\omega a/2\pi c = 0.232 - 0.290$, where c is the speed of light. (c) By removing a single cylinder, a structural defect is introduced in the previous structure. (d) The band diagram corresponding to the structure having the defect shows the existence of new electromagnetic states with frequencies ω_{01} and ω_{02} within the complete TE photonic band gap (see black horizontal lines). These states correspond to light spatially localized around the defect.

Because the periodic structure possesses a complete photonic band gap, we may be able to localize light within the structure. *The localization of light can be obtained by purposely introducing structural defects in the previously perfect periodic structure with a complete photonic band gap.*

One way to introduce a suitable structural defect is to vary the radius of a single cylinder in the perfectly periodic structure. To show how light can be localized within this specific two-dimensional structure, we choose, in particular, to introduce a structural defect by removing one of the air cylinders (i.e. set $r = 0$) from the perfectly periodic structure (Figure 9.4c). The band diagram corresponding to the

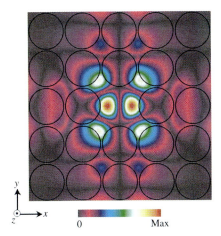

Fig. 9.5 The spatial localization of light around the defect created by removing a single cylinder in a periodic structure made of air cylinders in the square lattice. The figure shows the spatial distribution of the magnitude of the electric field vector (which is normal to the plane of the figure) for a TE electromagnetic wave with frequency ω_{01} within the two-dimensional structure shown in Figure 9.4c. Note the positions of the air cylinders surrounding the defect region. For this particular frequency, the electromagnetic wave is spatially localized around the defect created by removing a single cylinder.

structure *having the structural defect* is shown in Figure 9.4d.[14] After introducing the defect, the band diagram shows that the structure allows for new electromagnetic states with frequencies ω_{01} and ω_{02} within the complete photonic band gap (see horizontal black lines within the complete TE photonic band gap $\Delta\omega$). These electromagnetic states were not present in the previous perfectly periodic structure. Also note that after introducing the defect, the structure still possesses the complete photonic band gap $\Delta\omega$ for TE electromagnetic waves with frequencies within the range $\omega a/2\pi c = 0.232 - 0.290$, except for the particular frequencies ω_{01} and ω_{02}.

The newly allowed electromagnetic states, with frequencies ω_{01} and ω_{02} within the complete photonic band gap, correspond to TE electromagnetic waves spatially localized around the defect. This is shown in Figure 9.5, where we display the distribution of the electric field vector within the structure for the new electromagnetic state having a frequency ω_{01}. The color scheme shows the magnitude of the electric field vector along the z direction. We can see that the TE electromagnetic wave is localized around the defect, that is, the electric field vector has considerable

---

14) The large difference between the number of bands in the band diagrams shown in Figures 9.4b and d is due to the size of the unit cell considered for the numerical calculations. In the first case, a unit cell of size $a \times a$ is sufficient to describe the perfectly periodic structure. On the other hand, in the case of the structure with the defect, a supercell of size $5a \times 5a$ is considered for numerical calculations using periodic boundary conditions. This is a type of approximant to the actual structure made of a perfect photonic crystal containing a single isolated defect. This difference determines the number of bands that appear in the corresponding band diagrams. The physical response of the periodic structure (e.g. the photonic band gap) does not depend on the size of the unit cell considered for numerical calculations.

magnitude only in the spatial region around the defect. In this electromagnetic state, the electric field vector oscillates along the z direction with frequency ω_{01}, but the electromagnetic wave does not propagate within the photonic crystal. The electromagnetic state is 'bound' to the defect, forming a standing electromagnetic wave.

Note that the periodic structure with the structural defect has a ternary response. It allows the propagation of TE electromagnetic waves with frequencies outside of the complete photonic band gap, it prevents the propagation of TE waves with frequencies within the complete photonic band gap (except for the specific frequencies ω_{01} and ω_{02}), and it localizes electromagnetic waves with frequencies ω_{01} or ω_{02}. That is, depending on its frequency, a TE electromagnetic wave will be allowed to propagate within the structure, will be forbidden to propagate, or will be localized around the defect within the structure.

In this particular case, we chose to create a structural defect by removing one of the cylinders from the perfectly periodic structure. However, a structural defect can alternatively be created by varying the radius of a single cylinder in the perfectly periodic structure. Importantly, the frequencies ω_{01} and ω_{02} at which the localized electromagnetic waves oscillate depend on the specific radius of the structural defect. This means that the frequencies ω_{01} and ω_{02} of the localized states can be controlled by varying the radius (or alternatively the dielectric constant) of the structural defect. Therefore, a purposeful defect in a photonic crystal is really a type of device that allows the formation of a standing electromagnetic wave within the photonic crystal and is called an *optical microcavity* or *optical resonator*.

Next, we consider a further case where a *row* of cylinders is removed from a perfectly periodic structure. To study this, we consider an inverse periodic dielectric structure made of dielectric cylinders arranged in the triangular lattice (Figure 9.6), where the cylinders are made of silicon ($\varepsilon = 13\varepsilon_0$) and the background material is air ($\varepsilon = \varepsilon_0$). In this case, the radius of the cylinders is $r = 0.20a$, where a is the lattice constant of the triangular lattice. This two-dimensional photonic crystal possesses a complete TE photonic band gap $\Delta\omega$ for electromagnetic waves propagating in the plane of the structure with frequencies within the range $\omega a/2\pi c = 0.265 - 0.434$.

Instead of removing a single cylinder as we did in the previous case, we now remove a row of cylinders from the perfectly periodic structure to form a defect channel as shown in Figure 9.6. In this case, the resultant photonic crystal with a linear defect in the form of a *channel* allows for a new electromagnetic state (with frequency ω_0 within the photonic band gap $\Delta\omega$) that can freely propagate along the channel. For example, Figure 9.6 shows the instantaneous distribution of the electric field vector for a TE electromagnetic wave with frequency ω_0, which is incident on the surface of the photonic crystal from the left. The color scheme shows the value of the electric field vector, which is perpendicular to the plane of the structure (i.e. along the z axis). It can be seen that the incident electromagnetic wave is essentially *guided* through the channel created in the periodic dielectric structure and emerges on the opposite side of the photonic crystal. This is because the wave is allowed to propagate within the channel as it cannot propagate anywhere else inside of the photonic crystal because of the existence of the complete photonic band gap.

This unconventional mechanism to guide light is exclusive to periodic dielectric structures with complete photonic band gaps. In modern optical communication,

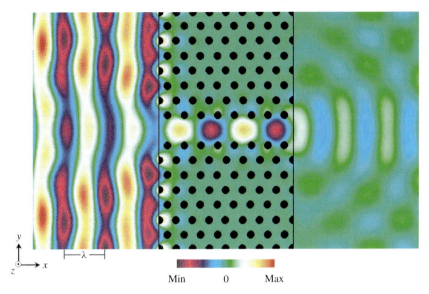

Fig. 9.6 Waveguides in photonic crystals. The figure shows the spatial distribution of the electric field vector (which is normal to the plane of the figure) for an incident TE electromagnetic wave on the surface of a periodic dielectric structure possessing a complete TE photonic band gap and a defect channel created by removing a row of cylinders from the perfectly periodic structure. It can be seen how the electromagnetic wave, which is incident from the left, is guided through the structure.

light is guided using optical fibers, which are made of an inner core dielectric cylinder with a high refractive index n_2 surrounded by a concentric low refractive hollow cylinder having a low refractive index n_1. In these optical fibers, light is confined within the central dielectric cylinder and guided along the axis of the cylinder by total internal reflection. That is, when a ray of light within the central dielectric cylinder strikes the interface between the inner cylinder and the outer hollow cylinder with an angle θ larger than the critical angle $\theta_c = \arcsin(n_1/n_2)$, the ray is reflected back to the central high dielectric cylinder. On the other hand, if the angle θ is smaller than the critical angle θ_c, the ray is refracted into the outer hollow cylinder and then into the air and escapes. (Note: the angle θ is measured with respect to the normal direction to the interface.) Those rays of light, within the central dielectric cylinder, for which $\theta > \theta_c$, will be reflected back to this dielectric region every time they strike the interface between the inner cylinder and the outer hollow cylinder. As a result, these rays are confined within the central dielectric cylinder and guided along the axis of the fiber. However, it is important to note that if at some point along the optical fiber, the curvature of the fiber is considerable, light can escape since in this case the ray of light may strike the interface between the inner cylinder and the outer hollow cylinder with an angle of incidence $\theta < \theta_c$.

Note the remarkable differences between this standard guiding mechanism and the unconventional mechanism shown in Figure 9.6. First, when light is guided through a defect channel created within a periodic dielectric structure having a complete photonic band gap, light can propagate through air instead of a dielectric material. This is an important consideration, because it greatly reduces the energy losses associated with the propagation of light in dielectric materials, such as those used in standard optical fibers. In addition, light guided within periodic dielectric structures *can even bend along sharp corners* (e.g. 90°) if a channel with this shape is created within the periodic structure (see Figure in the first page of this chapter). In this case, the reason behind the guiding mechanism is the existence of the *complete* photonic band gap, which determines that the propagating light cannot propagate within the photonic crystal irrespective of the direction of propagation.

Now imagine a three-dimensional photonic crystal possessing a complete photonic band gap. In this case, we can guide light in three dimensions simply with a continuous defect. Construction of such devices is only in its infancy, but the potential of such devices is a great driver for research in this important area.

In summary, periodic dielectric structures possessing complete photonic band gaps enable both light localization and wave guiding at microscopic length scales. As a result, these structures can be used as the basic platform for miniaturization and integration of optical components, circuits, and devices in all optical microchips. These microchips are the optical counterpart of electronic microchips used in computers, cell phones, and cameras having the additional benefits of increased signal speeds and minimal power dissipation.

9.3
Simultaneous Localization of Light and Sound in Photonic–Phononic Crystals: Novel Acoustic–Optical Devices

The localization and guiding of light is a physical phenomenon that is not exclusive to electromagnetic waves. In fact, all the effects shown in the previous section can similarly be obtained for sound waves. In the acoustic case, however, the periodic structures are made of appropriate elastic materials and possess complete *phononic* band gaps (Chapter 7).

An interesting and novel research area is to design periodic structures that can localize both light and sound in the same spatial region at the same time. This could in turn enhance the interaction of photons and phonons and create nonconventional acoustic–optical devices that can integrate the simultaneous control of electromagnetic and elastic waves. We previously mentioned that a prerequisite for the existence of localized electromagnetic states is the presence of a complete photonic band gap. Therefore, to obtain the simultaneous localization of light and sound, it is essential to have a periodic structure possessing both complete photonic and phononic band gaps.

The existence of a complete *photonic* band gap in a periodic structure does not necessarily imply that the structure also possesses a complete *phononic* band gap and vice versa. Because the simultaneous existence of complete photonic and phononic band gaps depends on several factors such as the intrinsic geometry of the structure, the physical properties of the materials comprising the structure, and the relative volume fraction of the materials, a comprehensive search for periodic structures possessing both gaps is needed. *Following the identification of a periodic structure exhibiting both complete photonic and phononic band gaps, the simultaneous localization of light and sound within the structure can in principle be obtained by purposely placing structural defects within the periodic structure.*

To illustrate the search for these "photonic–phononic" crystals, we consider two-dimensional periodic dielectric–elastic structures made of cylinders arranged in the square and triangular lattices. A judicious choice of materials is needed to increase the possibility of simultaneous complete photonic and phononic band gaps, and hence the likelihood of simultaneous localization of light and sound within these periodic structures. For example, many complete *phononic* gaps have been demonstrated in periodic structures made of solid cylinders in a solid background. However, these solid–solid structures are less likely to have complete *photonic* gaps. On the other hand, periodic structures made of a high dielectric solid material and air (or alternatively vacuum) help form complete *photonic* gaps and also promote the likelihood of complete *phononic* gaps due to the large density contrast between the solid material and air. As a result, we concentrate our study on these types of two-dimensional periodic structures.

Within the subset of air–solid structures, air connected (i.e. solid cylinders in air) structures show large complete *photonic* band gaps when the radius of the solid cylinders is small (Figures 9.7a and b). Unfortunately, these air connected structures only show large complete *phononic* band gaps when the radius of the solid cylinders is large (Figures 9.7c and d). As a consequence of this opposing requirement of the solid cylinder radius for large complete photonic and phononic band gaps, periodic structures made of solid cylinders in air are not good candidates for the simultaneous localization of light and sound. In fact, solid connected structures (such as vacuum cylinders in a solid background) appear to be the appropriate choice since, in contrast to structures made of solid cylinders in air, in this case both complete photonic and phononic band gaps are favored when the radius of the air cylinder is large (Figures 6.9 and 7.8).

For example, consider a periodic structure made of vacuum cylinders ($r = 0.48a$) arranged in the square lattice in a solid background, where the solid material is assumed to be silicon. Vacuum and silicon have dielectric constants $\varepsilon_2 = \varepsilon_0$ and $\varepsilon_1 = 13\varepsilon_0$, respectively. According to Figures 6.9 and 7.8, when the radius of the cylinders is $r = 0.48a$, the structure possesses a complete TE *photonic* band gap for electromagnetic waves with frequencies ω in the range $\omega a/2\pi c = 0.23 - 0.29$ and a complete *phononic* band gap for elastic waves (both in- and out-of-plane) with frequencies ω in the range $\omega a/2\pi c_T = 0.38 - 0.61$. Note that silicon has $\rho = 2330$ kg m^{-3} and velocities $c_T = 5360$ m s^{-1} and $c_L = 8950$ m s^{-1}, and its $c_L/c_T = 1.67$ ratio is very similar to that of epoxy, which is $c_L/c_T = 1.72$. Therefore, the complete phononic band gaps shown in Figure 7.8 are an excellent approximation for vacuum cylinders in a silicon background material.

If the periodic structure is fabricated in silicon with a lattice constant $a = 150$ nm, it has a complete TE *photonic* band gap for electromagnetic waves with visible frequencies $f = \omega/2\pi$ between 4.64×10^{14} and 5.80×10^{14} Hz and a complete *phononic* band gap for elastic waves with hypersonic frequencies $f = \omega/2\pi$

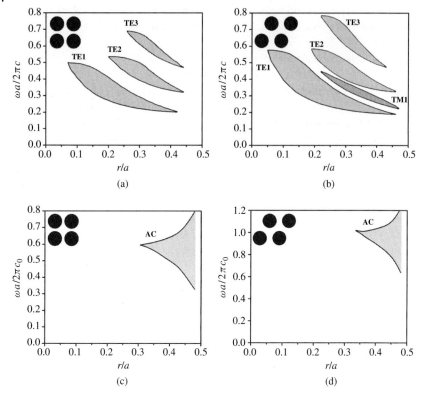

Fig. 9.7 Photonic and phononic band gap maps for solid cylinders in air background arranged in the square and triangular lattices. The dielectric/elastic material comprising the cylinders (see insets) is chosen as silicon ($\varepsilon = 13\varepsilon_0$). (a and b) Nondimensional frequencies $\omega a/2\pi c$ for *complete* transverse electric (TE) and transverse magnetic (TM) photonic band gaps as a function of the dimensionless radius r/a of the cylinders, where a is the lattice constant of the structures and c is the speed of light. (c and d) Nondimensional frequencies $\omega a/2\pi c_0$ for complete acoustic (AC) phononic band gaps as a function of the dimensionless radius r/a of the cylinders, where a is the lattice constant of the structure and c_0 is the speed of sound in air.

between 13.6×10^9 and 21.6×10^9 Hz. It is important to note that the frequencies of the complete photonic and phononic band gaps have different orders of magnitude. This is because, as mentioned earlier, the wavelengths λ of the electromagnetic and elastic waves that correspond to the band gap frequencies need to be of the order of the lattice constant of the periodic structure (i.e. $\lambda_{photon} \sim \lambda_{phonon} \sim a$). Since electromagnetic and elastic waves travel at velocities that are five orders of magnitude different from each other, the frequencies of the complete photonic and phononic band gaps are therefore different by the same order of magnitude.

We now introduce a structural defect within this photonic–phononic crystal in an attempt to localize electromagnetic and elastic waves in the same region at the same time. We choose to create a defect by removing a single vacuum cylinder from the perfectly periodic structure. The localization of *light* around this purposely introduced defect was demonstrated in Section 9.2 (Figures 9.4 and 9.5). (Note:

vacuum and air have nearly the same dielectric constant.) Here, we show that this periodic structure containing a purposeful defect can also localize *sound* around the defect.

Analogously to the electromagnetic case, after introducing the defect, the structure allows for new elastic states with frequencies within the complete phononic band gap. These elastic wave states were not present in the previous perfectly periodic structure. The newly allowed elastic states with frequencies within the complete phononic band gap correspond to elastic waves spatially localized around the defect. This is shown in Figure 9.8, where we display the distribution of the displacement field vector **u** for the new elastic states within the periodic structure containing the defect. The color scheme in Figure 9.8a shows the magnitude of the out-of-plane (transverse) displacement field vector u_z, which is along the z direction, while Figure 9.8b shows the magnitude $u = \sqrt{[(u_x)^2 + (u_y)^2]}$ of the in-plane (mixed) displacement field vector $\mathbf{u}_{//} = (u_x, u_y)$, which is in the xy plane. We can see that elastic waves are localized around the defect, that is, the displacement vector **u** has considerable magnitude only in the spatial region around the defect. In this localized state, the displacement field vector oscillates in the region around the defect but the elastic wave does not propagate within the phononic crystal forming a standing elastic wave.

This shows that *by introducing a structural defect within the perfectly periodic structure, we can localize both TE electromagnetic and elastic waves around the same spatial region at the same time.* The introduction of the defect allows for the simultaneous formation of standing electromagnetic and elastic waves within the periodic structure, and it therefore creates simultaneous optical and acoustical microcavities.

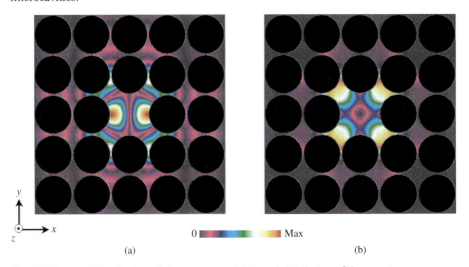

Fig. 9.8 The spatial localization of elastic waves around the defect created by removing a single cylinder in a periodic structure made of vacuum cylinders in the square lattice. (a) Spatial distribution of the magnitude of the out-of-plane displacement field vector u_z. (b) Spatial distribution of the magnitude of the in-plane displacement field vector $\mathbf{u}_{//} = (u_x, u_y)$. The figure shows that elastic waves can be spatially localized around the defect created by removing a single cylinder.

In conclusion, we showed that electromagnetic and elastic waves can be simultaneously localized around a structural defect introduced in a square periodic array of dielectric/elastic material. The localization of photons and phonons *in the same spatial region at the same time* could bring innovative research areas for combined management of electromagnetic and elastic waves. The coexistence of localized states for *light* and *sound* can influence photon–phonon interactions and open the possibility of dual acoustic–optical devices.

For example, by the use of the acousto-optical cavities presented in this chapter, the coupling between photons and phonons can be altered in such a way so as to open the possibility of the coherent generation of phonons or to influence acousto-optics effects that occur, for example, in the optical cooling of materials.

9.4
Negative Refraction and Superlenses

One of the most exciting (and recently discovered) features of periodic structures is their ability to create *negative* refraction of light. To understand this concept, consider Figure 9.9, where the refraction of light is shown in two different situations. In the first case, an electromagnetic wave is incident from air (which has $\varepsilon_1 > 0$ and $\mu_1 > 0$) onto the surface of a homogeneous material (with $\varepsilon_2 > 0$ and $\mu_2 > 0$) at an angle of incidence θ. As a consequence of the change in the refractive index, the incident electromagnetic wave is refracted within the homogeneous material as shown in Figure 9.9a. This case, which was explained in detail in Section 5.6.1, corresponds to "positive" or "normal" refraction of light.

On the other hand, Figure 9.9b shows an electromagnetic wave that is incident from air on the surface of a periodic dielectric structure (or a photonic crystal). For a certain range of frequencies, specific geometries, and constituent materials of the photonic crystal, it is possible that the incident electromagnetic wave undergoes negative refraction. That is, the angle of the refracted electromagnetic wave within

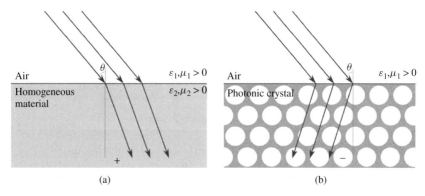

Fig. 9.9 (a) Positive refraction of light. (b) Negative refraction of light.

the photonic crystal, measured with respect to the normal direction, is reversed or *negative* (Figure 9.9b).

This negative refraction property exhibited by specific photonic crystals provides a significant practical application: photonic crystals can be used as lenses. This is shown in Figure 9.10, where light beams from a point source are incident on the surface of a photonic crystal that exhibits negative refraction for a certain range of frequencies $\Delta\omega$. If the frequency of the incident light beams is outside of the frequency range $\Delta\omega$, the beams undergo positive refraction at the interfaces between the air and the photonic crystal (Figure 9.10a) and a virtual image of the point source is generated within the photonic crystal. On the other hand, if the frequency of the incident light beams is inside of the frequency range $\Delta\omega$, the beams undergo negative refraction at the interfaces and a *real* image of the point source is generated (Figure 9.10b). In this latter case, the photonic crystal acts as a lens, with the particular feature that, in contrast to normal lenses, the photonic crystal is a *flat* lens. Importantly, photonic crystal lenses are able to resolve objects smaller than the wavelength of light under consideration, providing means to obtain "superlenses". In addition, these novel photonic crystal lenses can be fabricated at small length scales, bringing the opportunity for on-chip integration of these optical devices.

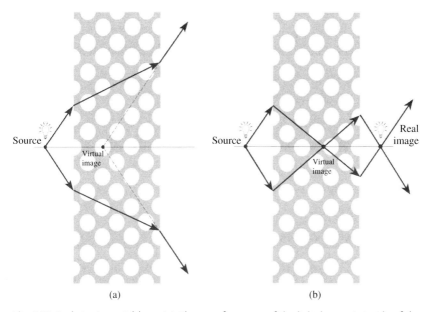

Fig. 9.10 A photonic crystal lens. (a) The frequency of the light beams coming from the point source is outside of the frequency range for negative refraction. Positive refraction occurs and a *virtual* image is formed within the photonic crystal. (b) The frequency of the light beams is inside of the frequency range for negative refraction. Negative refraction occurs and a *real* image is formed on the other side of the photonic crystal. In this case, the photonic crystal acts as a lens.

To have an efficient photonic crystal lens, it is important that negative refraction occurs *irrespective of the angle of incidence* of the light beams (which are incident from air into the surface of the photonic crystal). This property is called *all-angle negative refraction* and it is analogous to omnidirectional reflection, where the light beams are totally reflected by the photonic crystal irrespective of their angle of incidence. The range of frequencies $\Delta\omega$ for which a photonic crystal exhibits all-angle negative refraction can be identified by calculating the dispersion relation $\omega = \omega(\mathbf{k})$ for electromagnetic waves propagating within the photonic crystal. The details associated with the identification of the frequency ranges for which all-angle negative refraction occurs is beyond the scope of this book. The reader interested in this topic can consult the references at the end of this chapter.

We next show that the three-dimensional photonic crystal described by the simple periodic function $f_2^{6,I}(x, y, z)$ (Equation 2.40) exhibits all-angle negative refraction for a certain range of frequencies. As discussed in Chapter 4, this photonic crystal can be fabricated by interference lithography at small length scales and it can therefore exhibit all-angle negative refraction for electromagnetic waves with frequencies in the visible range, if fabricated in a high dielectric material. The photonic crystal is represented by the function

$$f_2^{6,I}(x, y, z) = +\cos\left[\frac{2\pi}{a}(x+y)\right] + \cos\left[\frac{2\pi}{a}(x-y)\right] + \cos\left[\frac{2\pi}{a}(x+z)\right]$$
$$+ \cos\left[\frac{2\pi}{a}(x-z)\right] + \cos\left[\frac{2\pi}{a}(y+z)\right] + \cos\left[\frac{2\pi}{a}(y-z)\right] + 1.5$$
(9.1)

and the laser beam parameters needed to fabricate this simple periodic structure by interference lithography were given in Table 4.4.

We want to note that the presence of all-angle negative refraction in this photonic crystal was first theoretically studied in Reference 14 based on a previous model introduced in Reference 12. Figure 9.11 shows the unit cell of the photonic crystal together with the dispersion relation $\omega = \omega(\mathbf{k})$ for electromagnetic waves propagating within the periodic structure. Note that in this case we consider the inverse case with respect to the $f_2^{6,I}$ structure shown in Figure 2.8 (i.e. the air and solid regions are interchanged). The solid material (shown in gray) is assumed to have a dielectric constant $\varepsilon = 18\varepsilon_0$, and its volume fraction is $f = 0.20$. As discussed in Chapter 2, the Bravais lattice that corresponds to this photonic crystal is the body-centered-cubic Bravais lattice. The graph of the dispersion relation $\omega = \omega(\mathbf{k})$ shows the range of frequencies $\Delta\omega$ for which the photonic crystal exhibits all-angle negative refraction. This means that an electromagnetic wave (which is incident from air on the surface of the photonic crystal) with frequency ω within the range $\Delta\omega$, is negatively refracted within the photonic crystal irrespective of the angle of incidence. In particular, for this photonic crystal, the frequency range $\Delta\omega$ extends from $\omega a/2\pi c = 0.396$ to 0.430. The dashed lines in Figure 9.11b correspond to the dispersion relation for electromagnetic waves propagating in air and they are needed to identify the upper-limit frequency of the all-angle negative refraction frequency range $\Delta\omega$.

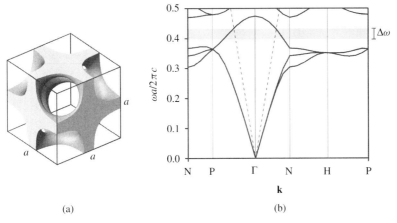

Fig. 9.11 (a) The level-set I-WP structure, which is given by the formula $f_2^{6,l}(x,y,z)$. The dielectric solid material (gray) has dielectric constant $\varepsilon_1 = 18\varepsilon_0$ and the background material is air ($\varepsilon_2 = \varepsilon_0$). The volume fraction of the solid material is 20%. (b) The dispersion relation $\omega = \omega(\mathbf{k})$ for electromagnetic waves propagating within the I-WP structure. The graph shows the frequency range $\omega a/2\pi c = 0.396$–0.430 (see the gray region) for which the structure exhibits all-angle negative refraction.

As mentioned earlier, because the photonic crystal exhibits all-angle negative refraction, it can in principle be used as a lens. This is demonstrated in Figure 9.12, where a point dipole source that oscillates with frequency $\omega a/2\pi c = 0.42$ is located at a distance a from the surface of the photonic crystal. Note that the frequency of the dipole source is chosen to be within the all-angle negative refraction frequency range. As a consequence of the negative refraction phenomenon, the image of the dipole is formed on the other side of the photonic crystal (Figure 9.12).

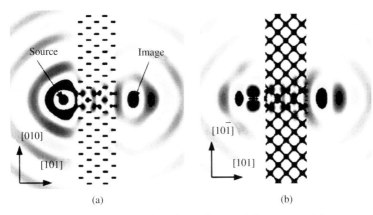

Fig. 9.12 Focusing the image of an electrical dipole (source) with a frequency $\omega a/2\pi c = 0.42$ polarized along the $(1, 0, -1)$ direction using a flat lens. The thickness of the photonic crystal is $3.6a$. The figure shows the spatial distribution of the electric field along $(1, 0, -1)$ on (a) the $(1, 0, 1)$–$(0, 1, 0)$ plane and (b) on the $(1, 0, 1)$–$(1, 0, -1)$ plane. [Reprinted with permission from [14].]

In summary, the recently discovered property of negative refraction brings unconventional means to control the propagation of light that may lead, for example, to the small-scale integration of lenses. Simple periodic structures exhibiting all-angle negative refraction and fabricated at small length scales by interference lithography can be employed as a lens that operates in the visible range of frequencies. The existence of all-angle negative refraction in photonic crystals is not trivial and the identification, fabrication, and experimental testing of three-dimensional photonic crystals exhibiting this property are currently an area of intense research. Significantly, the phenomenon of negative refraction is not exclusive to electromagnetic waves. In complete analogy, periodic elastic structures (i.e. phononic crystals) can exhibit negative refraction for sound waves and they can therefore be used as a platform for acoustical lenses. The development of superlenses for acoustic waves would have a strong impact on the production of images by ultrasound techniques.

9.5
Multifunctional Periodic Structures: Maximum Transport of Heat and Electricity

Throughout this book, we considered periodic structures made of two different constituent materials, which are called *binary composite structures*. We have shown many examples in which the proper distribution of the constituent materials in space can generate effective physical properties that are drastically different from the physical properties corresponding to the constituent materials (e.g. photonic crystals, phononic crystals, and cellular solids). In fact, in contrast to homogeneous materials, binary composite structures offer an extensive variety of desired physical properties. This, in part, is due to the degree of freedom that we have in distributing the different constituent materials in space.

As an additional example of the practical applications of these periodic structures, we consider the simultaneous transfer of *heat* and *electricity* through a three-dimensional binary composite structure. To study this, we suppose that one of the constituent materials is a good thermal conductor but a poor electrical conductor, and that the other constituent material is a good electrical conductor but a poor thermal conductor. We also assume that the constituent materials exist in equal proportions within the periodic composite structure (i.e. their volume fractions are the same). By requiring that the sum of the effective thermal and electrical transfer through the three-dimensional binary composite structure be maximized, a competition between the two effective properties is established [15].

Since the physical properties and the volume fractions of the constituent materials are fixed, the goal is to obtain a spatial distribution of the constituent materials (within the unit cell of the binary periodic composite structure) such that the sum of the effective thermal k_{eff} and electrical σ_{eff} conductivities of the composite structure is maximized. To obtain such a spatial distribution of materials, a numerical technique called *topology optimization* is commonly used. By using this technique, the unit cell of the periodic composite structure is divided into a sufficiently large number of volume elements, each of which is filled with one of the two constituent materials. The spatial distribution of materials within the unit cell of the structure changes as the objective function $k_{\text{eff}} + \sigma_{\text{eff}}$ is being maximized by the topology optimization algorithm, which conserves volume fraction composition and symmetry. A final distribution of the two materials, and consequently a final periodic composite structure, is obtained when the sum of the effective thermal and electrical transfer is maximized. This numerical technique allows us to obtain binary composite materials with optimized multifunctional characteristics. In the case of the optimization of two transport properties, connectedness of the materials is key (indeed, the optimus structures are bicontinuous) since isolated, discrete regions of a material do not contribute to the transport.

In our particular case, if we assume the spatial periodicity of the periodic composite structure is consistent with the simple cubic Bravais lattice, the topology optimization technique determines that the binary composite structure that maximizes the sum of the effective thermal and electrical transfer is the simple periodic structure $f_1(x, y, z)$ (Figure 2.4a). Note that the optimized structure has both the materials connected precisely in the same way with the same shape and topology (the P structure is its own inverse structure at $f = 0.50$ volume fraction).

On the other hand, if we assume the periodicity of the binary composite structure is consistent with the symmetry of the face-centered-cubic Bravais lattice, the binary composite structure that maximizes the sum of the effective thermal and electrical transfer is the diamond structure $f_3^{4,\text{II}}(x, y, z)$ shown in Figure 2.6 ($f = 0.50$ was assumed).

With this example, we demonstrate that the simple periodic structures introduced in this book can also be used for multifunctional materials such as composite materials exhibiting maximum transfer of heat and electricity.

9.6
Microfluidics

An additional useful application of periodic structures is the control of the flow of fluids. In recent years, fabrication techniques to miniaturize and integrate complex devices have been considerably enhanced via nanotechnology. In particular, the fabrication of devices that control the flow of fluids at small length scales gave rise to a novel research area called *microfluidics*. The aim of this research area is to provide a platform where the flow of fluids performs specific tasks under controlled conditions. One of the biggest challenges in this area is the development of a lab-on-a-chip device, where multiple laboratory functions (e.g. biological and chemical separation and sensing of solid particles or fluids) are miniaturized and integrated onto a single chip.

To achieve such a challenge, it would be interesting to have a control over the fluid permeabilities of the materials comprising the device. The fluid permeability is basically a material property that determines the rate at which a fluid flows through a specific material. In this regard, the simple periodic structure presented in this book can, in principle, be used to control the fluid permeabilities.

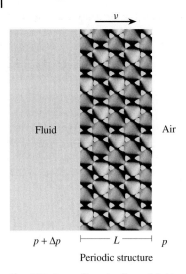

Fig. 9.13 Controlling the flow of fluids at small length scales by simple periodic structures. The rate at which the fluid crosses the periodic structure from left to right is determined by the effective fluid permeability of the structure.

Figure 9.13 shows how a simple periodic structure (such as those introduced in Chapter 2) made of solid material and air can be used to control the flow of a fluid. It is important to note that the structure needs to be bicontinuous, since the continuous solid network makes the structure self-supporting and stable while the continuous air region provides the interconnected channels for fluid flow. A pressure difference Δp exists over the thickness L of the structure, such that the fluid flows from left to right with velocity v. In addition, it is assumed that the fluid completely fills the air region within the simple periodic structure.

The fluid permeability k of the structure is given by Darcy's law

$$v = -\frac{k \Delta p}{\eta L} \qquad (9.2)$$

where v is the velocity of the fluid (averaged over the cross section of the material), η is the viscosity of the fluid, Δp is the pressure difference, and L is the thickness of the structure.

For a given pressure drop Δp, viscosity η, and structure thickness L, Equation 9.2 establishes that the fluid permeability k of the simple periodic structure determines the rate at which the fluid flows through the structure. The fluid permeability k depends on the shape, volume fraction, and length scale of the simple periodic structure. It would be certainly interesting to experimentally study the fluid permeabilities of the simple periodic structures introduced in Chapter 2 since these structures can be fabricated at small length scales by interference lithography. The understanding of the relationship between geometry and effective permeabilities in these structures can help to miniaturize and integrate devices that can control the flow of fluids.

9.7
Thermoelectric Energy

Periodic structures can also be useful for the management of heat in thermoelectric materials. Thermoelectricity is basically a physical phenomenon in which electricity is converted into heat and vice versa (i.e. heat is converted into electricity). This physical phenomenon allows for thermoelectric devices, which are based on the three following basic effects.

9.7.1
Peltier Effect

Consider two different conducting materials A and B, which are in contact with each other as shown in Figure 9.14a. If an electric current I flows from one material to the other, it turns out that heat is generated or absorbed around the junction between the two materials depending on the direction of the electric current. For example, if the current flows from left to right, heat is generated around the junction and this region becomes hot. On the other hand, if the current flows from right to left, heat is absorbed around the junction and the region becomes cold. This is because the specific heat of the electrons is different in the two materials

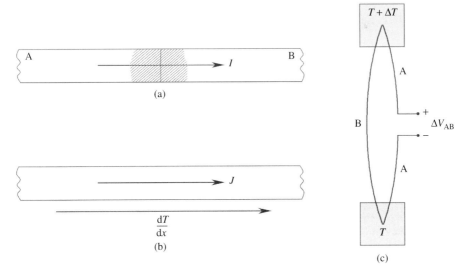

Fig. 9.14 (a) The Peltier effect. (b) The Thomson effect. (c) The Seebeck effect. [Adapted from Reference 18.]

and the electrons give out/absorb the excess/deficiency of heat as they pass from one material to the other.

It is important to note that the amount of heat generated or absorbed around the junction depends on the bulk properties of the two materials, the magnitude I of the electrical current, and the common temperature T of the materials. The amount of heat generated or absorbed around the junction per unit time is given by

$$\dot{Q} = \Pi_{AB}(T)I \qquad (9.3)$$

where $\Pi_{AB}(T)$ is called the *Peltier coefficient* between materials A and B. This coefficient depends on the temperature T of the materials and has units of volts.

Simply stated, the Peltier effect allows one to locally generate or absorb heat by changing the direction of the electrical current I flowing between two different materials. This means that we can heat or cool the region around the junction *reversibly*.

9.7.2
Thomson Effect

An additional thermoelectric effect, known as the *Thomson effect*, is created when an electric current I flows through a homogeneous conducting material subject to a temperature gradient dT/dx (Figure 9.14b). In this case, heat is generated or absorbed by the homogeneous conducting material depending on the relative direction between the electric current and the temperature gradient. Again the moving electrons either absorb or give off heat. The amount of heat generated or absorbed per unit volume per unit time is given by

$$\dot{q} = -\mu(T)J\frac{dT}{dx} \qquad (9.4)$$

where $\mu(T)$ is the Thomson coefficient of the material and J is the current density.

Simply stated, the Thomson effect allows one to generate or absorb heat by changing the direction of the electrical current I flowing through an homogeneous conducting material subject to a temperature gradient. Therefore, we can heat or cool the conducting material *reversibly*.

The heat generated in the Peltier and Thomson effects needs to be distinguished from the Joule heat generated in conducting materials as a result of an electric current flowing through them. Joule heat is *irreversible*, is always generated (never absorbed), and is due to the resistance of the conducting material to the flow of electrons. On the other hand, the heat in Peltier and Thomson effects is *reversible*; it can be generated or absorbed depending on the direction of the electric current. This is because the heat in the Peltier and Thomson effects is related to the specific heat of the electrons.

9.7.3
Seebeck Effect

The Seebeck effect allows us to directly create electricity as a result of a temperature gradient. Consider, for example, the open circuit in Figure 9.14c, which is made of two different conducting materials A and B with their ends at temperatures T and $T + \Delta T$, respectively. As a consequence of the temperature gradient, a voltage V_{AB} is created in the open circuit such that

$$\frac{dV_{AB}}{dT} = S_A(T) - S_B(T) \quad (9.5)$$

where $S_A(T)$ and $S_B(T)$ are the Seebeck coefficients (or thermoelectric power) of the conducting materials. These coefficients basically determine the voltage generated between two points in a conducting material as a consequence of the temperature gradient existing between those points. The Seebeck coefficient of a material depends on the temperature and has units of volt per kelvin.

The origin of the Seebeck effect is based on the diffusion of electrons from the hot region to the cold region. In addition, when phonon–electron interactions are strong, the diffusion of heat (or thermal phonons) from the hot region to the cold region tends to "sweep" or "drag" electrons, and there is an additional contribution to the diffusion of electrons from this effect. The Peltier, Thomson, and Seebeck thermoelectric effects are related by the Kelvin (Thomson) relations, which state that $\Pi(T) = S(T)T$ and $\mu(T) = TdS(T)/dT$.

An efficient thermoelectric material should have a large Seebeck coefficient, since this determines that the material develops a large voltage ΔV per degree of temperature difference between two different points in the material. In addition, an efficient thermoelectric material should have high electrical conductivity and low thermal conductivity. High conductivity is desired to generate a considerable electric current and to minimize Joule heating, which produces irreversible heat that can flow to the cold point and decrease the temperature gradient that generates the voltage ΔV. Low thermal conductivity is also a desired feature since it avoids the flow of heat from the hot point to the cold point, which decreases the temperature gradient. When all these aspects are considered, the efficiency of thermoelectric materials is determined by the figure of merit

$$Z = \frac{\sigma S^2}{k} \quad (9.6)$$

where S is the Seebeck coefficient, σ the electrical conductivity, and k is the thermal conductivity of the thermoelectric material. Because the figure of merit Z has units of 1/Kelvin, we can multiply Z by the temperature T of the thermoelectric material to obtain a dimensionless figure of merit ZT, which is often used to compare the efficiency of thermoelectric materials.

We now discuss the thermoelectric efficiency of different types of materials such as insulator, metallic, and semiconductor materials. Insulator materials have very low electrical conductivity σ, and they are, therefore, not suitable for thermoelectric materials. Metallic materials have large electrical conductivity σ, but they also have large thermal conductivity k since in metals the transport of thermal energy is mainly carried out by electrons. In these materials, the ratio between the electrical conductivity σ and the thermal conductivity k is constant, meaning that if we increase the electrical conductivity, the thermal conductivity is increased as well, and the figure of merit Z remains unchanged. In addition, Seebeck coefficients for metallic materials are small ($\sim 10\,\mu$V K^{-1}). As a result, metallic materials have small figures of merits Z. On the other hand, semiconductor materials can be relatively good electrical conductors. In this case, the transport of thermal energy is carried out by electrons and phonons, but the major contribution to the transport of thermal energy is provided by the flow of phonons. This means that if we could somehow alter the propagation of phonons in these materials, we could reduce the thermal conductivity k while keeping the electrical conductivity σ unmodified. This will result in thermoelectrical materials with increased figure of merits Z. In addition, Seebeck coefficients for semiconductor materials are comparatively large ($>100\,\mu$V K^{-1}). In fact, thermoelectric materials with large figure of merits are obtained by mixing different semiconductor materials.

We have shown that periodic elastic structures with complete two- and three-dimensional phononic band gaps, such as those introduced in Chapter 7, allow us to control the propagation of phonons. Importantly, for certain range of frequencies, the flow of phonons is prohibited irrespective of their propagation direction. As a result, these periodic structures provide a material platform for thermoelectric materials with increased figures of merit, since the effective thermal conductivity of these periodic materials can be in principle significantly decreased by the existence of the complete phononic band gap. However, to control the propagation of thermal phonons at room temperatures, the lattice constants a of the periodic structures must be extremely small, that is, $a \sim$ nm. Although the increase in the electrical conductivity in thermoelectric materials has been considerably studied, the reduction in the thermal conductivity by means of phononic crystals is still a largely unexplored research area.

Further Reading

Controlling the Spontaneous Emission of Light

1 Randall, G.H., Hilfer, E.S. and Kleppner, D. (1985) "Inhibited spontaneous emission by a Rydberg atom". *Physical Review Letters*, **55**, 2137–40.
2 Yablonovitch, E. (1987) "Inhibited spontaneous emission in solid-state physics and electronics". *Physical Review Letters*, **58**, 2059–62.
3 Yablonovitch, E. (1993) "Photonic band gap crystals". *Journal of Physics: Condensed Matter*, **5**, 2443–60.

Localization of Light: Microcavities and Waveguides

4 John, S. (1987) "Strong localization of photons in certain disordered superlattices". *Physical Review Letters*, **58**, 2486–89.
5 Meade, R.D., Brommer, K.D., Rappe, A.M. and Joannopoulos, J.D. (1991) "Photonic bound-states

in periodic dielectric materials". *Physical Review B*, **44**, 13772–74.
6 Joannopoulos, J.D., Villeneuve, P.R. and Fan, S. (**1997**) "Photonic crystals: putting a new twist on light". *Nature*, **386**, 143–49.

Simultaneous Localization of Light and Sound in Photonic-Phononic Crystals: Novel Acousto-Optical Devices

7 Maldovan, M. and Thomas, E.L. (**2006**) "Simultaneous complete elastic and electromagnetic band gaps in periodic structures". *Applied Physics B*, **83**, 595–600. band gap
8 Maldovan, M. and Thomas, E.L. (**2006**) "Simultaneous localization of photons and phonons in two-dimensional periodic structures". *Applied Physics Letters*, **88**, 251907.

Negative Refraction and Superlenses

9 Kosaka, H., Kawashima, T., Tomita, A., Notomi, M., Tamamura, T., Sato, T. and Kawakami, S. (**1998**) "Superprism phenomena in photonic crystals". *Physical Review B*, **58**, R10096–99.
10 Notomi, M. (**2000**) "Theory of light propagation in strongly modulated photonic crystals: Refraction-like behavior in the vicinity of the photonic band gap". *Physical Review B*, **62**, 10696–705. band gap
11 Luo, C., Johnson, S.G., Joannopoulos, J.D. and Pendry, J.B. (**2002**) "All-angle negative refraction without negative refractive index". *Physical Review B*, **65**, 201104.
12 Luo, C., Johnson, S.G. and Joannopoulos, J.D. (**2002**) "All-angle negative refraction in a three-dimensionally periodic photonic crystal". *Applied Physics Letters*, **81**, 2352–54.
13 Cubukcu, E., Aydin, K., Ozbay, E., Foteinopolou, S. and Soukoulis, C.M. (**2003**) "Negative refraction by photonic crystals". *Nature*, **423**, 604–5.
14 Ao, X. and He, S. (**2004**) "Three-dimensional photonic crystal of negative refraction achieved by interference lithography". *Optics Letters*, **29**, 2542–44.

Maximum Transport of Heat and Electricity

15 Torquato, S., Hyun, S. and Donev, A. (**2002**) "Multifunctional composites: optimizing micro-structures for simultaneous transport of heat and electricity". *Physical Review Letters*, **89**, 266601.
16 Torquato, S., Hyun, S. and Donev, A. (**2003**) "Optimal design of manufacturable three-dimensional composites with multifunctional characteristics". *Journal of Applied Physics*, **94**, 5748–55.

Microfluidics

17 Jang, J.-H., Dendukuri, D., Hatton, T.A., Thomas, E.L. and Doyle, P.S. (**2007**) "A route to three-dimensional structures in a microfluidic device: stop flow interference lithography". *Angewandte Chemie*, **46**, 9027–31.

Thermoelectricity

18 MacDonald, D.K.C. (**1962**) *Thermoelectricity: An Introduction to the Principles*, John Wiley & Sons, Ltd, New York.

Appendix A
MATLAB Program to Calculate the Optimal Electric Field Amplitude Vectors for the Interfering Light Beams

A.1
Two-dimensional Periodic Structures

In Chapter 4, we have seen that the superposition of $N = 3$ noncoplanar laser beams, with wave vectors $\mathbf{k}_1$, $\mathbf{k}_2$, and $\mathbf{k}_3$ and electric field amplitude vectors $\mathbf{E}_{01}$, $\mathbf{E}_{02}$, and $\mathbf{E}_{03}$, generates a two-dimensional spatial distribution of intensity of light $I(\mathbf{r})$ given by the formula

$$
\begin{aligned}
I(\mathbf{r}) = &\ |\mathbf{E}_{01}|^2 + |\mathbf{E}_{02}|^2 + |\mathbf{E}_{03}|^2 \\
&+ 2\,\mathrm{Re}(\mathbf{E}_{01}\cdot\mathbf{E}_{02}^*)\cos[(\mathbf{k}_1-\mathbf{k}_2)\cdot\mathbf{r}] - 2\,\mathrm{Im}(\mathbf{E}_{01}\cdot\mathbf{E}_{02}^*)\sin[(\mathbf{k}_1-\mathbf{k}_2)\cdot\mathbf{r}] \\
&+ 2\,\mathrm{Re}(\mathbf{E}_{01}\cdot\mathbf{E}_{03}^*)\cos[(\mathbf{k}_1-\mathbf{k}_3)\cdot\mathbf{r}] - 2\,\mathrm{Im}(\mathbf{E}_{01}\cdot\mathbf{E}_{03}^*)\sin[(\mathbf{k}_1-\mathbf{k}_3)\cdot\mathbf{r}] \\
&+ 2\,\mathrm{Re}(\mathbf{E}_{02}\cdot\mathbf{E}_{03}^*)\cos[(\mathbf{k}_2-\mathbf{k}_3)\cdot\mathbf{r}] - 2\,\mathrm{Im}(\mathbf{E}_{02}\cdot\mathbf{E}_{03}^*)\sin[(\mathbf{k}_2-\mathbf{k}_3)\cdot\mathbf{r}]
\end{aligned}
\quad (\mathrm{A.1})
$$

We also showed that the interference pattern (Equation A.1) can be used to fabricate two-dimensional periodic structures. From an experimental perspective, it is convenient to have a spatial distribution of intensity $I(\mathbf{r})$ for which the contrast is maximal (Section 4.2) because this reduces constraints on the structure reproduction. In fact, the contrast C of the intensity distribution $I(\mathbf{r})$ is maximal when the constant term $|\mathbf{E}_{01}|^2 + |\mathbf{E}_{02}|^2 + |\mathbf{E}_{03}|^2$ in Equation A.1 is minimal.

In this section, we introduce a program in version 7.2 of MATLAB that calculates (for given wave vectors $\mathbf{k}_1$, $\mathbf{k}_2$, and $\mathbf{k}_3$) the electric field amplitude vectors $\mathbf{E}_{01}$, $\mathbf{E}_{02}$, and $\mathbf{E}_{03}$ of the interfering beams that minimize

$$|\mathbf{E}_{01}|^2 + |\mathbf{E}_{02}|^2 + |\mathbf{E}_{03}|^2 \quad (\mathrm{A.2})$$

subject to the constraints

$$
\begin{aligned}
+2\,\mathrm{Re}(\mathbf{E}_{01}\cdot\mathbf{E}_{02}^*) &= A, & -2\,\mathrm{Im}(\mathbf{E}_{01}\cdot\mathbf{E}_{02}^*) &= D \\
+2\,\mathrm{Re}(\mathbf{E}_{01}\cdot\mathbf{E}_{03}^*) &= B, & -2\,\mathrm{Im}(\mathbf{E}_{01}\cdot\mathbf{E}_{03}^*) &= E \\
+2\,\mathrm{Re}(\mathbf{E}_{02}\cdot\mathbf{E}_{03}^*) &= C, & -2\,\mathrm{Im}(\mathbf{E}_{02}\cdot\mathbf{E}_{03}^*) &= F
\end{aligned}
\quad (\mathrm{A.3})
$$

Periodic Materials and Interference Lithography. M. Maldovan and E. Thomas
Copyright © 2009 WILEY-VCH Verlag GmbH & Co. KGaA, Weinheim
ISBN: 978-3-527-31999-2

$$\mathbf{k}_1 \cdot \mathbf{E}_{01} = 0$$
$$\mathbf{k}_2 \cdot \mathbf{E}_{02} = 0 \qquad (A.4)$$
$$\mathbf{k}_3 \cdot \mathbf{E}_{03} = 0$$

where A, B, C, D, E, and F are arbitrary real numbers. As discussed in Chapter 4, these real numbers are indeed the Fourier coefficients of the function $f(\mathbf{r})$ representing the periodic structure one wants to fabricate.

A.1.1
Real Electric Field Amplitude Vectors

For simplicity, we assume in this section that the electric field amplitude vectors $\mathbf{E}_{01}$, $\mathbf{E}_{02}$, and $\mathbf{E}_{03}$ are real vectors. This means that $D = E = F = 0$ in Equation A.3. The Cartesian components of the electric field amplitude vectors are written as

$$\mathbf{E}_{01} = (x_1, x_2, x_3)$$
$$\mathbf{E}_{02} = (x_4, x_5, x_6) \qquad (A.5)$$
$$\mathbf{E}_{03} = (x_7, x_8, x_9)$$

where $x_i (i = 1, \ldots, 9)$ are real numbers.

By substituting the real numbers in Equation A.5 into Equations A.2–A.4, the optimization problem is converted to calculate (for given wave vectors $\mathbf{k}_1$, $\mathbf{k}_2$, and $\mathbf{k}_3$) the $x_1, x_2, x_3, x_4, x_5, x_6, x_7, x_8,$ and x_9 values that minimize

$$x_1^2 + x_2^2 + x_3^2 + x_4^2 + x_5^2 + x_6^2 + x_7^2 + x_8^2 + x_9^2 \qquad (A.6)$$

subject to the constraints

$$x_1 x_4 + x_2 x_5 + x_3 x_6 - \frac{A}{2} = 0$$
$$x_1 x_7 + x_2 x_8 + x_3 x_9 - \frac{B}{2} = 0 \qquad (A.7)$$
$$x_4 x_7 + x_5 x_8 + x_6 x_9 - \frac{C}{2} = 0$$

$$k_{1x} x_1 + k_{1y} x_2 + k_{1z} x_3 = 0$$
$$k_{2x} x_4 + k_{2y} x_5 + k_{2z} x_6 = 0 \qquad (A.8)$$
$$k_{3x} x_7 + k_{3y} x_8 + k_{3z} x_9 = 0$$

This nonlinear optimization problem can be solved using the MATLAB program shown below, which calculates the x_i values (Equation A.5) that minimize Equation A.6 subject to the constraints from Equations A.7 and A.8.

Step 1: Write the following M-file (which includes the constraints given by Equations A.7 and A.8) and save it as "constraints.m".

```
function [c1, c2] = constraints(x)
c1 = [ ];
c2 = [x(1)*x(4)+x(2)*x(5)+x(3)*x(6)-(A/2);
      x(1)*x(7)+x(2)*x(8)+x(3)*x(9)-(B/2);
      x(4)*x(7)+x(5)*x(8)+x(6)*x(9)-(C/2);
              K1X*x(1)+K1Y*x(2)+K1Z*x(3);
              K2X*x(4)+K2Y*x(5)+K2Z*x(6);
              K3X*x(7)+K3Y*x(8)+K3Z*x(9)];
```

Important note: Replace the variables A, B, C, K1X, K1Y, K1Z, K2X, K2Y, K2Z, K3X, K3Y, K3Z by their corresponding numerical values, where A, B, C are the Fourier series coefficients of the periodic structure one wants to fabricate and K1X, K1Y, K1Z, K2X, K2Y, K2Z, K3X, K3Y, K3Z are the components of the wave vectors of the three interfering beams.

Step 2: Run the following M-file.

```
%Line1: Objective function to be minimized (Equation A.6)
f=@(x) x(1)^2+x(2)^2+x(3)^2+x(4)^2+x(5)^2+x(6)^2+x(7)^2+x(8)^2+x(9)^2;

%Line2: Initial Solution. Use any solution to Equations A.8.
x0 = [ X01 X02 X03 X04 X05 X06 X07 X08 X09];

%Line3: Optimization
options = optimset('LargeScale','off');
[x,fval] = fmincon(f,x0,[],[],[],[],[],[],'constraints', options);

%Line4: Solution for the electric field components
x

%Line5: Value of the minimized objective function (Equation A.6)
fval
```

Important note: Replace the variables X01 X02 X03 X04 X05 X06 X07 X08 X09 in Line 2 by their corresponding numerical values. These variables are the initial guess for the solution to our problem. Any trivial solution to Equations A.8 can be used. (Note that the wave vectors in Equations A.8 are known quantities.)

A.1.2
A Numerical Example

To show how to use the program, we next consider a particular numerical example. Suppose that we want to fabricate the periodic structure represented by the function

$f_1(\mathbf{r})$ given by Equation 4.11. By considering Equations 4.13, we have

$$\begin{aligned} A &= 1, & D &= 0 \\ B &= 1, & E &= 0 \\ C &= 0, & F &= 0 \end{aligned} \tag{A.9}$$

According to Equation 4.15, the wave vectors $\mathbf{k}$ of the light beams required to fabricate this structure are given by

$$\begin{aligned} k_{1x} &= 1, & k_{1y} &= 1, & k_{1z} &= 1 \\ k_{2x} &= -1, & k_{2y} &= 1, & k_{2z} &= 1 \\ k_{3x} &= 1, & k_{3y} &= -1, & k_{3z} &= 1 \end{aligned} \tag{A.10}$$

where the common factors π/a in Equation 4.15 need not be considered because the constraints from Equation A.8 are equal to zero.

By considering the wave vectors in Equation A.10, a trivial solution to the system given by Equation A.8 is

$$\begin{aligned} x_{01} &= 1, & x_{02} &= 1, & x_{03} &= -2 \\ x_{04} &= 1, & x_{05} &= 1, & x_{06} &= 0 \\ x_{07} &= 1, & x_{08} &= -1, & x_{09} &= -2 \end{aligned} \tag{A.11}$$

which can be used as the initial condition for the minimization problem.

To find the optimal electric field amplitude vectors $\mathbf{E}_{01}, \mathbf{E}_{02}$, and $\mathbf{E}_{03}$, which are determined by the variables $x_1, x_2, x_3, x_4, x_5, x_6, x_7, x_8$, and x_9, all we need to do is to replace the numerical values (Equations A.9–A.11) into the above MATLAB program. This results in the following program, which is ready for use.

Step 1: Write the following M-file (which includes the constraints from Equations A.7 and A.8) and save it as "constraints.m".

```
function [c1, c2] = constraints(x)
c1 = [ ];
c2 = [x(1)*x(4)+x(2)*x(5)+x(3)*x(6)-(1/2);
      x(1)*x(7)+x(2)*x(8)+x(3)*x(9)-(1/2);
      x(4)*x(7)+x(5)*x(8)+x(6)*x(9)-(0/2);
              1*x(1)+1*x(2)+1*x(3);
             -1*x(4)+1*x(5)+1*x(6);
              1*x(7)-1*x(8)+1*x(9)];
```

Step 2: Run the following M-file.

```
%Line1: Objective function to be minimized (Equation A.6)
f=@(x) x(1)^2+x(2)^2+x(3)^2+x(4)^2+x(5)^2+x(6)^2+x(7)^2+x(8)^2+x(9)^2;
```

```
%Line2: Initial Solution. Use any solution to Equations A.8.
x0 = [ 1 1 -2 1 1 0 1 -1 -2];


%Line3: Optimization
options = optimset('LargeScale','off');
[x,fval] = fmincon(f,x0,[],[],[],[],[],[],'constraints',options);


%Line4: Solution for the electric field components
x


%Line5: Value of the minimized objective function (Equation A.6)
fval
```

After running the program, the solution in the MATLAB command window should read

$$x = 0.5720\ 0.2188\ -0.7908\ 0.3539\ 0.5720\ -0.2180\ 0.2186\ -0.3533$$
$$-0.5719$$
$$\text{fval} = 2.0000$$

This means that the solution to the minimization problem is

$$\begin{aligned} x_1 &= 0.5720, & x_2 &= 0.2188, & x_3 &= -0.7908 \\ x_4 &= 0.3539, & x_5 &= 0.5720, & x_6 &= -0.2180 \\ x_7 &= 0.2186, & x_8 &= -0.3533, & x_9 &= -0.5719 \end{aligned} \qquad (A.12)$$

and that the value of the minimized function (Equation A.6) is 2.0000.

By substituting the values of Equation A.12 into Equation A.5, the optimal electric field amplitude vectors $\mathbf{E}_0$ to fabricate the periodic structure represented by Equation 4.11 are thus obtained.

A.1.3
Complex Electric Field Amplitude Vectors

In general, the electric field amplitude vectors $\mathbf{E}_{01}$, $\mathbf{E}_{02}$, and $\mathbf{E}_{03}$ are complex vectors. In this general case, the Cartesian components of the electric field amplitude vectors can be written as

$$\begin{aligned} \mathbf{E}_{01} &= (x_1 + ix_2, x_3 + ix_4, x_5 + ix_6) \\ \mathbf{E}_{02} &= (x_7 + ix_8, x_9 + ix_{10}, x_{11} + ix_{12}) \\ \mathbf{E}_{03} &= (x_{13} + ix_{14}, x_{15} + ix_{16}, x_{17} + ix_{18}) \end{aligned} \qquad (A.13)$$

where $x_i (i = 1, \ldots, 18)$ are real numbers.

By substituting these values of Equation A.13 into Equation A.2–A.4, the optimization problem is converted to calculate (for given wave vectors $\mathbf{k}_1, \mathbf{k}_2$, and $\mathbf{k}_3$) the values $x_1, \ldots, x_{18}$ that minimize

$$x_1^2 + x_2^2 + x_3^2 + x_4^2 + x_5^2 + x_6^2 + x_7^2 + x_8^2 + x_9^2 + x_{10}^2 + x_{11}^2 + x_{12}^2 + x_{13}^2$$
$$+ x_{14}^2 + x_{15}^2 + x_{16}^2 + x_{17}^2 + x_{18}^2 \quad (A.14)$$

subject to the constraints

$$x_1 x_7 + x_2 x_8 + x_3 x_9 + x_4 x_{10} + x_5 x_{11} + x_6 x_{12} - \frac{A}{2} = 0$$

$$x_1 x_{13} + x_2 x_{14} + x_3 x_{15} + x_4 x_{16} + x_5 x_{17} + x_6 x_{18} - \frac{B}{2} = 0$$

$$x_7 x_{13} + x_8 x_{14} + x_9 x_{15} + x_{10} x_{16} + x_{11} x_{17} + x_{12} x_{18} - \frac{C}{2} = 0$$

$$-x_1 x_8 + x_2 x_7 - x_3 x_{10} + x_4 x_9 - x_5 x_{12} + x_6 x_{11} + \frac{D}{2} = 0 \quad (A.15)$$

$$-x_1 x_{14} + x_2 x_{13} - x_3 x_{16} + x_4 x_{15} - x_5 x_{18} + x_6 x_{17} + \frac{E}{2} = 0$$

$$-x_7 x_{14} + x_8 x_{13} - x_9 x_{16} + x_{10} x_{15} - x_{11} x_{18} + x_{12} x_{17} + \frac{F}{2} = 0$$

$$k_{1x} x_1 + k_{1y} x_3 + k_{1z} x_5 = 0$$
$$k_{1x} x_2 + k_{1y} x_4 + k_{1z} x_6 = 0$$
$$k_{2x} x_7 + k_{2y} x_9 + k_{2z} x_{11} = 0 \quad (A.16)$$
$$k_{2x} x_8 + k_{2y} x_{10} + k_{2z} x_{12} = 0$$
$$k_{3x} x_{13} + k_{3y} x_{15} + k_{3z} x_{17} = 0$$
$$k_{3x} x_{14} + k_{3y} x_{16} + k_{3z} x_{18} = 0$$

This nonlinear optimization problem can be solved by using the MATLAB program shown below, which finds the x_i values (Equation A.13) that minimize Equation A.14 subject to the constraints from Equations A.15 and A.16.

Step 1: Write the following M-file (which includes the constraints from Equations A.15 and A.16) and save it as "constraints.m".

```
function [c1, c2] = constraints(x)
c1=[ ];
c2=[x(1)*x(7)+x(2)*x(8)+x(3)*x(9)+x(4)*x(10)+x(5)*x(11)+x(6)*x(12)-(A/2);
    x(1)*x(13)+x(2)*x(14)+x(3)*x(15)+x(4)*x(16)+x(5)*x(17)+x(6)*x(18)-(B/2);
    x(7)*x(13)+x(8)*x(14)+x(9)*x(15)+x(10)*x(16)+x(11)*x(17)+x(12)*x(18)-(C/2);
    -x(1)*x(8) +x(2)*x(7)-x(3)*x(10)+x(4)*x(9)-x(5)*x(12) +x(6)*x(11)+(D/2);
    -x(1)*x(14)+x(2)*x(13)-x(3)*x(16)+x(4)*x(15)-x(5)*x(18) +x(6)*x(17)+(E/2);
    -x(7)*x(14)+x(8)*x(13)-x(9)*x(16)+x(10)*x(15)-x(11)*x(18)+x(12)*x(17)+(F/2);
```

```
          K1X*x(1)  +K1Y*x(3)  +K1Z*x(5)   ;
          K1X*x(2)  +K1Y*x(4)  +K1Z*x(6)   ;
          K2X*x(7)  +K2Y*x(9)  +K2Z*x(11)  ;
          K2X*x(8)  +K2Y*x(10)+K2Z*x(12)   ;
          K3X*x(13)+K3Y*x(15)+K3Z*x(17)    ;
          K3X*x(14)+K3Y*x(16)+K3Z*x(18)]   ;
```

Important note: Replace the variables A, B, C, D, E, F, K1X, K1Y, K1Z, K2X, K2Y, K2Z, K3X, K3Y, K3Z by their corresponding numerical values, where A, B, C, D, E, F are the Fourier series coefficients of the periodic structure one wants to fabricate and K1X, K1Y, K1Z, K2X, K2Y, K2Z, K3X, K3Y, K3Z are the components of the wave vectors of the three interfering beams.

Step 2: Run the following M-file.

```
%Line1: Objective function to be minimized (Equation A.14)
f=@(x) x(1)^2+x(2)^2+x(3)^2+x(4)^2+x(5)^2+x(6)^2+x(7)^2+x(8)^2+x(9)^2
+x(10)^2+x(11)^2+x(12)^2+x(13)^2+x(14)^2+x(15)^2+x(16)^2+x(17)^2+x(18)^2;

%Line2: Initial Solution. Use any solution to Equations A.16
x0=[ X01 X02 X03 X04 X05 X06 X07 X08 X09 X10 X11 X12 X13 X14 X15 X16 X17 X18];

%Line3: Optimization
options = optimset('LargeScale','off');
[x,fval] = fmincon(f,x0,[],[],[],[],[],[],'constraints',options);

%Line4: Solution for the electric field components
x

%Line5: Value of the minimized objective function (Equation A.14)
fval
```

Important note: Replace the variables X01 ... X18 in Line 2 by their corresponding numerical values. These variables are the initial guess for our problem. Any trivial solution to Equations A.16 should be used. (Note that the wave vectors in Equations A.16 are known quantities.)

A.2
Three-dimensional Periodic Structures

In Chapter 4, we have seen that the superposition of $N = 4$ noncoplanar laser beams with wave vectors $\mathbf{k}_1, \mathbf{k}_2, \mathbf{k}_3$, and $\mathbf{k}_4$ and electric field amplitude vectors $\mathbf{E}_{01}, \mathbf{E}_{02}, \mathbf{E}_{03}$, and $\mathbf{E}_{04}$ generates a three-dimensional spatial distribution of intensity of light $I(\mathbf{r})$ given by the formula

$$\begin{aligned}
I(\mathbf{r}) = &|\mathbf{E}_{01}|^2 + |\mathbf{E}_{02}|^2 + |\mathbf{E}_{03}|^2 + |\mathbf{E}_{04}|^2 \\
&+ 2\operatorname{Re}(\mathbf{E}_{01}\cdot\mathbf{E}_{02}^*)\cos[(\mathbf{k}_1-\mathbf{k}_2)\cdot\mathbf{r}] - 2\operatorname{Im}(\mathbf{E}_{01}\cdot\mathbf{E}_{02}^*)\sin[(\mathbf{k}_1-\mathbf{k}_2)\cdot\mathbf{r}] \\
&+ 2\operatorname{Re}(\mathbf{E}_{01}\cdot\mathbf{E}_{03}^*)\cos[(\mathbf{k}_1-\mathbf{k}_3)\cdot\mathbf{r}] - 2\operatorname{Im}(\mathbf{E}_{01}\cdot\mathbf{E}_{03}^*)\sin[(\mathbf{k}_1-\mathbf{k}_3)\cdot\mathbf{r}] \\
&+ 2\operatorname{Re}(\mathbf{E}_{01}\cdot\mathbf{E}_{04}^*)\cos[(\mathbf{k}_1-\mathbf{k}_4)\cdot\mathbf{r}] - 2\operatorname{Im}(\mathbf{E}_{01}\cdot\mathbf{E}_{04}^*)\sin[(\mathbf{k}_1-\mathbf{k}_4)\cdot\mathbf{r}] \\
&+ 2\operatorname{Re}(\mathbf{E}_{02}\cdot\mathbf{E}_{03}^*)\cos[(\mathbf{k}_2-\mathbf{k}_3)\cdot\mathbf{r}] - 2\operatorname{Im}(\mathbf{E}_{02}\cdot\mathbf{E}_{03}^*)\sin[(\mathbf{k}_2-\mathbf{k}_3)\cdot\mathbf{r}] \\
&+ 2\operatorname{Re}(\mathbf{E}_{02}\cdot\mathbf{E}_{04}^*)\cos[(\mathbf{k}_2-\mathbf{k}_4)\cdot\mathbf{r}] - 2\operatorname{Im}(\mathbf{E}_{02}\cdot\mathbf{E}_{04}^*)\sin[(k_2-\mathbf{k}_4)\cdot\mathbf{r}] \\
&+ 2\operatorname{Re}(\mathbf{E}_{03}\cdot\mathbf{E}_{04}^*)\cos[(\mathbf{k}_3-\mathbf{k}_4)\cdot\mathbf{r}] - 2\operatorname{Im}(\mathbf{E}_{03}\cdot\mathbf{E}_{04}^*)\sin[(\mathbf{k}_3-\mathbf{k}_4)\cdot\mathbf{r}]
\end{aligned} \quad (A.17)$$

The interference pattern (Equation A.17) can be used to fabricate three-dimensional periodic structures. In the three-dimensional case, however, we want to calculate (for given wave vectors $\mathbf{k}_1, \mathbf{k}_2, \mathbf{k}_3$, and $\mathbf{k}_4$) the optimal electric field amplitude vectors $\mathbf{E}_{01}, \mathbf{E}_{02}, \mathbf{E}_{03}$, and $\mathbf{E}_{04}$ that minimize

$$|\mathbf{E}_{01}|^2 + |\mathbf{E}_{02}|^2 + |\mathbf{E}_{03}|^2 + |\mathbf{E}_{04}|^2 \quad (A.18)$$

subject to the constraints

$$\begin{aligned}
+2\operatorname{Re}(\mathbf{E}_{01}\cdot\mathbf{E}_{02}^*) &= A, & -2\operatorname{Im}(\mathbf{E}_{01}\cdot\mathbf{E}_{02}^*) &= G \\
+2\operatorname{Re}(\mathbf{E}_{01}\cdot\mathbf{E}_{03}^*) &= B, & -2\operatorname{Im}(\mathbf{E}_{01}\cdot\mathbf{E}_{03}^*) &= H \\
+2\operatorname{Re}(\mathbf{E}_{01}\cdot\mathbf{E}_{04}^*) &= C, & -2\operatorname{Im}(\mathbf{E}_{01}\cdot\mathbf{E}_{04}^*) &= I \\
+2\operatorname{Re}(\mathbf{E}_{02}\cdot\mathbf{E}_{03}^*) &= D, & -2\operatorname{Im}(\mathbf{E}_{02}\cdot\mathbf{E}_{03}^*) &= J \\
+2\operatorname{Re}(\mathbf{E}_{02}\cdot\mathbf{E}_{04}^*) &= E, & -2\operatorname{Im}(\mathbf{E}_{02}\cdot\mathbf{E}_{04}^*) &= K \\
+2\operatorname{Re}(\mathbf{E}_{03}\cdot\mathbf{E}_{04}^*) &= F, & -2\operatorname{Im}(\mathbf{E}_{03}\cdot\mathbf{E}_{04}^*) &= L
\end{aligned} \quad (A.19)$$

$$\begin{aligned}
\mathbf{k}_1 \cdot \mathbf{E}_{01} &= 0 \\
\mathbf{k}_2 \cdot \mathbf{E}_{02} &= 0 \\
\mathbf{k}_3 \cdot \mathbf{E}_{03} &= 0 \\
\mathbf{k}_4 \cdot \mathbf{E}_{04} &= 0
\end{aligned} \quad (A.20)$$

where A, B, C, D, E, F, G, H, I, J, K, and L are arbitrary real numbers. Indeed, these numbers are the Fourier coefficients of the function $f(\mathbf{r})$ representing the three-dimensional periodic structure one wants to fabricate.

By following the procedure discussed in Section A.1, the reader can modify the MATLAB programs provided above to calculate the optimal electric field amplitude vectors for the three-dimensional case.

Appendix B
MATLAB Program to Calculate Reflectance versus Frequency for One-dimensional Photonic Crystals

B.1
Input Data

The objective of this program in version 7.2 of MATLAB is to calculate the reflectance R as a function of $\omega a/2\pi c$ for normal incident electromagnetic plane waves on a one-dimensional photonic crystal consisting of homogeneous dielectric layers (Figure B.1), where ω is the frequency of the incident electromagnetic plane wave, a is a characteristic length of the photonic crystal, and c is the speed of light (Chapter 6).

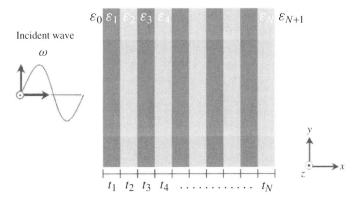

Fig. B.1 A one-dimensional photonic crystal consisting of N homogeneous dielectric layers with dielectric constants $\varepsilon_1, \varepsilon_2, \ldots, \varepsilon_N$ and thicknesses $t_1, t_2, \ldots, t_N$. The dielectric materials in contact with the photonic crystal have dielectric constants ε_0 (left-hand side) and ε_{N+1} (right-hand side). The layers and the surrounding materials are assumed to be nonmagnetic and the homogeneous dielectric layers extend infinitely in the yz plane. An electromagnetic plane wave with frequency ω is normally incident on the one-dimensional photonic crystal from the left. The figure shows the input data required to calculate the reflectance of the normal incident electromagnetic plane wave.

Periodic Materials and Interference Lithography. M. Maldovan and E. Thomas
Copyright © 2009 WILEY-VCH Verlag GmbH & Co. KGaA, Weinheim
ISBN: 978-3-527-31999-2

B MATLAB Program to Calculate Reflectance versus Frequency for One-dimensional Photonic Crystals

Because the materials are assumed to be nonmagnetic, the input data required to describe the one-dimensional photonic crystal comprise the following: the number of layers N; the dielectric constants ε_1, ε_2, ε_3, ..., ε_N and thicknesses t_1, t_2, t_3, ..., t_N of the homogeneous layers; and the dielectric constants ε_0, and ε_{N+1} of the homogeneous materials on both sides of the phononic crystal.

Input data to describe the photonic crystal

The number of layers N
The dielectric constants ε_0, ε_1, ε_2, ε_3, ..., ε_N, ε_{N+1}
The thicknesses t_1, t_2, t_3, ..., t_N

For example, consider a one-dimensional photonic crystal made of $N = 10$ alternating homogeneous layers with dielectric constants $\varepsilon_1 = 11\varepsilon'_0$ and $\varepsilon_2 = 2\varepsilon'_0$ (where ε'_0 is the dielectric constant of vacuum) and thicknesses $t_1 = 0.24\mu\text{m}$ and $t_2 = 0.56\mu\text{m}$. The materials on both sides of the structure are chosen to be air, that is, $\varepsilon_0 = \varepsilon_{N+1} = \varepsilon'_0$. The input data for the MATLAB program (which is given at the end of this chapter) are therefore

```
% Input Data for the Structure
N=10;                                         % USER
e0=     1.00;                                 % USER
e(1)= 11.00; t(1)= 0.24E-6;                   % USER
e(2)=  2.00; t(2)= 0.56E-6;                   % USER
e(3)= 11.00; t(3)= 0.24E-6;                   % USER
e(4)=  2.00; t(4)= 0.56E-6;                   % USER
e(5)= 11.00; t(5)= 0.24E-6;                   % USER
e(6)=  2.00; t(6)= 0.56E-6;                   % USER
e(7)= 11.00; t(7)= 0.24E-6;                   % USER
e(8)=  2.00; t(8)= 0.56E-6;                   % USER
e(9)= 11.00; t(9)= 0.24E-6;                   % USER
e(10)= 2.00; t(10)=0.56E-6;                   % USER
eN1=    1.00;                                 % USER
```

If the number of layers forming the one-dimensional photonic crystal is reduced to four, the input data for the MATLAB program should be changed to

```
% Input Data for the Structure
N=4;                                          % USER
e0=     1.00;                                 % USER
e(1)= 11.00; t(1)= 0.24E-6;                   % USER
e(2)=  2.00; t(2)= 0.56E-6;                   % USER
e(3)= 11.00; t(3)= 0.24E-6;                   % USER
e(4)=  2.00; t(4)= 0.56E-6;                   % USER
eN1=    1.00;                                 % USER
```

Important note: In our particular example, the one-dimensional photonic crystal consists of *alternating* homogeneous layers immersed in air. However, the MATLAB program can manage arbitrary real values of dielectric constants and thicknesses, and, therefore, arbitrary one-dimensional photonic crystal structures.

We now introduce three additional parameters that the MATLAB program needs to calculate the reflectance R as a function of $\omega a/2\pi c$. These parameters must be specified by the user.

Input data for reflectance calculations

The parameter RANGE
The parameter RESOLUTION
The characteristic distance a

The parameter RANGE determines the maximum value of $\omega a/2\pi c$. For example, if RANGE = 1.0, then the reflectance R is calculated for normal incident electromagnetic waves with frequencies ω within the range $0 < \omega a/2\pi c < 1.0$.

The parameter RESOLUTION determines the number of points used in the plot of the reflectance R as a function of $\omega a/2\pi c$. The larger the resolution, the more accurate the graph of the reflectance R versus $\omega a/2\pi c$. A typical value for the parameter RESOLUTION is 1000.

The parameter a is a characteristic length of the structure, which is arbitrary and must be defined by the user. In the particular case of a photonic crystal made of alternating homogeneous layers, it is common to define $a = t_1 + t_2$.

B.2
Running the Program

Below is the MATLAB program that calculates and plots the reflectance R as a function of $\omega a/2\pi c$ for normal incident electromagnetic plane waves on a one-dimensional photonic crystal consisting of N homogeneous dielectric layers (Figure B.1). In the example shown below, the photonic crystal is made of $N = 10$ alternating homogeneous layers with $\varepsilon_1 = 11\varepsilon_0'$, $\varepsilon_2 = 2\varepsilon_0'$, $t_1 = 0.24\mu m$, $t_2 = 0.56\mu m$, and $\varepsilon_0 = \varepsilon_{N+1} = 1\varepsilon_0'$. Users can obviously change these numbers according to their specific photonic crystal designs.

```
% Input Data for the Structure
N=10;                                    % USER
e0=    1.00;                             % USER
e(1)= 11.00;  t(1)= 0.24E-6;             % USER
e(2)=  2.00;  t(2)= 0.56E-6;             % USER
e(3)= 11.00;  t(3)= 0.24E-6;             % USER
e(4)=  2.00;  t(4)= 0.56E-6;             % USER
e(5)= 11.00;  t(5)= 0.24E-6;             % USER
```

```matlab
e(6)=  2.00; t(6)= 0.56E-6;                  % USER
e(7)= 11.00; t(7)= 0.24E-6;                  % USER
e(8)=  2.00; t(8)= 0.56E-6;                  % USER
e(9)= 11.00; t(9)= 0.24E-6;                  % USER
e(10)= 2.00; t(10)=0.56E-6;                  % USER
eN1=   1.00;                                 % USER
% Input Data for Reflectance Calculations
Range=1.0;                                   % USER
Resolution=1000;                             % USER
a=t(1)+t(2);                                 % USER
% Speed of light
c=3.0E8;
% Data Storage
ep(1)=e0;
for ii=2:N+1
   ep(ii)=e(ii-1);
end
ep(N+2)=eN1;
% Location of the Interfaces
d(1)=0.0;
for ii=2:N+1
   d(ii)=d(ii-1)+t(ii-1);
end
% Loop in angular frequencies w
for iw=1:Resolution
   wa2pc=(iw/Resolution)*Range;
   w=wa2pc*(2.0*pi*c)/a;
% Wave-vectors
   for jj=1:N+2
      kx(jj)=(w/c)*sqrt(ep(jj));
   end
% Matrices
   for ii=1:N+1
      a11= cos(kx(ii)*d(ii))+ i*sin(kx(ii)*d(ii));
      a12= cos(kx(ii)*d(ii))- i*sin(kx(ii)*d(ii));
      a21= cos(kx(ii)*d(ii))+ i*sin(kx(ii)*d(ii));
      a22= cos(kx(ii)*d(ii))- i*sin(kx(ii)*d(ii));
      A(ii,1)=              a11;
      A(ii,2)=              a12;
      A(ii,3)= sqrt(ep(ii))*a21;
      A(ii,4)=-sqrt(ep(ii))*a22;
      b11= cos(kx(ii+1)*d(ii))+ i*sin(kx(ii+1)*d(ii));
      b12= cos(kx(ii+1)*d(ii))- i*sin(kx(ii+1)*d(ii));
```

```
    b21= cos(kx(ii+1)*d(ii))+ i*sin(kx(ii+1)*d(ii));
    b22= cos(kx(ii+1)*d(ii))- i*sin(kx(ii+1)*d(ii));
    B(ii,1)=                  b11;
    B(ii,2)=                  b12;
    B(ii,3)= sqrt(ep(ii+1))*b21;
    B(ii,4)=-sqrt(ep(ii+1))*b22;
  end
  % Inverse Matrices
  for ii=1:N+1
    Ai(ii,1)= A(ii,4)/(A(ii,1)*A(ii,4)-A(ii,2)*A(ii,3));
    Ai(ii,2)=-A(ii,2)/(A(ii,1)*A(ii,4)-A(ii,2)*A(ii,3));
    Ai(ii,3)=-A(ii,3)/(A(ii,1)*A(ii,4)-A(ii,2)*A(ii,3));
    Ai(ii,4)= A(ii,1)/(A(ii,1)*A(ii,4)-A(ii,2)*A(ii,3));
  end
% Product of Matrices
  for ii=1:N+1
    At(2*ii-1,1)=Ai(ii,1);
    At(2*ii-1,2)=Ai(ii,2);
    At(2*ii-1,3)=Ai(ii,3);
    At(2*ii-1,4)=Ai(ii,4);
    At(2*ii,1)=B(ii,1);
    At(2*ii,2)=B(ii,2);
    At(2*ii,3)=B(ii,3);
    At(2*ii,4)=B(ii,4);
  end
  Ap(1,1)=At(1,1);
  Ap(1,2)=At(1,2);
  Ap(2,1)=At(1,3);
  Ap(2,2)=At(1,4);
  Bp(1,1)=At(2,1);
  Bp(1,2)=At(2,2);
  Bp(2,1)=At(2,3);
  Bp(2,2)=At(2,4);
  Cp=Ap*Bp;
  for ii=3:2*(N+1)
    Ap(1,1)=Cp(1,1);
    Ap(1,2)=Cp(1,2);
    Ap(2,1)=Cp(2,1);
    Ap(2,2)=Cp(2,2);
    Bp(1,1)=At(ii,1);
    Bp(1,2)=At(ii,2);
    Bp(2,1)=At(ii,3);
```

```
    Bp(2,2)=At(ii,4);
    Cp=Ap*Bp;
  end
% Reflectance
  Rx=Cp(2,1)/Cp(1,1);
  R(iw)=Rx*conj(Rx);
  X(iw)=wa2pc;
end
% Output
fid=fopen('matlab.out','wt');
for i=1:Resolution
  fprintf('%8.4f %8.4f\n',X(i),R(i))
  fprintf(fid,'%8.4f %8.4f\n',X(i),R(i));
end
plot(X,R,'k','linewidth',2)
axis([0, Range, 0, 1.05])
title('Plot of Reflectance vs wa/2Pic')
xlabel('wa/2Pic'); ylabel('Reflectance'); grid
clear
```

After running the above program in the MATLAB command window, the values of the frequencies $\omega a/2\pi c$ and the corresponding values of the reflectance R are written in two columns on the screen. These values are also stored in the output text file "matlab.out". For example, the first values for $\omega a/2\pi c$ and the reflectance R in the case of the above example are

$\omega a/2\pi c$	R
0.0010	0.0034
0.0020	0.0133
0.0030	0.0291
0.0040	0.0501
0.0050	0.0752
0.0060	0.1032
0.0070	0.1331
0.0080	0.1638
0.0090	0.1945
0.0100	0.2244
0.0110	0.2529
0.0120	0.2798
0.0130	0.3045
0.0140	0.3271
0.0150	0.3473

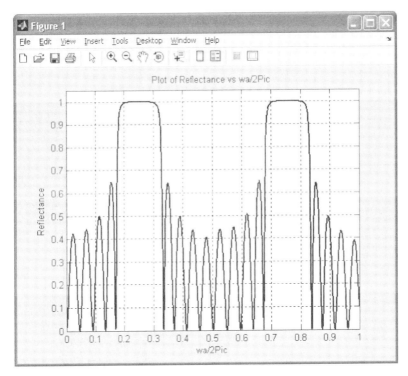

Fig. B.2 A plot of the reflectance R versus $\omega a/2\pi c$ for normal incident electromagnetic plane waves on a one-dimensional photonic crystal.

In addition, after running the program, the graph of the reflectance R versus $\omega a/2\pi c$ automatically appears on the screen (Figure B.2). From this graph, one can identify the frequency ranges for which the reflectance R of normal incident electromagnetic waves is large (e.g. $R > 0.99$). These high reflectance frequency ranges are called the *photonic band gaps* and can alternatively be identified from the output text file "matlab.out".

Appendix C
MATLAB Program to Calculate Reflectance versus Frequency for One-dimensional Phononic Crystals

C.1
Input Data

The objective of this program in version 7.2 of MATLAB is to calculate the reflectance R as a function of $\omega a / 2\pi c_T$ for normal incident elastic plane waves on a one-dimensional phononic crystal consisting of homogeneous elastic layers (Figure C.1), where ω is the frequency of the incident elastic plane wave, a is a characteristic length of the phononic crystal, and c_T is the transverse velocity of elastic waves in one of the homogeneous layers (Chapter 7).

The input data required to describe the one-dimensional phononic crystal comprise the following: the number of layers N; the densities $\rho_1, \rho_2, \ldots, \rho_N$; transverse velocities $c_{T1}, c_{T2}, \ldots, c_{TN}$; and longitudinal velocities $c_{L1}, c_{L2}, \ldots, c_{LN}$ of the homogeneous layers; and the densities and transverse and longitudinal velocities $\rho_0, c_{T0}, c_{L0}, \rho_{N+1}, c_{TN+1}$, and c_{LN+1} of the elastic materials on both sides of the phononic crystal.

Input data to describe the phononic crystal

The number of layers N
The densities $\rho_0, \rho_1, \rho_2, \rho_3, \ldots, \rho_N, \rho_{N+1}$
The transverse and longitudinal velocities $c_{T0}, c_{L0}, c_{T1}, c_{L1}, c_{T2}, c_{L2},$
$\ldots, c_{TN}, c_{LN}, c_{TN+1}, c_{LN+1}$
The thicknesses $t_1, t_2, t_3, \ldots, t_N$

For example, consider a one-dimensional phononic crystal made of $N = 10$ alternating homogeneous layers with densities $\rho_1 = 10\,760\,\text{kg}\,\text{m}^{-3}$ and $\rho_2 = 1140\,\text{kg}\,\text{m}^{-3}$; velocities $c_{T1} = 850\,\text{m}\,\text{s}^{-1}$, $c_{T2} = 1300\,\text{m}\,\text{s}^{-1}$, $c_{L1} = 1960\,\text{m}\,\text{s}^{-1}$ and $c_{L2} = 2770\,\text{m}\,\text{s}^{-1}$; and thicknesses $t_1 = t_2 = 0.5\,\mu\text{m}$. These values correspond to the choice of lead and epoxy for the homogeneous layers, respectively. Also consider that the material on both sides of the phononic crystal is nylon, i.e. $\rho_0 = \rho_{N+1} = 1110\,\text{kg}\,\text{m}^{-3}$, $c_{T0} = c_{TN+1} = 1100\,\text{m}\,\text{s}^{-1}$, and $c_{L0} = c_{LN+1} = 2600\,\text{m}\,\text{s}^{-1}$. The input

Periodic Materials and Interference Lithography. M. Maldovan and E. Thomas
Copyright © 2009 WILEY-VCH Verlag GmbH & Co. KGaA, Weinheim
ISBN: 978-3-527-31999-2

C MATLAB Program to Calculate Reflectance versus Frequency for One-dimensional Phononic Crystals

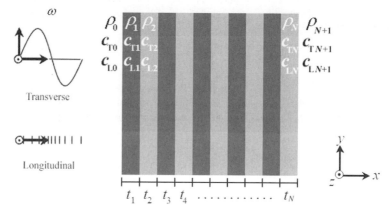

Fig. C.1 A one-dimensional phononic crystal consisting of N homogeneous elastic layers with densities $\rho_1, \rho_2, \ldots, \rho_N$; transverse velocities, $c_{T1}, c_{T2}, \ldots, c_{TN}$; longitudinal velocities $c_{L1}, c_{L2}, \ldots, c_{LN}$; and thicknesses $t_1, t_2, \ldots, t_N$. The elastic materials in contact with the phononic crystal have parameters ρ_0, c_{T0}, and c_{L0} (left-hand side) and ρ_{N+1}, c_{TN+1}, and c_{LN+1} (right-hand side). The homogeneous elastic layers are assumed to extend infinitely in the yz plane. An elastic plane wave (either transversal or longitudinal) with frequency ω is normally incident on the one-dimensional phononic crystal. The figure shows the input data required to calculate the reflectance of the normal incident elastic plane waves.

data for the MATLAB program (which are given at the end of this chapter) are therefore

```
% Input Data for the Structure
N=10;                                                       % USER
rho0=     1.11;  ct0=   1.10;  cl0=   2.60;                 % USER
rho(1)= 10.76;  ct(1)= 0.85;  cl(1)= 1.96;  t(1)= 0.50E-6;  % USER
rho(2)=  1.14;  ct(2)= 1.30;  cl(2)= 2.77;  t(2)= 0.50E-6;  % USER
rho(3)= 10.76;  ct(3)= 0.85;  cl(3)= 1.96;  t(3)= 0.50E-6;  % USER
rho(4)=  1.14;  ct(4)= 1.30;  cl(4)= 2.77;  t(4)= 0.50E-6;  % USER
rho(5)= 10.76;  ct(5)= 0.85;  cl(5)= 1.96;  t(5)= 0.50E-6;  % USER
rho(6)=  1.14;  ct(6)= 1.30;  cl(6)= 2.77;  t(6)= 0.50E-6;  % USER
rho(7)= 10.76;  ct(7)= 0.85;  cl(7)= 1.96;  t(7)= 0.50E-6;  % USER
rho(8)=  1.14;  ct(8)= 1.30;  cl(8)= 2.77;  t(8)= 0.50E-6;  % USER
rho(9)= 10.76;  ct(9)= 0.85;  cl(9)= 1.96;  t(9)= 0.50E-6;  % USER
rho(10)= 1.14;  ct(10)=1.30;  cl(10)=2.77;  t(10)=0.50E-6;  % USER
rhoN1=    1.11;  ctN1=  1.10;  clN1=  2.60;                 % USER
```

If the number of layers forming the one-dimensional phononic crystal is reduced to four, the input data for the MATLAB program should be changed to

```
% Input Data for the Structure
N=4;                                                        % USER
rho0=     1.11;  ct0=   1.10;  cl0=   2.60;                 % USER
```

```
rho(1)= 10.76;   ct(1)= 0.85;   cl(1)= 1.96;   t(1)= 0.50E-6;      % USER
rho(2)=  1.14;   ct(2)= 1.30;   cl(2)= 2.77;   t(2)= 0.50E-6;      % USER
rho(3)= 10.76;   ct(3)= 0.85;   cl(3)= 1.96;   t(3)= 0.50E-6;      % USER
rho(4)=  1.14;   ct(4)= 1.30;   cl(4)= 2.77;   t(4)= 0.50E-6;      % USER
rhoN1=   1.11;   ctN1=  1.10;   clN1=  2.60;                       % USER
```

Important note: Although we consider the particular case of a phononic crystal made of *alternating* homogeneous layers immersed in nylon, the densities, velocities, and thicknesses in the MATLAB program can have arbitrary values for solid materials.

We next introduce four additional parameters that the MATLAB program needs to calculate the reflectance R as a function of $\omega a/2\pi c_T$. These parameters must be specified by the user.

Input data for reflectance calculations

The parameter RANGE
The parameter RESOLUTION
The characteristic distance a
The transverse velocity c_T

The parameter RANGE determines the maximum value of $\omega a/2\pi c_T$. For example, if RANGE $= 1.0$, then the reflectance R is calculated for normal incident elastic waves with frequencies within the range $0 < \omega a/2\pi c_T < 1.0$.

The parameter RESOLUTION determines the number of points used in the calculation of the reflectance R versus $\omega a/2\pi c_T$. The larger the resolution, the more accurate the graph of the reflectance R versus $\omega a/2\pi c_T$. A typical value for the parameter RESOLUTION is 1000.

The parameter a is a characteristic length of the structure, which is arbitrary and must be defined by the user. In the particular case of a phononic crystal made of alternating homogeneous layers, it is common to define $a = t_1 + t_2$.

The parameter c_T is an arbitrary velocity and must be defined by the user. In the particular case of a one-dimensional phononic crystal made of alternating homogeneous layers, it is common to define c_T as the transverse velocity of elastic waves in one of the homogeneous layers (e.g. $c_T = c_{T2}$).

C.2
Running the Program

Below is the MATLAB program that calculates and plots the reflectance R as a function of $\omega a/2\pi c_T$ for normal incident elastic plane waves on a one-dimensional phononic crystal consisting of N homogeneous elastic layers (Figure C.1). In the example shown below, the phononic crystal comprises $N = 10$ alternating homogeneous layers made of lead ($\rho_1 = 10\,760\,\text{kg m}^{-3}$, $c_{T1} = 850\,\text{m s}^{-1}$, and $c_{L1} = 1960\,\text{m s}^{-1}$) and epoxy ($\rho_2 = 1140\,\text{kg m}^{-3}$, $c_{T2} = 1300\,\text{m s}^{-1}$, and $c_{L2} = 2770\,\text{m s}^{-1}$)

C MATLAB Program to Calculate Reflectance versus Frequency for One-dimensional Phononic Crystals

with thicknesses $t_1 = t_2 = 0.50\,\mu\text{m}$. The surrounding material is assumed to be nylon ($\rho_0 = \rho_{N+1} = 1110\,\text{kg m}^{-3}$, $c_{T0} = c_{TN+1} = 1100\,\text{m s}^{-1}$, and $c_{L0} = c_{LN+1} = 2600\,\text{m s}^{-1}$). Users can obviously change these numbers according to their specific phononic crystal designs.

```
% Input Data for the Structure
N=10;                                                              % USER
rho0=      1.11;   ct0=    1.10;   cl0=   2.60;                    % USER
rho(1)= 10.76;  ct(1)= 0.85;  cl(1)= 1.96;  t(1)= 0.50E-6;         % USER
rho(2)=  1.14;  ct(2)= 1.30;  cl(2)= 2.77;  t(2)= 0.50E-6;         % USER
rho(3)= 10.76;  ct(3)= 0.85;  cl(3)= 1.96;  t(3)= 0.50E-6;         % USER
rho(4)=  1.14;  ct(4)= 1.30;  cl(4)= 2.77;  t(4)= 0.50E-6;         % USER
rho(5)= 10.76;  ct(5)= 0.85;  cl(5)= 1.96;  t(5)= 0.50E-6;         % USER
rho(6)=  1.14;  ct(6)= 1.30;  cl(6)= 2.77;  t(6)= 0.50E-6;         % USER
rho(7)= 10.76;  ct(7)= 0.85;  cl(7)= 1.96;  t(7)= 0.50E-6;         % USER
rho(8)=  1.14;  ct(8)= 1.30;  cl(8)= 2.77;  t(8)= 0.50E-6;         % USER
rho(9)= 10.76;  ct(9)= 0.85;  cl(9)= 1.96;  t(9)= 0.50E-6;         % USER
rho(10)= 1.14;  ct(10)=1.30;  cl(10)=2.77;  t(10)=0.50E-6;         % USER
rhoN1=     1.11;   ctN1=   1.10;   clN1=  2.60;                    % USER
% Input Data for Reflectance Calculations
Range=2.0;                                                         % USER
Resolution=2000;                                                   % USER
a=t(1)+t(2);                                                       % USER
c0=ct(2);                                                          % USER
% Data Storage
rhop(1)=rho0; ctp(1)=ct0; clp(1)=cl0;
for ii=2:N+1
   rhop(ii)=rho(ii-1); ctp(ii)=ct(ii-1); clp(ii)=cl(ii-1);
end
rhop(N+2)=rhoN1; ctp(N+2)=ctN1; clp(N+2)=clN1;
% Location of the Interfaces
d(1)=0.0;
for ii=2:N+1
   d(ii)=d(ii-1)+t(ii-1);
end
% Loop in angular frequencies w
for iw=1:Resolution
   wa2pc=(iw/Resolution)*Range;
   w=wa2pc*(2.0*pi*ct(2))/a;
   % Loop for transverse and longitudinal waves
   for iT=1:2
      % Wave-vectors
      if (iT==1)
         for jj=1:N+2
            kx(jj)=w/ctp(jj);
         end
      end
      if (iT==2)
```

C.2 Running the Program

```
    for jj=1:N+2
      kx(jj)=w/clp(jj);
    end
  end
  % Matrices
  for ii=1:N+1
    a11= cos(kx(ii)*d(ii))+ i*sin(kx(ii)*d(ii));
    a12= cos(kx(ii)*d(ii))- i*sin(kx(ii)*d(ii));
    a21= cos(kx(ii)*d(ii))+ i*sin(kx(ii)*d(ii));
    a22= cos(kx(ii)*d(ii))- i*sin(kx(ii)*d(ii));
    if (iT==1)
      AlfaA=rhop(ii)*ctp(ii);
    end
    if (iT==2)
      AlfaA=rhop(ii)*clp(ii);
    end
    A(ii,1)=       a11;
    A(ii,2)=       a12;
    A(ii,3)= AlfaA*a21;
    A(ii,4)=-AlfaA*a22;
    b11= cos(kx(ii+1)*d(ii))+ i*sin(kx(ii+1)*d(ii));
    b12= cos(kx(ii+1)*d(ii))- i*sin(kx(ii+1)*d(ii));
    b21= cos(kx(ii+1)*d(ii))+ i*sin(kx(ii+1)*d(ii));
    b22= cos(kx(ii+1)*d(ii))- i*sin(kx(ii+1)*d(ii));
    if (iT==1)
      AlfaB=rhop(ii+1)*ctp(ii+1);
    end
    if (iT==2)
      AlfaB=rhop(ii+1)*clp(ii+1);
    end
    B(ii,1)=       b11;
    B(ii,2)=       b12;
    B(ii,3)= AlfaB*b21;
    B(ii,4)=-AlfaB*b22;
  end
  % Inverse Matrices
  for ii=1:N+1
    Ai(ii,1)= A(ii,4)/(A(ii,1)*A(ii,4)-A(ii,2)*A(ii,3));
    Ai(ii,2)=-A(ii,2)/(A(ii,1)*A(ii,4)-A(ii,2)*A(ii,3));
    Ai(ii,3)=-A(ii,3)/(A(ii,1)*A(ii,4)-A(ii,2)*A(ii,3));
    Ai(ii,4)= A(ii,1)/(A(ii,1)*A(ii,4)-A(ii,2)*A(ii,3));
  end
% Product of Matrices
  for ii=1:N+1
    At(2*ii-1,1)=Ai(ii,1);
    At(2*ii-1,2)=Ai(ii,2);
    At(2*ii-1,3)=Ai(ii,3);
    At(2*ii-1,4)=Ai(ii,4);
    At(2*ii,1)=B(ii,1);
```

```
        At(2*ii,2)=B(ii,2);
        At(2*ii,3)=B(ii,3);
        At(2*ii,4)=B(ii,4);
      end
      Ap(1,1)=At(1,1);
      Ap(1,2)=At(1,2);
      Ap(2,1)=At(1,3);
      Ap(2,2)=At(1,4);
      Bp(1,1)=At(2,1);
      Bp(1,2)=At(2,2);
      Bp(2,1)=At(2,3);
      Bp(2,2)=At(2,4);
      Cp=Ap*Bp;
      for ii=3:2*(N+1)
        Ap(1,1)=Cp(1,1);
        Ap(1,2)=Cp(1,2);
        Ap(2,1)=Cp(2,1);
        Ap(2,2)=Cp(2,2);
        Bp(1,1)=At(ii,1);
        Bp(1,2)=At(ii,2);
        Bp(2,1)=At(ii,3);
        Bp(2,2)=At(ii,4);
        Cp=Ap*Bp;
      end
    % Reflectance
      Rx=Cp(2,1)/Cp(1,1);
      if (iT==1)
        RT(iw)=Rx*conj(Rx);
      end
      if (iT==2)
        RL(iw)=Rx*conj(Rx);
      end
      X(iw)=wa2pc;
    end
  end
% Output
fid=fopen('matlab.out','wt');
for i=1:Resolution
  fprintf('%8.4f %8.4f %8.4f\n',X(i),RT(i),RL(i))
  fprintf(fid,'%8.4f %8.4f %8.4f\n',X(i),RT(i),RL(i));
end
plot(X,RT,'k',X,RL,'--r','linewidth',1.5)
axis([0, Range, 0, 1.05])
title('Plot of Reflectance vs wa/2PicT')
xlabel('wa/2PicT'); ylabel('Reflectance'); grid
clear
```

After running the program in the MATLAB command window, the values of $\omega a/2\pi c_T$ and the corresponding values of the reflectance R for transverse and

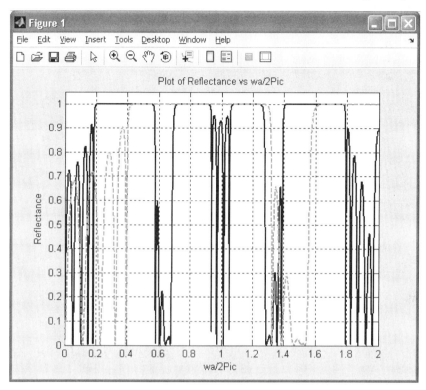

Fig. C.2 A plot of reflectance R versus $\omega a/2\pi c_T$ for normal incident transverse (solid line) and longitudinal (dotted line) elastic waves on a one-dimensional phononic crystal.

longitudinal incident elastic waves are written in three successive columns on the screen. These values are also stored in the output text file "matlab.out". For example, the first values of $\omega a/2\pi c_T$ and the reflectance R for the above example are given as

```
wa/2πcT  R(transverse) R(longitudinal)
0.0010   0.0083       0.0014
0.0020   0.0322       0.0057
0.0030   0.0694       0.0128
0.0040   0.1163       0.0225
0.0050   0.1692       0.0347
0.0060   0.2247       0.0491
0.0070   0.2802       0.0655
0.0080   0.3335       0.0836
0.0090   0.3835       0.1031
0.0100   0.4294       0.1239
0.0110   0.4710       0.1455
```

0.0120	0.5083	0.1678
0.0130	0.5415	0.1904
0.0140	0.5708	0.2133
0.0150	0.5966	0.2362

In addition, after running the program, the graph of the reflectance R versus $\omega a/2\pi c_T$ for transverse (black solid lines) and longitudinal (gray dashed lines) incident elastic waves automatically appears on the screen (Figure C.2). From this graph, one can identify the frequency ranges for which the reflectance of normal incident transverse and longitudinal elastic waves is high (e.g. $R > 0.99$). These high reflectance frequency ranges are called the *phononic band gaps* for transverse and longitudinal elastic waves, respectively, and they can also be identified from the output text file "matlab.out".

Index

a
acoustic waves, definition 184
 – in homogeneous fluid materials 187–188
acoustic–optical devices 264–268
air cylinders, photonic band gap 160–161
air spheres, structures made of 240–242
all-angle negative refraction, definition 270
arbitrary unit cells 25

b
backfilling 133–134
band diagrams
 – in three dimensions 168–170
 – in two dimensions 157–160
 – of photonic crystals 147
 – structures for localization of light 260–261
band gap
 – phononic *see* complete phononic band gap
 – photonic *see* complete photonic band gap
beam configuration, periodic structure fabrication 119–121
beam direction, preserving 128–130
beam parameters, periodic structure fabrication
 – three-dimensional 107–109
 – two-dimensional 102–103
beam splitter, polarizing 119
bicontinuous structures 93
binary composite structures 272
Bloch wave
 – definition 147–148
 – elastic 194
body-centered-cubic lattice 51–59
 – Bravais 242
 – primitive vectors 165

body-centered-cubic photonic crystals 173–176
body-centered unit cells, definition 14
Bravais lattices
 – body-centered-cubic 242
 – cubic 165
 – definition 12–14
Brillouin zones
 – definition 155
 – in three dimensions 164–167
 – in two dimensions 152–157
 – irreducible 156, 167
bulk modulus, cubic crystals 237–238

c
CAR *see* chemically amplified resists
cellular solids
 – closed-cell 216
 – definition 215–216
 – fabrication 249–250
 – open-cell 216
 – periodic *see* periodic cellular solids
channel, defect 262–263
characteristic distance, between atoms 184
chemical modification, photoresist materials 89–91, 122
chemically amplified resists (CAR) 124
circularly polarized waves
 – definition 72
 – plane waves 74–75
closed-cell cellular solids 216
CO_2, supercritical drying 133
collinearity, and interference 83
complete band gaps
 – in photonic–phononic crystals 264–267
 – phononic 201, 204–206
  – three-dimensional 209–210

- photonic
  - alternative definitions 159
  - localization of light 260
  - three-dimensional 163, 169
  - two-dimensional 158
  - wave guiding 264

compliance
- definition 219
- see also elastic constants

composite structures, binary 272

constitutive parameters, Maxwell's equations 66

contrast, definition 131–132

conventional unit cells 12
- definition 15

crystal systems, definition 13–14

crystals
- cubic see cubic crystals
- phononic see phononic crystals
- photonic see photonic crystals

cubic Bravais lattices, primitive vectors 165

cubic crystals
- elastic constants 232–238
- photonic
  - body-centered 173–176
  - face-centered 172–173
  - simple 172

cubic lattices
- body-centered 51–59
- Bravais 165
- face-centered 47–51
- simple 44–46

cubic periodic structures, mechanical properties 244

cylinders, photonic, elastic 189

d

Darcy's law 274

defect channel, in photonic crystals 262–263

defects, structural 267

deformation
- as strain plus rotation 227–228
- definition 221–222
- general 226–228

deformation tensor, definition 224

density, as a material parameter 187

developing of photoresist materials
- definition 89
- structure obtaining 93

diamond structure, definition 168–169

diazonaphthoquinone (DNQ) 126–127

dielectric constant, Maxwell's equations 66

dielectric-constants ratio, photonic band gaps 169–170

direct structure
- definition 34–35
- three-dimensional 45

dispersion relation
- definition 147
- in two dimensions 157–159, 199
- solid–fluid phononic crystals 203
- solid–solid phononic crystals 195–197

displacement vectors
- definition 222–223
- transverse elastic plane waves 186

distance, characteristic 184

DNQ see diazonaphthoquinone

drying, photoresists 132–133

e

elastic constants, cubic crystals 232–238

elastic cylinders 189

elastic layers 189, 191

elastic mechanical properties, linear 243–247

elastic spheres 189

elastic wave equations 186

elastic wave velocity, as a material parameter 187

elastic waves
- Bloch 194
- definition 184
- forbidden propagation 198
- in homogeneous solid materials 184–187

electric field vector, definition 64

electric permittivity see dielectric constant

electricity transport, maximum 272–273

electromagnetic (plane) waves 68–69
- basic description 64–65
- electromagnetic energy 76–81
- energy density 77
- energy flux 77
- forbidden propagation 162
- in-plane 157
- intensity 80–81
- interference 81–89
- out-of-plane 200
- phase properties 72
- polarization 69–75
- propagation in photonic crystals
  - three-dimensional 168–170
  - two-dimensional 157–160
- superposition 83
- time-averaged values 77–80
- transverse character 69–72
- wave equation 65–68

electromagnetic spectrum, classification 65
elliptical polarization, electromagnetic plane waves 75
emission of light, spontaneous 256–259
energy
– electromagnetic 76–81
– per unit area per unit time 78–80
energy conservation law 76
energy density
– electromagnetic (plane) waves 77
– instantaneous 76
– time-averaged 78
energy flux, electromagnetic plane waves 77
expansion, periodic cellular solids 225

f

F-RD structure 242–243
– inverse 247–248
face-centered-cubic lattice 47–51
– Bravais 242
– primitive vectors 165
face-centered unit cells, definition 14
figure of merit, efficiency of thermoelectric materials 277
finite element program, to calculate linear elastic mechanical properties 243
finite periodic structures 143–146
– phononic crystals 190–194
– versus infinite 150–151
fluid materials, acoustic waves 187–188
fluid permeability, definition 274
fluid–fluid phononic crystals 207
foaming, nonperiodic cellular solids 217
forbidden propagation
– elastic waves 198, 201
– electromagnetic waves 162
four-beam interference 116–117
four-beam technique
– periodic structure fabrication 119–120
– prism configuration 130–131
Fourier coefficients
– calculation of 21–23
– definition 21
– relative weight 33
Fourier series 20–26
– and interference 98–99
– for three-dimensional periodic functions 23–25
– for two-dimensional periodic functions 20–23
Fourier series expansion 98
– definition 21
– three-dimensional 42–43
– to create periodic structures 29–61
frequency, definition 64
frequency ranges, in photonic crystals 146, 148
fringe-to-fringe spacing, definition 86

g

gamma point, Brillouin zone 156
generalized Hooke's law
– definition 219
– matrix notation 230–232
– stress–strain relationship 229
geometry of the periodic structure 169–170
gold spheres, phononic crystal 208–209
guiding mechanisms, in photonic crystals 262–264

h

half-wave plate, polarization 120
heat transport, maximum 272–273
holographic interference lithography *see* interference lithography
homogeneity, material parameter 66
homogeneous materials
– fluid 187–188
– solid 184–187
– wave vector 147
Hooke's law
– generalized *see* generalized Hooke's law
– one-dimensional 218–219

i

I-WP structure 242
– negative refraction 271
in-plane waves
– definition 200
– electromagnetic 157
incident light beam, TE and TM components 129–130
infinite periodic structures 147–149, 194–198
– three-dimensional 5
– versus finite 150–151
inks, polyelectrolyte 239
intensity, definition 80–81
intensity distribution
– and interference parameters 94
– spatial 88–89
interference
– and Fourier series 98–99
– definition 83
– electromagnetic plane waves 81–89

interference lithography 89–94
- and periodic structures 97–110
- backfilling 133–134
- contrast 131–132
- fabrication technique 239
- obtaining structure by 92–94
- practical considerations 128–135
- simple cubic cellular solid fabrication 249–250
- technique 92–93
- three-dimensional periodic structures 104–110
- two-dimensional periodic structures 100–104
- volume fraction control 134–135
interference parameters, and intensity distribution 94
interference patterns
- definition 86
- three-dimensional 86–89
interference terms 98
inverse structures
- creation of 133–134
- definition 34–35
- three-dimensional 45
irradiance see intensity
irreducible Brillouin zone
- definition 156
- three-dimensional 167
isotropic solid material, mechanical properties 244
isotropy, material parameter 66

k
Kelvin (Thomson) relations 277

l
Lame coefficients
- fluid material 188
- homogeneous solid material 187
laser beams 115, 123, 135
- four-beam technique 131
- interference lithography 97
- negative refraction 270
- see also light beams
lattice constants
- three-dimensional 12
- two-dimensional 8
layers, elastic 189, 191
lenses, photonic crystals 269
level-set structures 240–242
- negative refraction 271
light
- localization of 259

- simultaneous localization with sound 264–268
- spontaneous emission of 256
light beams
- periodic structure fabrication 116–118
- see also laser beams
- TE and TM components 129–130
linear elastic mechanical properties 243–247
- calculated with finite element program 243
- periodic cellular solids 243–247
linear polarization, basics 73–75
linearity, material parameter 66
linearly polarized waves
- definition 72
- plane waves 73–74
liquid CO_2, supercritical drying 133
lithography
- interference see interference lithography
- phase mask 120–121
- road map 123
- two-photon 239
localization of light
- and sound 264–268
- definition 259
- microcavities and waveguides 259–264
- spatial 261
longitudinal plane wave 184–185

m
magnetic field vector, definition 64
magnetic permeability, Maxwell's equations 66
maximum transport of heat and electricity 272–273
Maxwell's equations 65–66
- curl 76
mechanical waves, propagation 184–187
methacrylate polymers (PMMA) 127
microcavities 259–264
microchips, optical 264
microfluidics 273–275
microframes, plastic deformation 250–252
monochromatic waves, definition 69
monomeric systems, negative photoresists 123
multifunctional periodic structures 272–273
multiple gratings, and the registration challenge 118

n

negative photoresists 124–125
- definition 92
- monomeric systems 123
- oligomeric systems 124

negative refraction
- and superlenses 268–272
- definition 268–269

noncollinearity, as prerequisite for interference 83, 86
nonconductivity, material parameter 66
noncoplanarity, and interference 86
nondimensional frequency
- phononic crystals 192–193
- photonic crystals 145

nondispersivity, material parameter 66
nonhomogenous material, wave vector 147
nonperiodic cellular solids, definition 216
nonperiodic structures, versus periodic structures 4–6
nonperpendicularity, as prerequisite for interference 83
nonprimitive unit cells 14–16
- definition 15

normal stress, definition 220
Novolac resin 126–127
numerical techniques, plane-wave method 148

o

oligomeric systems, negative photoresists 124
omnidirectional reflection, definition 150
one-dimensional Hooke's law 218–219
one-dimensional phononic crystals 190–198
- solid–solid 195–197

one-dimensional photonic crystals 143–151
open-cell cellular solids, definition 216
optical microcavity 262
optical microchips 264
optical resonators 262
organic–inorganic hybrids photoresists 128
out-of-plane waves, definition 200

p

P structures 249
PAG *see* photoacid generator
parameter d
- periodic functions in three dimensions 43
- square lattice 31–33
- triangular lattice 39

pattern transfer, periodic structure fabrication 122–128

Peltier effect 275–276
periodic cellular solids 215–252
- body-centered-cubic Bravais lattice 242
- definition 216
- expansion 225
- face-centered-cubic Bravais lattice 242
- general deformation 226–228
- linear elastic mechanical properties 243–247
- simple cubic 249–250
- strain tensor 221–228
- stress tensor 219–221
- topological design 238–243
- twelve-connected stretch-dominated 247–249

periodic functions
- and structures 29–61
- combination 59–61
- simple 33–38, 40–41
- three-dimensional 23–25, 41–59
- two-dimensional 20–23, 31–41

periodic structure fabrication 115–135
- beam configuration 119–121
- beam parameters 102–103, 107–109
- four-beam technique 119–120
- interference lithography 97–110
- light beams 116–118
- material platforms 122–128
- multiple gratings 118
- pattern transfer 122–128
- photoresists 122–128
- prism configuration 130–131

periodic structures
- creation of 133–134
- design of 93–94
- finite *see* finite periodic structures
- further applications 255–278
- infinite *see* infinite periodic structures
- level-set 240–242
- light-emission control 256–259
- mathematical description 16–20
- microfluidics 273–275
- multifunctional 272–273
- phononic band gaps 205–207
- photonic band gaps 160–162, 170–176
- thermoelectric energy 275–278
- three-dimensional 5, 104–110
- two-dimensional 100–104, 205–207
- versus nonperiodic structures 4–6
- wave vector prerequisites 99–103
- waveguides 259–264

periodicity, spatial 7
permeability, fluid 274
perpendicularity, and interference 83
phase control 118
phase mask lithography 120–121
phononic band gaps
– complete 201
– definition 193
– origin 189
– solid–solid 196–197
– two-dimensional 205–207
phononic crystals 183–210
– definition 188–190
– examples 189
– finite periodic structures 190–194
– infinite periodic structures 194–198
– one-dimensional 190–198
– solid cylinders 202–205
– solid spheres 208–210
– solid–fluid 198–207
– solid–solid 194–198
– solid–vacuum 199–200
– three-dimensional 207–210
– two-dimensional 198–207
– vacuum cylinders 198–202
PHOST see polyhydroxystyrene
photoacid generator (PAG) 124
photonic band gap maps
– as a function of the volume fraction 170
– definition 160
– for simple periodic structures 172–175
photonic band gaps 142
– calculation of 148–149
– complete 158–159
– definition 146
– three-dimensional 168–176
– two-dimensional 157–162
photonic crystals 141–175
– as lenses 269
– band diagram 147
– body-centered 173–176
– cubic 172
– definition 142
– face-centered cubic 172–173
– frequency ranges 146
– guiding mechanisms 262–264
– one-dimensional 143–151
– three-dimensional 162–176
– two-dimensional 151–162
photonic–phononic crystals
– complete band gaps 264–267

– simultaneous localization of light and sound 264–268
– structural defects 267
photoresists 89–92
– chemical modification 89–91
– definition 122
– developing 89
– drying 132–133
– negative 124–125
– negative and positive, definition 92
– organic–inorganic hybrids 128
– periodic structure fabrication 122–128
– shrinkage 133
– wave vectors in 128–129
– work done on 90
photosensitizers 124
plane-wave method, numerical technique 148
plane waves, electromagnetic 68–69
plastic deformation, of microframes 250–252
point lattices
– primitive and nonprimitive unit cells 14–16
– three-dimensional 10–16
– two-dimensional 6–10
– types 9
Poisson's ratio
– cubic crystals 233–235
– definition 234
polarization
– electromagnetic plane waves 69–75
– elliptical 75
– linear 73–75
– preserving of 128–131
polarizing beam splitter 119
polyelectrolyte inks 239
polyhydroxystyrene (PHOST) polymers 127
positive photoresists 126–127
– definition 92
positive refraction, definition 268
Poynting vector, instantaneous 76
primitive unit cells 14–16
primitive vectors
– cubic Bravais lattices 165
– three-dimensional 12
– two-dimensional 8
prism configuration, periodic structure fabrication 130–131

r
reciprocal lattice vectors 153
– definition 25

reciprocal lattices
- in three dimensions 164–167
- in two dimensions 152–157

reflectance 144–145
- as a function of the nondimensional frequency 145, 192–193

reflection, omnidirectional 150

refraction
- in photoresist 129
- negative 268–272
  - all-angle 270
  - definition 268–269

registration challenge 118
rod-connected model 240–242
rotation tensor, definition 224

S

scaling, in periodic structures 145
Seebeck effect 277–278
shear modulus, cubic crystals 235–237
shrinkage, photoresists 133
side-centered unit cells, definition 14
silicon, as background material 208–210
simple cubic lattice, primitive vectors 165
simple periodic functions
- combinations 59–61
- square lattice 33–38
- three-dimensional 41–59
  - body-centered-cubic lattice 51–55
  - face-centered-cubic lattice 47–49
  - simple cubic lattice 44
- triangular lattice 40–41

simple periodic structures, photonic band gap maps 172–175
simultaneous localization of light and sound 264–268
single beam technique, periodic structure fabrication 120–121
solid cylinders, in air 202–205
solid materials, elastic waves 184–187
solid spheres
- in solid background material 208–210
- structures made of 240–242

solid–fluid phononic crystals 198–207
- dispersion relation 203

solid–solid phononic crystals 194–198, 207
- dispersion relation 195–197

solid–vacuum phononic crystals 199–200, 205–207

solids, periodic cellular *see* periodic cellular solids
sound, simultaneous localization with light 264–268
spacing, fringe-to-fringe 86, 88
spatial period
- in three dimensions 42
- square lattice 31
- triangular lattice 39

spatial periodicity, definition 7
spheres
- elastic 189
- solid *see* solid spheres

spontaneous emission of light
- controlling 256–259
- definition 256

square lattice 31–38
- primitive vectors 154

stiffness constant, definition 219, 234
strain, definition 218
strain tensor
- definition 224
- periodic cellular solids 221–228

stress, definition 218
stress tensor
- periodic cellular solids 219–221
- symmetry 221

stress–strain relationship
- definition 222
- generalized Hooke's law 229
- *see also* Young's modulus

stretch-dominated periodic cellular solids 247–249
structural defects
- and localization of light 261
- photonic–phononic crystals 267

structural periodicity 3–25
- nonperiodic versus periodic structures 4–6

structures
- bicontinuous 93
- binary composite 272
- cubic periodic 244
- diamond 168–169
- direct *see* direct structure
- F-RD 242–243, 247–248
- finite periodic *see* finite periodic structures
- geometry of periodic 169–170
- I-WP 242, 271
- infinite periodic *see* infinite periodic structures
- inverse 34–35, 45
- level-set 240–242, 271

- multifunctional periodic 272–273
- nonperiodic versus periodic structures 4–6
- obtained by interference lithography 92–94
- P 241, 249
- periodic *see* periodic structures
- simple periodic 172–175
- three-dimensional 5, 104–110
- two-dimensional 100–104

SU-8 microframes 250–252
SU-8 photoresist 124–125
supercritical drying 133
superlenses, and negative refraction 268–272
superposition
- definition 83
- of two or more waves 86

t

TE and TM waves
- in two-dimensional photonic crystals 151–152
- photonic band gap 158–159
- *see also* transverse electric/magnetic

tensile stress, definition 218
thermoelectric energy, periodic structures 275–278
Thomson effect 276
three-dimensional band diagrams 168–170
three-dimensional Brillouin zones 164
three-dimensional fluid–fluid phononic crystals 207
three-dimensional interference patterns 86–89
three-dimensional periodic functions
- creation 41–59
- Fourier series 23–25
three-dimensional phononic crystals 207–210
- gold spheres 208–209
three-dimensional photonic crystals 162–176
three-dimensional point lattices 10–16
three-dimensional solid–fluid phononic crystals 207
three-dimensional solid–solid phononic crystals 207
three-dimensional structures
- infinite periodic 5
- nonperiodic 5
- periodic, via interference lithography 104–110

time-averaged values, electromagnetic waves 77–80
TM *see* transverse electric/magnetic
topological design, periodic cellular solids 238–243
topology optimization, definition 272
traction vectors, definition 219–220
translation in space, definition 221
transport of heat and electricity, maximum 272–273
transverse character, electromagnetic plane waves 69–72
transverse electric/magnetic (TE/TM) components
- incident light beam 129–130
- *see also* TE and TM waves
transverse plane wave, mechanical 184–185
transverse waves, definition 69
triangular lattice, primitive vectors 154
triaryl sulfonium salt 125
tubular P structures 241, 249
twelve-connected stretch-dominated periodic cellular solids 247–249
two-dimensional periodic functions
- creation 31–41
- Fourier series 20–23
two-dimensional periodic structures, via interference lithography 100–104
two-dimensional phononic crystals 198–207
two-dimensional photonic crystals 151–162
- Brillouin Zones 152–157
- dispersion relation 157–158
- reciprocal lattices 152–157
- TE and TM waves 151–152
two-dimensional point lattices 6–10
- types 9
two-photon lithography 239

u

umbrella configuration 120
unit cells
- arbitrary 25
- conventional 12
- primitive and nonprimitive 14–16
- three-dimensional 11, 46
- two-dimensional 8

v

vacuum cylinders, in a solid background 198–202
velocity, of longitudinal and tranverse elastic plane wave 187

volume fraction
- and linear elastic mechanical properties 243–247
- control of 134–135
- definition 35
- photonic band gaps 169–170

w

wave equation 65–68
- definition 67

wave vectors
- for three-dimensional periodic structures 105–109
- for two-dimensional periodic structures 99–103
- in photoresist 128–129
- interference in three dimensions 88
- of homogeneous/ nonhomogenous material 147

waveguides 259–264
- in photonic crystals 262–263

wavelength
- definition 64
- mechanical waves 184

work, done on photoresist material 90

y

Young's modulus
- cubic crystals 233–235
- definition 234
- *see also* stress–strain relationship